COMMUNICATION, TECHNOLOGY, AND POLITICS IN THE INFORMATION AGE

For Daniel and Jacqueline—
May their tribe increase

Communication, Technology, and Politics in the Information Age

Gerald Sussman

SAGE Publications
International Educational and Professional Publisher
Thousand Oaks London New Delhi

For information address:

SAGE Publications, Inc.
2455 Teller Road
Thousand Oaks, California 91320
E-mail: order@sagepub.com

SAGE Publications Ltd.
6 Bonhill Street
London EC2A 4PU
United Kingdom

SAGE Publications India Pvt. Ltd.
M-32 Market
Greater Kailash I
New Delhi 110 048 India

Printed in the United States of America

Library of Congress Cataloging-in-Publication Data

Sussman, Gerald.
    Communication, technology, and politics in the information age /
by Gerald Sussman.
        p. cm.
    Includes bibliographical references and index.
    ISBN 0-8039-5139-6 (acid-free paper). — ISBN 0-8039-5140-X (pbk.:
acid-free paper)
        1. Telecommunication—United States—History. 2. Technological
innovations—Economic aspects—United States—History.
    3. Information technology—United States—History. 4. Information
society—United States—History. I. Title.
    HE7775.S87 1997
    384'.0973—dc21   97-4788

97 98 99 00 01 02 03 10 9 8 7 6 5 4 3 2 1

Acquiring Editor:	Margaret Seawell
Editorial Assistant:	Renée Piernot
Production Editor:	Astrid Virding
Production Assistant:	Denise Santoyo
Typesetter/Designer:	Danielle Dillahunt
Cover Designer:	Candice Harman
Print Buyer:	Anna Chin

Contents

Preface

This book is about the political content of communication technology. It is intended to develop in its readers a deeper understanding of the nature of the "information society." The core idea is to show that technology is much more than an apparatus, a tool, a technique, a toy, a package of electronics, or an invention. It is also the embodiment of human decisions made at particular moments in particular political settings.

Technology and politics are conceptually inseparable because the rules under which technological enterprise is undertaken is essentially political in nature. Politics is precisely about what the rules are and who is favored by them, but most important, it is about who gets to make them. Where there is technology, there is embedded politics.

Society's rule makers get to determine the areas of public and private investment, as well as the laws concerning property ownership, tax obligations and waivers, legal protections and rights of workers, the statutory limits of natural resource exploitation, the funding of physical and educational infrastructure, and other policies that establish the boundaries of technology development. During the 20th century, the United States wavered between relatively active and inactive central government, but in the last half of the century, the big story was the rise of transnational corporations (TNCs) and their global dominions.

The biggest of these TNCs hold more assets than 90% of the world's nation states. They and the global institutions that support them (e.g., World Bank, International Monetary Fund, General Agreement on Tariffs and Trade, World

Trade Organization, United Nations, government aid agencies) have structured the world economy, including how it is wired for sound, text, and image.

Even before TNCs became dominant, private large-scale business enterprises and the federal government were the key players in building America's "information highway" of the air, which originally was based on telegraph and telephone. During the 19th century, relatively few citizens got to use that highway, but together with the railroad, it was a great boon to the consolidation of corporate power in the United States and the establishment of major business and urban centers. In the first half of the 20th century, radio and television were introduced through corporate and commercial sponsorship, and by the close of the century, it was apparent that computer-networked communications via the Internet and the World Wide Web would similarly be dominated by profit-seeking business organizations.

Modern communication and other technologies do not arrive on the scene by accident, nor are they primarily developed for the use of the general public. Rather, they are the product of careful business investment decisions involving control of existing technical know-how, substantial financial resources (including large public subsidies), research facilities, scientific talent, and most important, market data. These decisions ultimately translate into not only salable goods and services but also jobs, wages, salaries, stock options, educational priorities, environmental and health impacts, government policies of many kinds, welfare needs (for those displaced by restructuring), and quality-of-life issues. In short, technology affects the whole production infrastructure on which the economy is based.

There are definite societal trade-offs when governing institutions permit the private sector this much latitude over the lives of millions of citizens. Much of the excited rhetoric about the glorious future of the information society does not take into consideration the material and personal interests of the private institutions and governments that actually dominate it. Instead, communication technology is usually portrayed (e.g., in the mass media) as if it's all there for anyone's taking. Part of this book, particularly Chapters 5 and 6, charts just who the heavy truckers are on the information highway and what that means for the rest of us.

Most people operate household or office communication equipment (like the laptop computer used in writing this manuscript) without reflecting on the means and social costs involved in their making and how technology affects our society as a whole—and ultimately our individual lives. In this book, I want to reveal and demystify the social innards of a range of communication technologies— television, cable TV, radio, telephone, computers, the Internet, facsimile (fax), orbiting satellites, oceanic cable, and others—showing how they are historically and politically related to one another, where they originated, and what effects they have had on the people involved in their production and in their use.

In recent years, a flood of books and other media coverage has celebrated the "information revolution" and the emergence of the information society or "global village." The vision of technology taking over the direction of our everyday affairs is a very powerful narrative about social reality, but it is not an accurate one. Apart from its continuous *anthropomorphizing* of technology— that is, repeating the fallacy that inanimate objects are capable of human intentions and behavior and being the change agents of history—technological determinism embraces the dangerous ideological tendency to disregard the human actors (agency) involved in public and private decision making. Technological determinist readings of social change tend to either glorify or condemn technology, while personal, political, and institutional responsibility for their development, use, and consequences are concealed. The ultimate technological deterministic fantasy lies in the computer engineer's imagination: the creation of "artificial intelligence" (AI). AI is a technology whose possession would confer transcendent power, like that of Dr. Frankenstein, while reducing human reasoning to that of lower-order species.

To understand technology is not only to be aware of its electromechanical functions—for example, that a television set receives, amplifies, demodulates, and processes into images and sound electrical broadcast or cable signals—but also to recognize its social content, purposes, and ramifications. Here, one might consider what it is like to work in a television assembly plant, how the presence of television has changed the way people relate to one another, and what impact technology has on the natural environment that sustains us. These are not easy questions, and their answers should not be left to industry, its sponsors, and public relations officers, who have little incentive in disclosing their self-interests and actual agenda. To understand the politics of technology, one looks beyond the techno-functional aspects of product and process and focuses on the human agents and institutions involved in their making. In this way, we can begin to understand the real meaning of the information revolution.

Technological changes do not come about merely because certain technical blueprints make them theoretically feasible. Feasibility is itself the outcome of a complex set of nontechnological forces and considerations. Henry Ford had developed an automobile engine prototype in the 1930s that could run on a corn-based alcohol derivative, which, given the surplus corn in the United States, would have made engine combustion an inexpensive component of driving and would have created a windfall for American farmers and car makers. But that happened just before powerful Anglo-American petroleum interests gained leverage in the Middle East and began pumping mass quantities of inexpensive oil from that region. Alexander Graham Bell, a man paid homage in standard American history for having "invented the telephone," was among those who believed that, sometime within the 20th century, there would be an airplane in every garage. I can recall reading in a children's weekly magazine during my elementary school days in the 1950s that, by the year 1975, Americans would be routinely flying about their communities, powered by jet packs strapped to their backs. I could hardly wait for 1975!

It may very well be that had the political and industrial communities been committed to making jet packs available to almost every household, as they did in the case of automobiles, we might very well be buzzing about the lower atmosphere. But other political and economic considerations intervened, including the priority of the United States becoming a leading nuclear power, arms merchant, electronics manufacturer, and international police force. Americans also began to learn that polluting the atmosphere on that scale was not without costs to people's health, the quality of the physical environment, and ultimately to society as a whole. Economic costs alone made jet packs a questionable investment, given the private and public funds that would be required. Clearly, technology in and of itself was not in command. What was is the thrust of this book, which, I hope, will bring readers to a critical understanding of the politics of communication technology.

In applying the critical approach to an understanding of communication technology, the measuring stick is how it benefits ordinary people, the vast majority who are not engaged in great material or political power pursuits. Inasmuch as the mainstream literature all around us is filled with stories of such pursuits by the captains of industry, finance, and commerce, there is no room or reason to reiterate here all the glorified claims of conquest made about the advanced technological universe in which we live. A critical analysis demands a detailed rethinking of much of our received knowledge, and that is the pursuit toward which the pages of this book are engaged: What does the "information age" mean for most of us?

Communication technology can be and is used not only in making certain information more available but also in keeping secret dossiers on citizens, over which they have little or no control. Every day, commercial, police, and political interests transmit and read without permission of their subjects millions of biographical profiles, which, as a number of authors have pointed out (see Chapter 6), has turned American society into a massive surveillance operation and seriously eroded constitutional guarantees of rights to privacy. At the same time, people who have the most need of functional information in the form of good and affordable education are those most frequently denied access, particularly people of color and those disenfranchised by class and poverty. Women, too, continue to be victimized by sexual discrimination and harassment in the various channels of public information, including the mass media and the Internet.

In writing about scientific knowledge, the historian of science Thomas Kuhn speculated that revolutions in thinking occur following major *paradigm shifts*— that is, changes in core beliefs that fundamentally alter technical understanding.

In submitting information age ideology to critical investigation, we find that the theoretical approach that most thoroughly and coherently undermines its assumptions is political economy. This approach enables people to ask key questions about the material interests that back up the exercise of power in government and in the marketplace. In the tradition of political economic analysis influenced by marxian thinking, these two sets of empowered actors are interdependent. Some writers, not necessarily political economists, talk about the "revolving door" of political and corporate life—executives forming big businesses and moving over to administrative positions in government and vice versa and keeping policy discussion within a very narrow framework of safe, status quo actors, issues, and technical solutions.

In developing this critical approach, I start with theory (Chapters 1 and 2), move on to review the social history of communication technology (Chapters 3 and 4), and proceed to an analysis of contemporary issues in the U.S. context (Chapters 5 and 6). In Chapter 7, I bring in the global dimension of the information society. Here, we look at both the instruments of communication technology, where and how they are produced, and the content flowing through them that is targeted to worldwide audiences. This dimension helps us realize that the information society involves billions of people who, especially from the Third World, are rarely given press or airtime in the American media, and yet without them the availability of an information society as we now know it would be more like the jet pack example cited above—theoretically possible but without the practical means of realization. The social cost of many countries' involvement in the wired world often has revolutionary or destabilizing consequences, as we have seen in South Africa, the Philippines, and the former Soviet Union.

In liberal democratic societies, many opportunities for social empowerment have been enhanced with the aid of new communication and information tools, especially with the formation of an alternative press, public access cable TV, and the Internet, and we discuss these options as ways of illustrating that the hegemony of political economic power is never complete and always subject to "paradigmatic shifts" and new ideological challenges and ways of seeing. As we look to the future (Chapter 8), the struggle will be to make education, science, justice, and opportunity the commonwealth of ordinary people. This struggle will employ some of the instruments of the communication system designed and intended for other purposes and will require new and creative forms of political education and political initiative. This book is written with such objectives in mind.

Acknowledgments

I wish to acknowledge the help of staff at Emerson College and Portland State University (PSU), particularly Katherine Belbin, Robin Panzer, Carleen Simmering, and Tim Smith, and my students, especially those in Communication, Technology, and Politics at Emerson and in Critical Theories of Mass Communications at PSU. Colleagues and friends, including Sy Adler, Ibrahim and Liana Aoude, Nancy Chapman, Beverly and Levon Chorbajian, Ed Clemmer, Dick Duprey, Flora Gonzalez, Steve Kosokoff, Oliver Lee, John and Rose Lent, John Marlier, Walt Littlefield, Kathy McCann, Catherine McKercher, Bella Mody, Vinny Mosco, Ed Oasa, Boone and Peggy Schirmer, Nohad Toulan, Janet Wasko, Bob White, Lillian Yamasaki, and family members Muriel and Paul Eagle, Edna and Tom Ozawa, Linda Sasaki, and Claire and Dave Sussman offered encouragement, valuable discussion, insights, and sustenance in helping me see this project to the end. Carl Abbott, my PSU colleague, read the entire manuscript and provided good critical insights, as did Douglas Kellner, Vinny Mosco, Dan Schiller, and Sage's anonymous reviewers. Sage editors Sophy Craze and Margaret Seawell were supportive, professional, and best of all, patient.

My children, Daniel and Jacqueline, are a constant and wonderful reminder of the simple joys of life and the questions that are really important. My companion, friend, spouse, colleague, and most appreciated critic, Connie Ozawa, who read and commented on the manuscript, inspired me to think about many of the ideas, especially related to the larger environmental and policy concerns, that I discuss in this book.

PART I

THE MEANING AND POLITICS OF THE "COMMUNICATIONS REVOLUTION"

Chapters 1 and 2 discuss the theoretical underpinnings of the "communications revolution" and look at various ways of interpreting this dramatic and unsettling period of history. They bring in the meaning of political economy as a way of demystifying many of the exaggerated claims made about communication and information technology, as well as of developing a related and critical understanding of how ideology is constructed and used to lend authority to certain interpretations of history and social change. One argument made in these chapters is that powerful institutions in society, in government, and in business have turned information into more of a salable commodity in recent decades, assigning marketplace values to it, and remaking the workplace and cultural institutions for the purpose of selling it.

CHAPTER 1

Ideology and Discourse on the "Information Society"

The more exactly he [the intellectual] is thus informed on his position in the process of [capitalist] production, the less it will occur to him to lay claim to "spiritual" qualities.

—*Walter Benjamin (1969), "The Author as Producer"*

THE POLITICS OF DISCOURSE

In asking critical (socially analytical) questions about the "information society," one might begin by examining how communication systems are socially organized. Most textbooks on communication studies, journalism, or mass media focus on the how-to aspects of communicating, on how "audiences" construct messages, or on how to be media professionals. Where the political economy analysis differs from mainstream approaches to communication studies, as one critical writer notes, is in "placing its emphasis on production or supply rather than consumption or demand as the determining moment" (Garnham, 1990, p. 7). He means that discourse, culture, and consumption are coordinated within the space, time, and system of production that is organized to carry out social, economic, and political activities. The political means of organizing and allocating resources and energy in society ultimately explain much, though certainly not all, of what a society is, including its communication needs and practices.

We conduct our day-to-day existence within laws, job and reward structures, cultural conventions, locations, norms of civic behavior within the built environment, achievement expectations, material values, social relationships, and so on that are all part of the political order. Society functions within boundaries governed or mediated by institutions of organized power (e.g., the legal system, corporations, business associations, Congress, police, prisons, mass media). Garnham (1990) sees organized power in a *capitalist society* primarily serving to defend the interests of private property and wealth. Constitutions and other charters of capitalist societies bestow power rights to the propertied classes, a fact acknowledged and defended by John Locke in the English tradition and by other classical writers. Locke wrote in *Two Treatises of Government* (1690): "The great and chief end, therefore, of men's uniting into commonwealths, and putting themselves under government, is the preservation of their property" (cited in Ebenstein & Ebenstein, 1992, p. 229).

Any system of institutional interests, whether democratic or authoritarian, is not static or immutable, but rather is inherently unstable over time. Therefore, organized power interests necessarily must be conscious of their own preservation, as would any individual member of the species. The preservation of power (the means to influence or control the outlook, behavior, and welfare of others) has among its requirements the need to influence the mode of discourse—to infuse language practices with concepts that protect the legitimacy of the existing organization of control and the particular institutions within it. The Italian political philosopher Gaetano Mosca (1858-1941) believed that elite interests inevitably dominate any society or organization, although there is need for a "circulation of the elites" (transfer of power) from time to time to prevent stagnation and the formation of radical alternatives. He wrote:

> In all societies—from societies that are very underdeveloped and have largely attained the dawnings of civilization, down to the most advanced and powerful societies—two classes of people appear—a class that rules and a class that is ruled. The first class, always the less numerous, performs all of the political functions, monopolizes power, and enjoys the advantages that power brings, whereas the second, the more numerous class, is directed and controlled by the first, in a manner that is now more or less legal, now more or less arbitrary and violent. (Mosca, 1939, p. 50)

Political economy as a way of understanding also focuses on the "ruling class" but, unlike Mosca, seeks to explain how that class is constituted through control of the production of the material and cultural filaments of society. As Garnham and others argue, communication in a capitalist society is increasingly permeated by considerations of market and money. Most people have little choice but to accept their wages and cultural sustenance from a system that governs the opportunities for work (through corporate management) and symbolic interaction (through the cultural industries). Few people have any direct voice in the marketing of mass culture, although some would argue that the act of consumption is a sufficient expression of participation. During the long history of capitalist development, now in a phase dominated by colossal industries, people have come to rely less on their natural rhythms, common sense, and communities and more on external and institutionalized sources of information and cultural processes (Garnham, 1990, p. 9).

Communication is continually being reinvented by the marketplace. The convention of calling the voice, message, and image transmissions in the mass media "communication" would have struck our pre-electronic-age ancestors as a very strange use of language. According to the German poet and essayist Hans Enzensberger (1974), with the few and limited exceptions of call-in radio and television programs, what we call the mass media permit "no reciprocal action between transmitter and receiver; technically speaking . . . [they reduce] feedback to the lowest point compatible with the system" (p. 97). Enzensberger observed, "In its present form, equipment like television or film does not serve communication but prevents it" (p. 97). In his analysis, it is not that electronic communications are inherently devoid of value, but that in the context of the capitalist form of ownership and use (appropriation), they have been turned into instruments of a "consciousness industry" designed for the preservation of ruling class values, and no less so in the hands of (former) Soviet bureaucrats.

The French sociologist Jean Baudrillard takes a different view, arguing that mass media are inherently repressive regardless of who owns or controls them. The importance that industrial society assigns to the media, he says, reveals how organized power centralizes and stabilizes authority even if television may occasionally be used in ways that momentarily weaken established power (e.g., inciting public disillusionment with Johnson's and Nixon's war in Indochina). Because of the media's "function of *habitual* social control," they stay "right on top of the action." There is no better way to stifle the momentum of popular resistance, Baudrillard says, "than to administer it a mortal dose of publicity" and transmute the original impulse toward popular revolt into filtered images for "exchange value" (profit) (1986, pp. 131-132). In other words, television edits and ultimately destroys spontaneity.

"Communities" of Captive Audiences

As a fixed ingredient in modern American society, television symbiotically links industry to citizen in a cycle of production and consumption. The private corporate control of American network television and the reluctance of government to regulate either the insatiable commercialism or tabloid excesses have effectively stripped the medium of public accountability and societal standards. Television's programming, advertising, and news products are seen by many as the educational and intellectual equivalent of a jelly doughnut: They tranquilize but do not nourish and sometimes contaminate their viewers (see the collected critical essays edited by Lazere, 1987).

One early critic of television described the networks' treatment of news as "events disassociated in time, space, and subject matter . . . [to the extent that] it is impossible for [viewers] to make any meaningful order out of the never-ending kaleidoscope" (Cirino, 1972, p. 134). Pharmaceutical advertisers, for example, do not warn viewers that frequent heartburn, indigestion, constipation, or body pain may be symptomatic of serious health problems or harmful lifestyles in need of attention; rather, they suggest their products as ways of hiding symptoms and living dangerously in denial. "The truth of the matter," say two knowledgeable media observers, "is that financial interests play a major role in determining what we see—and don't see—on television" (Lee & Solomon, 1990, p. 59).

In actuality, television programming, including its news presentations, is not all that haphazard, because the stations and networks do select and edit the content, often with "scientific precision," for drama, entertainment, and story lines that are consciously nonthreatening to commercial values or sponsors (Ben Bagdikian, cited in Lee & Solomon, 1990, p. 65). Media expose viewers to urban outlooks and the wider world, but they present a way of seeing that is carefully filtered and constructed for commercial and political ends, a kind of information management, or propaganda, that serves the interests of media's main patrons— industry, commerce, and the governing state apparatus. On the bottom line, news is not what enlightens, but what *sells*. Its worth is not its intrinsic educational, as much as its financial and ideological, enrichment—geared toward market-ability, profitability, and validation of the "free enterprise" system.

With a $134,000 receipt for each 30-second spot for prime-time sponsors ($900,000 per half-minute for the 1994 National Football League Super Bowl; $25 billion spent on television ads in 1988), why would a network want to offend its high-roller clients with stories about cars that fall apart, the contents of soft drinks, the profits and shady lifestyles of sports team owners, ripoffs in the insurance industry, the corporate bribes and kickbacks paid for overseas busi-ness deals, the toxic and oppressive working conditions in the Third World factories producing for U.S. consumer electronics and clothing corporations— or how media corporations manage the news? Although cigarette advertising is banned from television, Philip Morris, the third largest advertiser on television, gave $390 million in 1991 to the big three networks to sell its other wholesome products: Jell–O, Kool–Aid, and Miller Beer (Cohen & Solomon, 1993, p. 50).

News coverage of violence in society does little to educate viewers about the causes or prevention of antisocial behavior ("crime") but does play a big part in inducing support for more prisons, longer prison terms, and capital punishment. The United States already has the world's highest incarceration rate (5.1 million prisoners, 3% of the nation's adult population in 1995). George Gerbner, a professor at the University of Pennsylvania, found in a study that people who frequently watch television news are far more likely than infrequent viewers to feel threatened about their neighborhoods (cited in Postman & Powers, 1992, p. 23).

Mass media, in fact, give inordinate attention to violent crime even though that portion of the police ledger amounts to only 10% of the total, the rest registered largely as violations against property. A content analysis of *Time* magazine's crime reporting in the postwar period revealed that 73% of such stories focused on violent crime (Barlow, Barlow, & Chiricos, 1995, p. 11). The fear excited by the constant barrage of sensational news of random murder and violence, with its "adrenalin-pumping visuals" (Cohen & Solomon, 1993, p. 117), may actually help create a self-fulfilling prophecy by inducing excessive privacy, distrust of neighbors, and more time spent indoors—all of which contribute to the isolation of individuals from one another—indeed, the optimum environment for crime. And with some 38 million Americans watching network news each night, the captive market for fear-mongering is huge.

The implicit message is that we need more cops and prisons. Hence, the glut of police, detective, FBI, and other "crime buster" television formats—even though the evidence shows that the overwhelming number of crimes actually solved results from citizen initiative, not police detection. There is also abundant evidence that harsh prison environments and capital punishment do not deter, as much as beget, criminal outlooks. But in the television newsroom, the golden rule (or rule of gold) is: "If it bleeds, it leads."

Another utility of television-induced fear and anxiety is the association with consumption. Many people, anxious to rejuvenate feelings of well-being, find momentary alleviation by consumption fixes, often resulting in eating and shopping addictions. Understanding this well, advertisers attempt to associate gratification symbols (e.g., love, sex, status, power) with the act of spending and pay networks $400,000 per news show to get access to audiences. And with the average American exposed to more than 30,000 television ads per year, network television does deliver (Postman & Powers, 1992, pp. 117, 119).

From the advertiser's perspective, what could be a better arrangement for marketing goods than having a steady audience for fast and slick commercials and without having to answer the questions that buyers would ask in a true marketplace: Does the product do what the advertising images suggest? How long will it stay on the market before succumbing to early obsolescence? Is the product good for people who use it—and those who make it?

The media set an agenda for public discourse, and within this framework the citizen has little more input than opting/not opting for its menu of offerings. For Baudrillard, feedback loops such as letters to the editor, talk shows, and polls only reinforce the fiction of responsiveness without in any way altering the system of manipulation and control that is organized within such media structures. Only the elimination of mass media can restore the basis of real communication—that is, person-to-person communication. Whether or not this way of thinking is realistic or even desirable, it is evident that, in industrial societies, the media function as the principal information system through which money, economic exchange, the authority of government, and the dominant ideas in society are transmitted. To some extent, the youth rebellions in Western societies in the 1960s did reflect Baudrillard's rejection of official and organized structures of information and representation—a repudiation of corporate society, big government, and big media.

TRANSFORMATION OF CAPITALISM, COMMUNICATION, AND CULTURE: A PRIMER ON POLITICAL ECONOMY

The political economy approach to understanding society starts with the power context in which decisions about the accumulation and distribution of wealth take place. In the marxian tradition (following the critique of capitalism by Karl Marx and his successors, though without making a catechism of their arguments), political economy seeks to explain the historical and social changes from the focal point of capitalist enterprise and the private ownership of the means of production. In so doing, it also looks at the structural forms of exploitation practiced under capitalism on the direct producers of wealth—the working class (whom Marx called the proletariat). Technology, as a form of capital, does not come about of its own accord or accidentally, but rather through the logic of capitalist necessity.

In early (circa 16th century), small-scale capitalism, producers and potential purchasers most often had to deal with one another in a tangible, real-life marketplace, where long-term relationships required a greater degree of reliability and responsibility for product sales. The industrial revolution, starting in the late 1700s, went hand in hand with a new phase of capitalism in which large-scale technology, production, and markets began to form. Mass production took off in the early 19th century with the development of interchangeable parts for gun manufacture. A century later, under a more advanced stage of capitalism, Henry Ford perfected the system ("Fordism") organized on the basis of not only interchangeable parts but also a fragmented division of the work process, a specialized set of complex machine parts and machine designers, skilled systems managers, an abundance of disciplined unskilled and low-skilled workers, and a moving assembly line (Cohen, 1993, p. 109).

With the maturation of capitalism, the larger the market, the larger the need for coordination of the many complex factors that went into them: skilled labor, training centers, management, finance, transportation, telecommunications, legal services, and so on. In contemporary transnational capitalism, the abstraction of the borderless marketplace separates by thousands of miles the goods producers—themselves isolated by a global division of labor—and the purchasers. In most forms of modern production, this makes direct communication and accountability between producer and consumer nearly impossible. In the abstract market, an assortment of mediators (e.g., wholesalers, retailers, distributors, advertisers, managers, labor unions, government regulators, city and town administrators, media outlets) fill in the space between direct producers and consumers.

The owners and managers of the means of production, the capitalists, are the principal agents in sponsoring innovations in the methods, instruments, and supporting infrastructure of production. Labor constitutes the direct and main component in the production process; the state (the primary institution of societal control) provides supporting mechanisms (legal, coercive, ideological) in defense of private enterprise. In the contemporary stage of capitalism, political economy is concerned with, among other things, how owners and managers of large-scale corporate research and development centers, the state, and other players participate in the development and "social construction" of technology. *Technology* refers to the practical and applied uses of science in the various processes of production, wealth accumulation, and the organization of power, as well as in the fulfillment of social values.

The neoclassical (conservative) approach to political economy is essentially different and generally hostile to the marxian version. The former is principally concerned with so-called laws of economics and sees the market as the appropriate venue for decisions about determining the organization, allocation, and distribution of wealth. Interventionary state politics in this approach imposes impediments to the otherwise autonomous and self-regulating tendencies of the market (Adam Smith's "invisible hand") in maximizing profitable and fair commercial transactions. Therefore, conservative (also called *neoliberal*) political economists advocate laissez-faire (minimum state intervention) policies toward business so as to maximize the self-governing growth potential of market forces. Yet, even Smith, in his *Wealth of Nations,* recognized a tendency toward monopoly capitalism: "People of the same trade seldom work together," Smith wrote, but when they do, "the conversation ends in a conspiracy against the public, or in some diversion to raise prices" (cited in Heilbroner, 1992, p. 70).

Keynesian or liberal political economists also maintain a strong faith in the efficacy of the world capitalist structures but see a major role for government in regulating and investing in key areas of the economy to avoid the worst excesses of the market system and the cyclical trend of national and international recession. Marxians significantly differ from Keynesians in their view of the social class relations within capitalism and within the modern corporation, the former viewing such relations as essentially undemocratic. They also differ in their support of the idea of state and worker ownership and control of the strategic centers of economic activity, marxians favoring maximum public participation and strong regulation. Economic growth in the short term is not as important as laying the foundation of a public economy established on the basis of redistributed economic power and a political-legal structure that emphasizes greater equity, opportunity, social justice, and collective responsibility for the affairs of the state.

Capitalism is not one permanent or enduring political economic system, but a form of production that has changed radically since it became ascendant, around the year 1500, in the Western world. There have been different phases of capitalism, corresponding, according to the political economist Marx, to the state of development of the forces (material means) and social relations (class structure) of production. For Marx, all social relations ultimately derived from the system and ownership of production in a given society, so where early or mercantile capitalism existed (ca. 1400-1770), commercial (pre-industrial) interests prevailed. (This and the following time demarcations of the stages of capitalism are modified from Stavrianos, 1981.) *Mercantile* (or *commercial*) *capitalism* was still based heavily on the production and sale of raw materials, such as agricultural products, minerals, timber, and precious metals, relatively small markets and small-scale production of handicrafts—although with increasing importance of international exchanges of goods and services (and the early phase of Western colonialism).

Early European (commercial) capitalism was based on an *ideology* (a more or less coherent mental picture of the "real world" that underlies the rationale for governing institutions, the rules of citizenship, and the norms of appropriate public behavior) that, in opposition to official Church doctrine, emphasized the sanctity of private property, personal accumulation of wealth, and the legal, secular rights of the individual. These were revolutionary ideas, compared with the monopoly of privilege by the clerical, monarchical, and aristocratic classes of the feudal era that preceded capitalism. Politically, the rise of commercial capitalism ushered in new commercial and propertied interests that undermined the "divine right" ideological tenets of monarchy and the ecclesiastical hegemony of the Vatican. This eventually resulted in civil war in England and, later, the revolutionary movement in the American colonies. John Locke is the best source in defense of the values of early capitalist ideology.

Even an absolutist (divine right) monarch like Louis XV of France had to make concessions to the rise of scientific and secular thought, or at least appear to do so as a gesture of sophistication. The passing of electrical current was first artificially created in 1745 with von Kleist's Leyden jar, and it soon became the fashion in aristocratic circles to reproduce the experiment. On one occasion, Louis demonstrated before an audience the phenomenon of shock by ordering the palace guard to have current from a Leyden jar battery passed through them, causing them to jump in unison, much to the court's royal and electrifying amusement (Bernal, 1971a, pp. 601-602). As superficial an acknowledgement of the Age of Reason as it was, it represents an interesting moment of the intersection of declining feudal privilege and rising capitalist science.

Communication at this juncture was pre-electric, although with growing opportunities for the wealthy classes (explorers, merchants, bankers, speculators, landholders, royalty) for long-distance transactions through improved commercial shipping and equipment for personal travel. The need for control over territory induced the creation of different signaling systems (semaphore, optical telegraph), which in turn brought unexpected innovations, including an amusement spinoff, the "lovers' telegraph"—the still familiar string telephone. State-run postal systems were gradually developed, which in turn depended on state subsidies to road and waterway construction and, in the early years, were used primarily for government and commercial communications. State communications also brought unwanted communiques, such as to those conscripted into military service or coerced into emigration (e.g., Pilgrims, Puritans, Africans, Jews, Celts, Huguenots, impoverished farmers).

Capitalism's second phase, *industrial capitalism* (1770-1870), was distinctly different. This was the era of the industrial revolution, first in England, later in continental Europe and North America. Absolutist power of monarchs was further eroded with the American and French revolutions. Adam Smith, a Scottish political economist, published in 1776 his manifesto of capitalism, *The Wealth of Nations,* which called for entrepreneurial "free markets" and minimizing state supervision of economic activity (laissez-faire). David Ricardo, a British economist, published in 1817 (*Principles of Political Economy and Taxation*) an early explanation of prices based on labor cost theory. Ricardo made the important observation that labor is embodied both directly in products and indirectly in machines and other forms of capital, such as property and finance.

Major changes in the composition of *capital* (basically, forms of wealth that are used to create more wealth) occurred as the Western states and capitalist enterprises organized new explorations, colonial conquests, national competition, and internecine conflicts (e.g., the Napoleonic wars) that helped stimulate capital, technical innovation, and new opportunities for expansion and wealth. Britain, France, Holland, Spain, and Portugal were the main global colonizing powers, whereas the United States was in the process of resolving its own North-South conflict, pursuing extermination, internal colonization, or forced emigration of the Native American tribes and pushing its western boundaries through negotiated purchase, confiscation, and the spoils of war. The east-to-west physical expansion of capitalism proceeded apace: a government National Road starting in Cumberland, Maryland, in 1808 and reaching Vandalia, Illinois, in 1852; canal building; riverboats of various kinds along the great water channels; the railroads; and the telegraph.

During the industrial phase of capitalism, modern machinery and manufacturing came into existence, and the entrepreneurs of the age systematically began exploiting the patent laws for the private ownership, use, and control of knowledge. Social classes became more distinct, with greater wealth and grander lifestyles available for industrialists and financiers and increasing pauperism and harsh working conditions forming at the other end of the social spectrum. Hand-operated tools and artisanship were replaced, with the aid of steam power, by complex machines and automated systems of production, with greater separation of manual and mental forms of labor, and factories become the main sites of manufactured goods. Small cottage industries were virtually eliminated, including many in Third World regions, as modern large-scale industry captured almost all the internal markets for goods and services, with a rapid increase in hours demanded of workers, creating a market system that Marshall Sahlins called, "a life at hard labor" (Bottomore et al., 1983, p. 229: Schor, 1992, p. 7).

Increasingly, science was taking on applied industrial characteristics, especially in heat and energy physics, electricity and magnetism, engineering and metallurgy, chemistry, and biology. These changes were joined by new forms of art, music, ethnic culture, urban demography, agriculture, politics, business, travel, and other social interests. As the 19th-century observer Marx put it, "The revolution in the modes of production of industry and agriculture made necessary a revolution in the general conditions of the social process of production, i.e., in the means of communication and transport" (Marx, 1967, p. 384). The most "revolutionary" communication device in this period was the telegraph, followed by early experiments in telephony (see Chapter 3). This phase represented great leaps in development of control technologies—designs for managing the growing complexities of industrial production, management, and attendant social life changes, which included Charles Babbage's early prototypes of computer-type calculators, in practical terms conceived too early.

In the third phase, *national monopoly capitalism* (1870-1945), industry became highly concentrated, resulting in the near elimination or circumscribing of competition in such areas as steel, chemicals, oil drilling and refining, the automobile, the railroads, sugar production and refining, and telegraph and telephone. It was an age of social Darwinism in which business tycoons, appropriating the observations of the 19th-century naturalist Charles Darwin, rationalized a "survival of the fittest" ethos to justify aggressive and ruthless behavior in the hot pursuit of capital. National monopoly capitalism induced intensive state competition among European rivals, which led to the Franco—Prussian War (1871) and the two world wars of the 20th century. Mass production and mass destruction were two sides of the expanding worldwide capitalist industrial system. For Joseph Schumpeter (1942), capitalism in the 20th century, with its emphasis on scientific, economic, and environmental management, the resistance of unions and intellectuals, and the increased role of public policy intervention, began to lose the entrepreneurial character on which it was based and would inevitably lead to the formation of socialist states.

Mass production employed more intensive levels of automation and required manufacturers to develop ways to increase productivity by extracting more output from workers through redesign of equipment and work patterns and the use of low-wage immigrants, women, and children. In redesigning tools, skills are transferred from worker to machine, and workers are converted into easily replaceable, standardized, interchangeable parts. Machines make machine tools, which make other machines. Henry Ford's motor plant becomes the apotheosis of the automated assembly line, and Ford himself was proud to claim in his memoirs that 43% of jobs in his plant required a mere day's training, and 79% less than a week (cited in Huberman, 1970, p. 313). The 1920s and 1930s were periods of growing labor militancy.

The current and fourth phase of capitalism, *transnational capitalism* (beginning around 1945), has been marked by a postwar global industrial and political order; a growing scale and spread of corporate production of goods and services; new types of productive equipment, especially computer and digital based; and convergence of the forms and instruments of financial control, management, technology, and markets. Chastened by the destructive impact of national competition, economic depression, and two world wars, the leading industrial powers in Western Europe, North America (United States and Canada), Asia (Japan), and the South Pacific (Australia and New Zealand) formed military and economic alliances with a collaborative approach to development and peaceful coexistence.

Transnational corporations (TNCs), enterprises with subsidiaries in many (at least seven) other countries, became the dominant actors in the world economy and began to overwhelm the traditional role of national governments in domestic and international policy making. Currently, the scale of TNCs is such that the assets of the 300 largest of them together constitute a quarter of the world's productive assets, a number of them larger in wealth than the majority of nation states (cited in Barnet & Cavanagh, 1994, pp. 14-15). If corporations and nation states were ranked as such, two thirds of the biggest "countries" would be TNCs (Lairson & Skidmore, 1993, pp. 252-253).

In the 1990s, TNCs continue to expand and dominate the world's material and human resources. By 1988, TNCs made up 67% of U.S. merchandise trade exports and 41% of imports, and intrafirm (between TNC branches) merchandise trade was 30% of U.S. export and 20% of U.S. import totals. On a worldwide basis, the largest 50 TNCs alone grew from $540 billion in sales in 1975 to $2.1 trillion by 1990, with profits reaching $82 billion in 1989 (Carnoy, 1993, pp. 45n, 49). In the 1970s, U.S.-based TNCs overwhelmingly dominated global economic activity, but by the 1980s the country leading both the industrial and banking sectors was Japan. Partnerships of various types among TNCs are so common (e.g., in the automobile industry, the world's two largest corporations—General Motors and Ford—have many joint ventures with Japanese and European companies), however, it is hard to give a realistic calculation of "country" domination in any meaningful way (especially with Japan providing an average of 40% of "American" automobile components). TNCs have, in many respects, superseded the significance of the nation state.

In the United States, the most market centered of the capitalist countries, the average family income in 1992 of the top 5% of the population was more than 15 times the lowest 20% (recalculated to 24 times for 1994), and their respective shares of national income were 17.4% versus 4.6% (the top 20% had 44.3%, recalculated to 46.9% for 1994). These numbers reflected the widest recorded income gap between the richest and the rest of Americans since World War II (Holmes, 1996b).

Measured in terms of assets, in 1983 (prior to a phase of further wealth redistribution to the rich), the top 1% of American households owned 53% of all income-producing wealth, the top 10% had 83%, and the lowest 90% had 17% of total capital (Du Boff, 1989, p. 181). The trend

in the United States has been toward increasing polarization of wealth, the poorest fifth receiving a decreasing share of the national pie, and the country consistently ranking at or near the bottom of the leading capitalist industrial countries in such quality-of-life indicators as job security, health and unemployment protection, leisure time for workers, women's earnings compared with men's, life expectancy, infant mortality, teenage pregnancy, air pollution, voter turnout, violence, and poverty (Du Boff, 1989, pp. 183-184).

Among the leading industrial countries, the United States is also the most class divided, a distinction that formerly belonged to France. Of the new employment created in 1994, the overwhelming number of relatively good paying jobs (above the national average of $15.50 an hour) were in managerial and professional positions. One quarter of American workers, 30 million employees, live below or slightly above the official poverty line of $11,500 for a family of three, $14,800 for a family of four. Wages continue to slide for the lower half of the workforce even though they are working more hours and enjoying less leisure. Real unemployment (including those looking for full-time work but not counted because they either are no longer collecting unemployment insurance or have part-time work) in the early 1990s averaged about 13%, and of the 124 million employed Americans, 20 million are part-timers (McDermott, 1994, pp. 576-577; Schor, 1992, p. 5). A related statistic is that the United States has the lowest rate of unionized workers, less than 16% of the workforce, down from more than 35% just after World War II.

Capitalism started by freeing up labor from the constraints of feudal landlord-tenant relations of the medieval era and giving rise to wage labor. In the current phase of capitalism, labor is increasingly devalued and replaced by technologies that make no demands on income, health and retirement benefits, worktime reduction, job safety, affirmative action, collective bargaining, or severance pay. Labor-intensive industries (having relatively many workers) still exist, but much of this work is farmed out to Third World countries. This reality made the 1993 North American Free Trade Agreement (NAFTA) particularly threatening to American workers and to industries not large enough to organize overseas labor operations in low-wage countries such as Mexico.

The television industry has been hit by severe layoffs in recent years, as have the electronics and telephone industries. Rupert Murdoch, who owns Fox Broadcasting Network, among many other media holdings, built his American television empire on the strength of low wages, layoffs, and job reclassification (after sacking 5,000 union employees from his London newspapers; Downing, 1990, p. 39). In Britain, television employment has been falling drastically. At the British Broadcasting Corporation (BBC), the number of jobs dropped by 17% between 1987 and 1993, and at the ITV

stations, employment slumped 40% from 1988 to 1993 (Cornford & Robins, in press).

It is not only the scale, the "global reach," of capitalism that has changed during the transnational phase but also the modalities and rhythms of everyday life. The industrial centers have seen a decentralization of cultural collectivities, and as a result popular identification with cultural symbols has been fragmented. Compared with the pre-World War II and pre-transnational era, a multitude of lifestyles now engage people either by choice or by circumstance. Government commands far less respect for the majority of people (most do not vote in local and state elections and cannot even identify their congressional representative). Organized religion, the traditional nuclear family, stable neighborhoods and local business districts, identity by trade or occupation or company, family proprietorships, long-term single-institution employment, standard school curricula, the common reading experience, and respect for literature in general are among the social norms that no longer serve as the benchmarks in the lives of most Americans.

Subsidiary and overseas manufacturing and services and corporate conglomerates have broken down the identity of businesses. General Electric and its subsidiaries, for example, produce nuclear power plants, nuclear weapons, missile guidance systems, "Star Wars" electronics, radar systems, communication satellites, naval defense electronics, airplane engines, commercial electrical goods, consumer lighting equipment, electronics and appliances, and medical equipment; own television, radio, publishing, and advertising enterprises, cable programming, a transportation company, a Wall Street brokerage firm, a financial company with $80 billion in assets (one of the two "largest U.S. funding sources for industrial investments," the other being another electronics and media conglomerate, Westinghouse); plus have numerous other operations (Carnoy, 1993, p. 79). Moreover, some of these businesses are very fluid and can change hands from one year to the next. When they do, it means instant layoffs, loss of income, and disorientation for thousands of people.

Automation of the shop floor and the office, flexible and smaller-scale batch-production techniques, "just-in-time" manufacturing, "quality circle" work methods, and the mobility of corporations have largely broken the power of labor unions. Organized labor no longer has the means to broker business decisions and political agendas the way it did in the 1950s and 1960s. Corporate flexibility and mobility mean that most U.S. workers can expect to have to switch jobs often, relocate frequently during the course of their lives, and mortgage their futures with debt bondage to maintain the same material standards as their parents' generation.

From the usually conservative bourgeois (comfortably invested upper middle class) perspective, the breakdown of the traditional nuclear family structure represents moral decay. From the view of most working people coping with the regular uprooting and instability of everyday life, the avoidance of long-term family commitment represents rational choice. Cultural industries and their sponsors target their prospective audience/ consumer markets between the interstices of the employment and class structure and remake time around spectacles and light entertainment, such as the sports seasons, blockbuster movies, TV miniseries, sitcoms, soap operas and "specials," sensational news stories, talk shows, lifestyle fashions, and staged political dramas. The mass media devote little time to understanding the complexities of everyday social life; avoiding them is simply far too easy and too profitable.

MASS MEDIA ECONOMICS: THE SEGMENTED MARKET

In the post-World War II era, media and communications have been "demassified" into various and specialized, segmented user markets. These include cable TV, radio, broadcast TV, film, newspapers, magazines, Walkmans, camcorders, VCRs, word processors, computer software, handheld home and arcade video games, audiotape, compact disc, CB radio, shortwave and single sideband radio, fax, e-mail, electronic "chat groups," electronic bulletin boards, "virtual reality" networks, digital and long-distance and various other telephone services, teleconferencing, photocopying, airplane travel, voice mail, private automatic branch exchanges, answering machines, overnight post, mailing lists and junk mail, automated banking, videotex, teletext, and satellite (Robins & Webster, 1988).

Where once there were unified communication systems that catered to diverse sets of users, such as business executives and workers using the post office's first-class mailing system or middle and lower classes watching or listening to the same broadcast programs or different occupational groups reading the same newspapers, common standards provided the privileged and unprivileged classes relatively high-quality communication services. The coming of corporate-driven technological change and government deregulation enabled the privileged and business strata to segregate themselves in terms of access to communication services, with fewer trickle-down subsidies going to the working class.

The newer services cater to those with the ability to pay. What was once a commonly used public postal service has greatly declined, except for one of its

expensive spinoffs, overnight delivery. Sports, musical events, and movies carried on cable TV are not within financial reach of the poor, and neither are the prices charged in arenas and theaters. For the middle class, cable TV is the vicarious alternative to "being there." And because communication systems and information are so fragmented, no holistic picture of change that is communicated by the media is relevant to people's lives.

The mass media separate, isolate, and atomize individuals into receptacles for consumption. And where political protest creates new ideological communities, the media are there to domesticate the lexicon of dissent into the vernacular of advertising (the rap, the soul, the feminist, the gay and lesbian, the human rights, the peace, and so forth, images into messages of lifestyle and product association). Rebellion is appropriated and turned into a marketing campaign (e.g., the feminist movement transformed into advertising images of women obsessed with dieting, "dressed for success," sexually "liberated," or smoking Virginia Slims). Janis Joplin's lyrics of blues and class rebellion, "Oh, Lord, won't you buy me a Mercedes Benz," is converted by the blueblood car manufacturer, 25 years after the singer's tragic death, into a promotional appeal to the more affluent greyheads of the 1960s generation. The larger purpose of communication and information systems is not to sell social change but to absorb people's incomes and deliver them to advertisers.

Under transnational capitalism, the success of demobilizing the public into many publics in the leading industrial countries is reproduced in less developed and Third World countries. Thus, corporations need not worry about the saturation of markets in their countries of origin when they can expand into unchartered waters like the conquistadors of old. Unlike the early colonizers, TNCs do not have to arm themselves with Bibles and bullets in the Third World, but with automobile franchises, cellular telephones, and commercial high tech. *Free trade,* meaning no barriers to foreign investment, is the new catechism, and those who do not fully accept this Western religion may be visited by the U.S. military directly (Cuba, Vietnam, Guatemala, Grenada), by U.S.–armed proxies closer to the border (Iraq vs. Iran, Israel vs. Lebanon, South Africa vs. Angola, the Nicaraguan *Contras* vs. the Sandinistas), or by U.S.–equipped internal insurrectionists (the militaries in Chile, Indonesia, and Brazil; Afghan fundamentalists against the Soviet-backed Kabul government).

APPROPRIATING SCIENCE
AND TECHNOLOGY

Technology is the study and conversion of the natural, the built, or the conceptual world, leading to the creation or refinement of socially useful methods, tools, or products. It is the applied use of scientific principles. The deployment of a tree

branch for striking prey or for feeding fire are rudimentary forms of technology. The early domestication of cattle as draft animals and the development of the telescope for astronomical or navigational information or of programmed computer instructions to monitor space flight are later adaptations of scientific principles. Over the millennia, humans have applied their innate and expanding intelligence to gain control over the natural environment so as to alter the quality of their existence, for better and for worse. This evolution accelerated with the redefinition of land use as private property—the foundation of capitalism.

Technology is often thought of as simply equipment or technique, but this level of understanding misses its most important elements. Embedded within technology is a system of know-how that disposes of former knowledge and introduces a new set of rules and opportunities. Switching from an electric typewriter to a microcomputer means having to learn new skills and behavior. If a company has an electronic mail (e-mail) utility, for example, this might discourage the use of telephone voice exchanges and require learned methods for the sending of interoffice messages by computer. Automated automobile assembly moves many workers off the line and possibly into a white-collar job—or the unemployment office. Along with eliminating skilled labor, new technologies may be designed to reduce the power of labor unions and increase the control of company management. Electronic monitoring of telephone operators by their supervisors makes it difficult for them to carry on unauthorized communication with fellow workers or to take breaks from their work stations without supervisory approval.

The development of technology involves not only technical choices but also political and economic ones. Medical research incorporates decisions, for example, about what diseases to study; such studies require government, industrial, and private financial backing. This will involve, in turn, calculations of political and commercial returns and decisions about constituencies they are obligated to serve. Cutbacks in medical assistance for the poor in the 1990s, disproportionately affecting African Americans, single mothers, and Latinos, is one indication of those priorities. To use a different example, when chemists work on the production of color film, one implicit consideration is how well African American skin tones will show up. The choice of what kind of dyes to use in production of film was historically determined, in fact, by an interest in bringing out the best possible exposure of Caucasian skin tones. That was a political economic (and, in effect, racist) determination, not a technical one (Winston, 1995, pp. 72-73).

Regulation and control of the workplace, one of the resident features of many new technologies, relates to the corporation's larger interest in economic "efficiency" and profit. Efficiency, as it is often defined in business circles, does not have such a positive ring among non-executive ranks. Workers are far less likely to feel sanguine about the prospects of robots and automation in the workplace as substitutes for human labor. In the mid-1980s, a Louis Harris survey found that corporate executives are "more optimistic about factory automation than are the people who work in factories" by a difference of 39 percentage points, by 41% over skilled and unskilled labor in general, and by 54% measured against public perceptions (cited in Noble, 1986, p. 26). The continued depletion of better-paid skilled labor into the 1990s and manufacturing jobs in general could only have added to public skepticism about robotics.

The rapid changes in the organization and scale of capitalist enterprise brought a shift from the "backyard" craft quality of technical innovation typical of the early 19th century to a much more concentrated form of factory-based, professional, applied scientific research by the end of the century. Calling it the "wedding of science to the useful arts," David Noble found that the existing captains of industry sponsored the formation of technical education schools and the beginnings of the engineering and chemistry professions. The Rensselaer School, created in 1823 by a wealthy capitalist for the study of applied science, became the Rensselaer Polytechnic Institute in 1849. About the same time, Yale and Harvard started up chemistry and engineering education, and the Massachusetts Institute of Technology began in 1861. Scientific education was introduced at Dartmouth, Cornell, the University of Pennsylvania, the University of Michigan, and several other schools during this same period. The end product was the engineer-manager and the engineer-entrepreneur (Noble, 1979, pp. 22-23).

Engineer-managers next turned their attention to the social engineering of the factory worker. "Scientific management" specialists did "time and motion" studies of skilled workers as a way of mechanizing assembly and further fragmenting the labor force for productivity gains, offering individual piecework pay incentives for cooperation and dismissal or discipline for non-cooperation, and eventually redesigning heavy equipment to make the less predictable human skills redundant. The work of Frederick W. Taylor (*Principles of Scientific Management*), a metallurgical engineer who later made a career as an industrial consultant for machine tool and weapons manufacturers, is most closely associated with this approach to industrial reorganization (ca. 1900-1920). It was a period in which technology was applied toward making labor more of a commodity and workers into interchangeable parts. Taylor believed that the modern workshop should secure all knowledge aspects of production in the hands of management: "All possible brain work should be removed from the shop and concentrated in the planning or laying out department" (cited in Braverman, 1974, p. 113).

Worker resistance to Taylorism, especially in unionized shops and after a protest walkout in 1911 by molders at a Watertown, Massachusetts, foundry, forced industry to rethink some of its harsher methods of control. The greater the degree of centralized management, the greater the worker's sense of lost control, which often resulted in lack of initiative and low productivity. Noble sees this leading to the development of industrial management as an area of university study and as a more sophisticated form of control involving psychologists and sociologists who studied the "laws" of human behavior the way physicists studied the law of gravity. The goal was the achievement of labor peace through corporatist management of workers' emotional and social-psychological needs (Noble, 1979, pp. 272, 297n-298n, 315).

The industrial revolution of the late 19th century thereby represented a threshold in American capitalism and the organization of work. The craftsperson and crafts guilds were largely eliminated by corporate capitalism, and control over the pace and conditions of work was also lost to the skilled artisan. Before the end of the century, the capitalist and management were setting the rules of work under the roof of the factory, continually redesigning the labor process so as to reduce the objective basis of labor power and cause the "separation of the worker from his tools" (Christopher Lasch, cited in Noble, 1979, p. xi). Taylor saw scientific management as a way of rationalizing and thereby modernizing the production process by increasing output and reducing the training, cost, and importance of labor.

In his book *Shop Management,* Taylor said the full development of the system of production "will not have been realized until almost all of the machines in the shop are run by men who are of smaller caliber and attainments, and who are therefore cheaper than those required under the old system" (cited in Braverman, 1974, p. 118). The more recent rationalization of American labor in the service industries, contrary to claims in much of the trade literature, has not been to upgrade the knowledge base of workers, but rather to further extend control over the production process. As crafts were degraded in their human technical requirements, capitalist production largely reduced labor to an undifferentiated element "adaptable to a large range of simple tasks, while as science grew, it would be concentrated in the hands of management" (Braverman, 1974, p. 121).

The modern history of technology is tied to the quest for markets, market power, and the ascendance of the capitalist over the means (material elements) of production that went hand in hand with the struggle for political control of the state and the community. As C. S. Lewis commented, "Man's power over Nature often turns out to be a power exerted by some men over other men with Nature as its instrument" (cited in Pacey, 1991, p. 74). Confident of their collective physical grip of the national economic infrastructure, industrialists in the 1920s turned to the commodification of popular culture as a way of transforming a society of great regional and social differences into a market of mass production and mass consumer consciousness (Ewen, 1976). The means of gathering information about consumers and about workers have allowed the modern corporation to essentially bypass supply and demand mechanisms that Adam Smith had seen as the counterbalancing principles within capitalism.

APPROPRIATING DISCOURSE:
TECHNOLOGICAL DETERMINISM
AND THE DISAPPEARING SUBJECT

For authority to be successfully exercised, power holders must on some level withdraw themselves from the light of public scrutiny, create some myth or ideology that shields and deflects public attention from their real agenda. The myth/ideology may be secular, such as "rule by the people" or a "dictatorship of the proletariat," or it may be religious, such as "divine right" of monarchs or "God's chosen people." Language is, after all, the spoken aspect of consciousness, what Marx and Engels called "practical consciousness" (Marx & Engels, 1970, p. 51). In the struggle for power, the defenders of the existing division of labor in society, as a practical matter of managing consciousness, are required to explain the world in language that makes the present arrangements seem normal, natural, and necessary.

Where people do not directly rule themselves—and it is safe to say there is not one nation state where they do—some form of "representational" government is standing in for them, a reality of great distaste to radically democratic philosophers, such as Jean Jacques Rousseau, Robert Owen, or Karl Marx. Elitists, autocrats, and anti-democrats, such as Machiavelli, Hobbes, Pareto, Mosca, and Burke, all recognized the need for rulers to make certain concessions to democracy, if only symbolic, without surrendering the many special privileges that belong to the ruling class. Fascists such as Mussolini and Hitler fostered the idea of submission of one's individual identity to a higher loyalty to the state and its leader. For each of these thinkers, the political uses of language are critical in maintaining the stability of the state and society.

To retain state power, it is necessary but not sufficient to control the various factors of production (e.g., factories, technology, land, labor recruitment); the *ideology* of production must be controlled as well. By this I mean that, in the absence of coercion, the various worker and professional groups will cooperate in their respective roles in the production process only as long as their belief in its legitimacy remains intact. To the wielders of power, discourse is not just talk but also the struggle for ideas and ideology. As apparatuses of control, communication technologies are infused with dual functions: as instruments of production and as tools for managing hearts and minds.

The overthrow of the communist parties in the Soviet Union and eastern Europe in the early 1990s showed how important ideological legitimacy is to the preservation of the means of production (in those cases, state control and ownership or state socialism). Within the state socialist countries, the official language of "scientific socialism" and the representation of the Communist Party as "the vanguard of the working class" were out of synch with the popular perception of reality. Even the ancient Roman rulers understood that public acquiescence in their control, and not force alone, was necessary to keep occupied peoples from rebelling against established authority. By the early 1990s, with hardly a shot, the war of ideas proved decisive, and the Western capitalist states claimed the spoils—the triumph, if only momentary, of market ideology over the command (state-centered) economies of the East.

The inadequate standards of state socialism (e.g., the high levels of theft and corruption; the breakdown of the official planning, circulation, and distribution systems; the poor quality of many consumer goods and services) gave rise to an underground language and ideology of capitalist and libertarian counterculture, which, coinciding with the reformist (less coercive) political practices of the Gorbachev era, undermined the Soviet system and Communist Party control. Various conspiracy theories have been proposed about what took place, and certainly the relentless Cold War initiatives of the United States contributed to the demise of the Soviet Union, but ultimately the breakdown of the economy *and* the ideological consensus determined the outcome. Given the instability of present-day Russian capitalism and its confusing quasi-democratic ideology, an early reversion to some form of socialist economy is also conceivable, if not likely.

Within the capitalist states, an ideological theme and language that gained currency in the 1980s is *technological determinism.* This is a construction of reality based on the idea that world history is the unfolding of a progression of technological achievements that ultimately "transcend political boundaries, language, religion, and local tradition" (Boorstin, 1978, p. xiii). In their use of language, technological determinists (technophiliacs) treat technology as the initiator of events, as the grammatical subject of action, as the driving force and agency of change. Hence, the American historian Daniel Boorstin (1978) writes: "While the Machine made man feel himself master of his world, it also changed the feeling of the world that he had mastered. The Machine was a homogenizing device. The Machine tended to make everything—products, times, places, people—more alike" (p. 4). And again addressing the cultural homogenization theme with respect to communication technology, he says: "The telegraph and the power press and the mass-circulating newspaper brought the same information and the same images to people thousands of miles apart" (p. 44).

Daniel Bell (1979), a Harvard sociologist, also treats technology as the source of social change and of what he calls the "post-industrial society," although he makes his argument in a more circumspect manner. For example, he writes: "Every society is tied together by three different kinds of infrastructure—transportation, energy grids and communications. . . . The industrial heartland of the United States, for example, was created by the interplay of resources and water transport" (pp. 21-22). He also argues that "new communications enlarge the arenas in which social action takes place" (p. 36). The use of the passive voice in his first two sentences is a linguistic mechanism for ignoring human agency, power configurations, or organized interests and implying that technology is the active force.

In the third sentence construction, he *anthropomorphizes* the subject "communications" (the technology of communication) by attributing animate capabilities ("enlarge the arenas") to inanimate instruments ("communications"). Does inanimate technology have the capability of "enlarging," or is that another way of evading the real actors—the corporate communication industries? Writing for the conservative *Harvard Business Review* in this case, it is understandable that he might not want to offend his sponsors (Bell, 1979, pp. 22, 36).

When the actor is ignored, the emphasis is placed on action—what happened—rather than on citing the parties responsible for the action, thereby concealing a potential controversy. For example, to say "John Doe was injured in a car accident" does not tell us who was responsible. It may have been the "respectable" and married town mayor, distracted while at the wheel by a passionate secret lover, who forced Doe's car off the road; or it may have been John Doe himself, a powerful political figure in his own right, a judge perhaps, driving while intoxicated. Human interests employ language for politically instrumental ends.

The passive voice construction "In the future, all homes will be equipped with interactive cable TV-telephones" does not identify the presumed actors, the *who* (e.g., monopoly telephone service companies merged with regional cable operators). People might be more skeptical if it did. It should not be hard to imagine how and why certain speakers avoid political and power issues through the use of the passive voice. Intentionally hiding the actor is itself a political act, of course. We can conclude, therefore, that there is no avoiding the fact that all language uses are, on some level, political and can never be fully neutral. In the service of empowered interests, language functions to either legitimate or challenge the status quo.

It is important to make a distinction, however, between uses of language, on the one hand, in which the writer or speaker occasionally employs the passive voice or inanimate subjects for the purpose of adding color, using metaphorical expression, or merely varying sentence patterns and, on the other hand, those more consistent uses in which an ideological reconstruction of political reality is intended. In the first case, we are talking about ordinary literary and spoken conventions; in the second, we mean the use of language as an explanatory device in presenting the world as the mysterious outcome of inanimate forces. Although in this text I now and then resort to those very written conventions as a kind of literary license, my intention is to argue for a human-centered and politically focused, rather than a technology-centered and apolitical, paradigm.

Among the technological determinists, Bell is more attuned to the idea that technology has social outcomes, and he acknowledges that the consequences of communications technology include the dangerous possibilities of increased surveillance, record keeping, and secrecy. He prefers to de-emphasize these tendencies, however, and to avoid discussing in any detail the real-life situation of secret information gathering in the United States, which is, in fact, very well documented (see Chapter 5). In general, his arguments about the post-industrial society tend to treat politics and ideology as largely irrelevant in a computer- and information-based society ruled by the rationality of a technocratic elite (scientists, engineers, technicians, professional administrators, and leading academics). Bell has consistently argued against the idea that a social-economic elite governs the United States (see his 1960 work, *The End of Ideology*). But as one author observed, even Bell "has become silent about the affluence and leisure he once associated with postindustrialism" (Lyon, 1988, p. 8).

Anthropomorphizing Technology

Technological determinist language employs a confident anthropomorphizing of inanimate instruments, assigning to computers, robots, and machine technology in general the animate attributes of consciousness, will, creativity, and spontaneity—often portraying such instruments as being "in rebellion against the human community" (Winner, 1977, pp. 30-31). Boorstin (1978) extends the revolutionary image, claiming, "Technology invents needs and exports problems" (p. 8). This way of ascribing power to the machine gives the impression that technological developments unfold as inevitable progressions, driven by their own momentum, as if economic conquest and expansion and the ruthless appropriations of property and wealth are irrelevant considerations. The political value of such language allows the technocracy within business and government to present itself as "proceeding without passion or favoritism; and, above all, with the neutrality toward partisan interests that leaves it free to select and implement the best technological solution" (Gouldner, 1976, pp. 250-251).

Another true believer in "the coming information age" is Wilson Dizard (1982), who confidently observed:

> In the communications and information fields, *new technologies no longer develop in a linear fashion,* separated by decades, with enough time in between for their implications to be sorted out and for them to be brought into active use. Now *we are dealing with* a wide range of converging technologies, *which forces us to make immediate choices* and leaves considerably less margin for error. (p. 5; italics added)

Consider how this syntax structures a way of thinking about technology and also anthropomorphizes the inanimate ("new technologies no longer develop in a linear fashion"). It were as if there is a genealogy of technology constantly giving birth, so to speak, to technological offspring and, until the "information age," reproducing in a rational progression ("linear fashion"). Technology is in charge, leaving captive humans with only the "forced" and "immediate choices" of "dealing with" new technologies and with "less margin for error." Humans are not actors, but objects. Technology has no economic or political interest, only a way of governing that is politically neutral, though demanding and unstoppable.

Another, rather utopian, formulation proposes that digital technology has nullified the hierarchical, dominating, and machine-based discipline of the Fordist era and introduced a more knowledge-based, liberating alternative in which intellect, and not mechanical power and its masters, governs. Others see the information society as little more than an intensive version of Fordism. Are politics (the struggle for representation and power) and ideology (competing worldviews based on conflicting ideas and interests) obsolete because the information society is based on neutral, politically detached, and scientific solutions for a better world? Or does such thinking represent only a clever subterfuge for the political and economic status quo masquerading in a new jargon of technical, value-free rules and rationality?

The anthropomorphism of assigning *agency* to technology (e.g., "computers bring about changes in society"), in effect, rationalizes the existing organization of power by moving discussions to technical solutions to social problems, thereby insulating power elites from scrutiny. Indeed, technology is so ubiquitous in Western societies that it is not difficult to foster illusions of its neutrality and objectivity. When an ATM disburses money and prints out a balance statement, it almost appears that this electronic service is working for its depositors, rather than assisting bank owners and management with their investment and profit strategies (while continually raising the threshold of penalty-free depositing). The omission of politics and human interests in explaining the function of machines, an unfortunate starting point when children are first introduced to technology by their teachers and elders, continues into adulthood when various established power groups (e.g., CEOs, military officers, media executives) take over the inculcation of citizenship values.

In an earlier age, it was easier to plainly recognize the political and economic interests embedded in technology. When Thomas Jefferson expressed reservations about the industrial revolution ("Those who labor in the earth are the chosen people of God"), he and his followers understood that bankers and industrialists were bent on seizing control of the natural environment for commercial profit and, in the process, destroying the agrarian (farming and frontier) way of life. Eventually, the Jeffersonian ideal was overwhelmed by the momentum of industrialism. By the mid-19th century, against the wishes of some writers and poets of the day, including Thoreau, the railroad was belching its way across a subjugated continent. Native Americans, up to and beyond the resistance of Sitting Bull and Geronimo, also understood the conquering logic within the design of European American technology.

The domination of society by the empires of wealth in the contemporary era is more complex and diffused, compared with the 19th century. Urbanization, corporatism, the destruction of agrarian society, and the management of labor and the working class have increased social dependency and deeply internalized the American ethos of capitalism in the consciousness of its citizens. At the same time, many skeptics, dissenters, and public intellectuals (among the best known are Ralph Nader, Barry Commoner, and Noam Chomsky) are using their knowledge and understanding to challenge industry's self-serving deceptions about the glorious future of technology. One of the more discerning skeptics on the subject of electronic communications exhorts his readers to develop a critical understanding of technology:

> Techno-skeptics need to build their own repositories of knowledge about technology, not in the data banks of the information czars but in the minds and experiences of ordinary people. The single greatest discomfort to those who design, develop, and apply technology is caused by the demand of ordinary people to exercise the right to know. There is at present so little questioning of the direction technology is taking that the technocrats and their allies have been allowed to pursue their objectives amid silent acquiescence. (Reinecke, 1984, p. 243)

Assigning creative social and historical agency to technology is an example of what in marxian analysis is called a "commodity fetish." Armand Mattelart (1978) appended the idea that:

> [The] communication fetish hides the repressive and manipulative character of the dominant technological power of the diffusion of information (a veritable new *productive force*) and presents it to those dominated by it as a force of liberation and good will. (p. 14)

The grammar of technological determinism hides the human agents precisely to disguise the political economic and repressive aspects and identities of empowered institutions and interests acting through their technocratic instruments. Albert Einstein, speaking on the social core of technology, commented that "science in the making, science as an end to be pursued, is as subjective and psychologically conditioned as any other branch of human endeavor" (cited in Smythe, 1984, p. 205). A critical theory of communications insists that the analysis consider the "unequal distribution of control . . . and the wider pattern of inequality in the distribution of wealth and power . . . especially the class structure and the unequal exchange between advanced and developing nations." It must also "unmask how communication systems maintain, reproduce and continually legitimate the prevailing structure of advantage and inequality as natural and inevitable" as well as the "sources of social dissent and political struggle" against domination (Robert White, cited in Smythe, 1984, p. 211).

Repudiating technological determinist and apolitical readings of history, including the evolution of communication technology, Raymond Williams (1975) argued that it is important to understand the centrality of human intention in the processes of research and development (p. 14). The crosscurrents of various technological developments, such as in electricity, engineering, and applied technology industries, were associated in the public mind with the momentum of industrialization in the 19th century. Industrialism itself was the outgrowth of private property laws and the aggressive accumulation of wealth under that stage of capitalism. Whatever doubt there may have been about the economic direction of the United States before the 1860s was settled in the war between the largely industrial and the largely agricultural regions of the country.

Although telegraphy was technically possible earlier than the 1830s, the spread of the railroad system, its freight, and passenger transfer made its development a practical necessity. The two technologies of long-distance transportation and communication contributed decisively to the outcome of the Civil War, as they did to the delayed industrial revolution that followed. Economic development itself has no purpose other than through the generation of "need," and the determinants of need in the case of communication devices like the telegraph, the telephone, and broadcasting, as we shall further discuss in Chapters 3 and 4, have been initiated primarily by the community of commercial and state interests and stakeholders, which organizes and furnishes the necessary resources and official permissions (Williams, 1975, pp. 18-19).

The twin forces of territorial and commercial conquest worked hand in hand to expand the frontiers of control: promotion of print media to announce everyday events and their meanings in the early centuries of capitalist society, wired electrical communication in the national-scale industrial phase, and wireless electronic communication in the transnational phase. In fascist countries, communication devices are defended as instruments of direct social control; in communist party-run states, the media are for political education in line with the party's agenda for the resocialization of its citizens; in liberal democratic societies, media controls are more subtle, less mobilized, and less direct, although no less pervasive. And although media and communication technologies (or any other technologies) may not be so explicit in their social control purposes in liberal democratic societies like the United States, they also do not have an impartial character in the practical sense of being accessible or adaptable for any purpose, person, or group. Although digital technologies are more flexible in certain ways than their mechanical predecessors, they are also circumscribed by political and economic interests that have fostered and imprinted particular designs and objectives, countercultural applications notwithstanding.

Although the social and commercial potential of broadcasting was not fully recognized in the earliest years of its existence, the mass scale of the audience and point-to-multipoint system of transmission soon brought out its broad-ranging and unifying political and economic utilities. Advertisers, politicians, ideologues, and entertainers by the 1920s and 1930s could reach thousands or even millions of radio listeners without having to face them. Heady realizations of this kind of reach gave rise to formula messages and program content packaged with "mass appeal"—amusing and interesting on some basic level to many; offensive, incomprehensible, and occasionally intellectually challenging to others. As the history of the medium would show, the design of radio content in the United States and elsewhere made it a powerful instrument of discourse and ideology and an important form of political and commercial capital.

Social and Cultural Mediation of Change

Although in its finer utterances American democracy promised elected government "of the people, by the people, and for the people," it never assured that economic life would be so egalitarian. Citizens do not go to the polls to vote on those economic policies that most affect them or to elect those who wield the most economic power, such as the executives of modern-day TNCs. Yet, political independence is intimately intertwined with economics, technology, and culture. The artifacts of technology take shape within the existing political economy, in which their purpose, design, and development are authorized by a small group of executives. Ordinary people come to define their lives in terms of these technologies (e.g., office buildings, the highway system, neighborhood design, prisons, mass communication structures) that "reflect who we are, what we aspire to be," and who both use and are dependent on them (Winner, 1993, p. 284).

In the last years of the 20th century, the leading industrial societies, but particularly the United States, came under the influence of a more strident business culture that gave a market definition to almost everything. Few institutions in public and private life escaped the purview of market enthusiasts and entrepreneurs in search of new business opportunities, in the process exposing material ambitions almost unthinkable a decade or two earlier: the flamboyant celebration of wealth and fame, the junk mail bonanza, the smashing of labor unions, massive layoffs, the flight of TNCs to cheap labor havens, attacks on public education, tabloid journalism, and voyeuristic talk shows. The new cultural diet contained massive doses of real and simulated television violence, wholesale attacks on and erosion of government services and democratic institutions by leading media pundits, a contemptuous indifference to the poor, cynical commercialization of television aimed at children, advertising styles capitalizing on and penetrating public space (e.g., works of art, elections, political movements, sports stadiums and events), the professionalization of amateur athletics, the cult of entertainment, and other profitable pursuits.

To be sure, there have been countervailing currents. Social activism in such areas as ecology, the natural environment, health, and racial, ethnic, gender, and human rights awareness have altered the thinking of millions of Americans and politicians. However, the difficulties of mobilizing such diverse though shared progressive interests, on the one hand, and the extremely well-financed organization of various conservative and right-wing groups opposed to social change, on the other, have led to enormous confusion and polarization in the way issues are publicly portrayed in the media and other forums. Most people apparently do not recognize the inconsistency of favoring protection of the environment while coveting carefree consumption, demanding high levels of moral character in government officials but tolerating avaricious behavior within business enterprises, or advocating the theory of democracy but withholding criticism when the government supports ruthless oppression among its right-wing foreign allies. Are people to blame for their lack of education, or do the government, business, and the media, for reasons of self-interest or self-defense, use their vast arsenal of resources to masterfully organize disinformation and ignorance?

Packaging the "Information Society"

Language, written and spoken (or signed), does not simply convey *objective* statements of fact or even expressions of intent. Language also transmits various semantic codes that have underlying narratives, stories, and ways of seeing the world. One's social and cultural experiences also shape the casual, the formal, and the instrumental uses of language, whether it be in addressing a waiter, a lover, or a pope. Clearly, language is used to establish or reinforce various representations of power, and those aspects of language are of interest in understanding politics.

Within contemporary transnational capitalism, few stones are left unturned in its search for new space and time to occupy. Beyond the control of resources, capitalists must be able to create legitimacy for themselves in their role of resource developers, and this requires the capture of public discourse. When the captains of industry are not capable of responding to people's real needs, it becomes necessary for them to change the popular perception and language of "need." Capitalism employs and rewards those who defend or do not challenge its governing ideology and ignores or marginalizes those who are critical and oppositional. And although many are dissatisfied with its reward and resource development system, the sedative of passive consumption seems to dull the inclination to rebel.

At a moment of declining consumption and deterioration of public life, the ideological offensive of American capitalism has targeted the public sector (condemning "tax and spend" politicians) while unabashedly celebrating privilege, social Darwinism, and business culture (yuppie lifestyles, racism, nationalism, tougher anti-crime measures, deregulation, corporate mergers, union busting, and hostility toward the homeless and Third World immigrants). The maintenance of power requires intensified forms of indoctrination to be internalized as public beliefs, especially through the mass media. Herbert Marcuse (1964), a critical philosopher of the Frankfurt School, ruefully conjectured:

> The mere absence of all advertising and of all indoctrinating media of information and entertainment would plunge the individual into a traumatic void where he would have the chance to wonder and to think, to know himself (or rather the negative of himself) and his society. Deprived of his false fathers, leaders, friends, and representatives, he would have to learn his ABC's again. (pp. 245-246)

"Post-Industrialism" as Ideology

"Post-industrialism" is another technological determinist construct depicting history as a progression of technological innovations that continually change the way people live. Human behavior is organized under the compelling requirements of machines. According to the economist and business writer Peter Drucker (1966), irrigation technology was the root of human civilization. "The irrigation city first established government as a distinct and permanent institution. It established an impersonal government with a clear hierarchical structure in which very soon there arose a genuine bureaucracy." It "developed a standing army," "first conceived of man as a citizen," and, in effect, "created the individual" (pp. 143-146).

To another noted academic, the printing press brought about modern democracy (Pool, 1983b)—not, as others presume, the Puritans and Presbyterians, the new commercial classes, artisans, gentry, yeomen farmers, and laborers who forced revolutionary parliamentary changes on an intractable English monarchy and aristocracy during and after their Civil War. Moreover, Pool overlooked the fact that printing was first developed, not in 15th-century Europe, but in 11th-century China, and that it certainly did not propel a social revolution against the ruling mandarins.

The communications theory of Marshall McLuhan (1964), full of exaggerated claims about the determining nature of technology, lacks almost any political dimensionality. He ascribed great powers to electronic communications, especially television, to mesmerize people into tribal assemblies, linking them to a "global village." McLuhan's formulations lack human accountability, just as other cult treatments of technology have assumed that machine tools relocated people to industrial centers, that the automobile created suburbia, and that the computer established service industries based on knowledge rather than on mechanical skills, thereby making factories and mechanical labor anachronistic: hence, post-industrialism.

In each case, technology, without conscious or purposeful intent, is seen as establishing the range of possibilities in people's lives. In the 19th and early 20th centuries, there was a cultivated belief that railroads would eliminate European wars, that steam power would eliminate the need for human labor, and that the radio would upgrade the level of culture (Mulgan, 1991, p. 23). Similarly, in the late 20th century, communication technologies are said to provide the means by which industrial society becomes obsolete in favor of one based on the production and distribution of information and knowledge. Toffler (1980) and others believed that computers would create the interactive world of the "electronic cottage," a decentralized system of economic life in which work was returned to the household, not unlike American society before the factory system.

One popular illusion is that computers with large-scale information processing capacities get to do the deciding, that institutional decision making can be accomplished by the complex, "non-partisan," "objective," multivariate logic of "thinking machines." Self-interested motives of human decision makers are substituted by the computer's cool calculations for maximizing efficiency. The volume of information replaces the centrality of money in creating "knowledge capital" and a new, decentralized basis of power.

Herbert Simon, who has long been a leading proponent of artificial intelligence (see Chapter 5), claimed in 1957 that "there are now in the world machines that think, that learn and that create. Moreover, their ability to do these things is going to increase rapidly until—in a visible future—the range of problems they can handle will be coextensive with the range to which the human mind has been applied" (cited in Dreyfus, 1992, p. 81). Critics, such as Joseph Weizenbaum, in contrast, reject the implication that the nature of the "problems" that computers can "solve" is akin to the reasoning capabilities of humans. Beyond that, Weizenbaum (1981b) criticizes such habits of speech:

> It surely reflects a habit of thought, that makes instruments responsible for events, leads directly to speaking and thinking of science and technology as autonomous forces and to the idea of technological inevitability. It leads finally to the proposition that man is, after all, impotent to struggle with powerful impersonal agencies of his own making over which he has lost control, and that he is therefore justified in abdicating his responsibility for the consequences of his acts. (p. 435)

The technological determinist's rendition of history has a superficial reasonableness and appeal, especially when it suggests a future without the drudgery of factory work and its associated environmental and health hazards. Large smokestack industries do seem to be less visible along America's urban skylines, often having given way to service-type employment. And certain job classification measures back up this observation, with services (including information-based work) constituting more than 70% of employment. John Naisbitt, author of *Megatrends* (1984), predicted a future in which "innovations in communications and computer technology will accelerate the pace of change by collapsing the *information float*" and "will succeed or fail according to the principle of high tech/high touch" (p. 19). Although buoyant about the future of digital technologies, Naisbitt at least recognizes that one result of the microprocessor is the potential for increased layoffs in sections of the economy where workers have developed a livable wage and that unions, as a result, "have been forced to fight automation in both industry and agriculture" (p. 28).

The major shortcoming of this kind of reasoning is, not the probability of specific predictions about technology actually occurring, but the ideological assumptions underlying such expectations. It would be just as reasonable to argue that we are living, not in a post-industrial age in which technology is in control, but in a new stage of capitalist industrialism, a super- or hyper-industrialism, a stage in which considerably more emphasis has been given to managing consumption. One writer has argued,

> What we are witnessing is not a transition towards a post-industrial society, where information would take the place of energy and raw materials. We are witnessing the industrialization of data, information, knowledge, even wisdom, in a process which tries to apply to this field of human activity the basic principles of industrial society: standardization, mass production, maximalization of output, synchronization of activities, concentration, centralization, etc. (Blanc, 1985, p. 78)

It is the age of what Enzensberger (1974) called "the industrialization of consciousness." People become undifferentiated from products and are standardized by the advertising industry into interchangeable units of consumption and are studied and segmented into computerized consumer profile samples according to "values, attitudes, and lifestyles" ("psychographics"). In this way, goods and services can be packaged, marketed, and sold in formats either consciously or subliminally appealing to target audiences. In Enzensberger's view, the production of information is not an end in itself, but rather subordinate to the production of salable commodities.

Another problem with the "post-industrial" thesis is that it universalizes the American or the Western experience. In actuality, many of the manufacturing industries have not disappeared but, still under North American, Japanese, or western European control, have simply been relocated to select Third World countries. Under this globalized production system, much of the work simply has been moved "offshore," remaining tied to the core capitalist economies. A few Third World countries are beginning to undergo extensive industrialization, but the majority are still largely pre-industrial and more economically marginalized than ever—hardly a case for a post-industrial world.

Moreover, even within the United States, the small percentage of "agricultural" employment, officially calculated as less than 3% of the workforce, greatly understates how important agricultural goods, compared with the "service" and "information" sectors, are to the economy. Apart from the fact that agriculture is the largest export sector, consider how many service and manufacturing areas are dependent on farm products. Here is a partial list: railroading, trucking, air carriage and shipping of farm goods, supermarkets, food and produce retailers and distributors, the alcohol industry, restaurants, bakeries, and fast-food places, advertisers of agricultural products, food processing plants, the packaging industries, clothing and tobacco industries, the underground drug trade, pharmaceuticals, the agriculture-based government sector, agricultural research centers and university departments, farm services (e.g., legal, medical, mechanical, finance, veterinary), the farm machinery and equipment industries, sections of the petroleum industry (e.g., pesticides, herbicides, fungicides), fuels, lubricants, industrial vegetable oils, paints, soybean-based inks, plastics, coatings, fabric softeners, hair rinses and conditioners, cosmetics, and other product cycle materials.

NEO-INDUSTRIALISM:
THE COMMODIFICATION OF EVERYTHING

Capitalism as a political economic system seems to be endlessly adaptable to opportunities for exploitation of the natural environment. In its earliest global form in the 1500s, capitalism involved a set of social and legal practices that freed up the use of land and water for greater commercial activity, most actively at the time in Britain and Holland. Entrepreneurs, encouraged and supported by state subsidies and the weight of official legal and property protection, embarked on new forms of commercial ventures, including exploration of the New World and areas of the globe now called the Third World. As new sources of wealth in the form of scarce minerals and agricultural goods became available, land progressively took on the shape of a capital commodity, something to be exchanged for profit; that is, land came to be seen and used not simply as a basic natural source of survival but as a central form of wealth that could be rented, sold, mined, or cultivated with human instruments for the production of marketable commodities. These market commodities could be directly exchanged, in turn, for currency, goods, gold, or silver and indirectly for luxuries, personal conveniences, or other forms of capital.

With the rise of the factory system in the 19th century, land assumed subsidiary importance in the production cycle as the source of raw materials (e.g., cotton, iron, silver, copper, trees) in support of new urban industrial, commercial, and residential centers. Steam power, later electricity, and the new machines of mass production were harnessed to turn out standardized and more economically efficient goods that no longer bore the signature of skilled workers, people who had been regarded as respectable artisans in the early part of that century. Technology was becoming more important relative to the talent, status, and reward structure for skilled labor. The craftsperson working with small-scale technology in her or his own home or village was a thing of the past. The most serious, though not the only, challenge to the rise of the factory system and the introduction of labor-saving technologies was made by the English Luddites, craftspeople despairing of their losses in pay, status, and gainful employment with the rise of the machine.

With the coming of the factory system, conditions of work were set by capitalists and managers who directly oversaw the production of goods, often demanding 12-hour days, child labor, and machinery that frequently caused workers bodily injuries and even death. The goods produced by worker and machine were generally of little use or otherwise unavailable to the workers themselves and were destined for remote, not local, consumption. This new relationship broke contact between the direct producer and the buyer. Luddite rebellions in early 19th-century England, especially their assaults on machines, which in the pre-union years represented a form of "collective bargaining by riot" (Slack, 1984, p. 42), have usually been historically misrepresented as phobic reactions to new technology. A more careful reading suggests an anxiety borne of technology-induced permanent unemployment at a time of severe economic conditions in England. Hence, the modern-day opprobrium of *Luddite* directed toward critics of technology policy is founded on an original misreading of the machine-busting worker revolt (Watson, 1993).

The inherent aspects of manufactured commodities indeed were not very well understood until Karl Marx (1818-1883), the author of *The Communist Manifesto, Das Kapital* (*Capital*), and other well-known discourses, applied thorough sociological and political economic analyses in interpreting the characteristics and evolution of capitalism. In his dissection, Marx tried to show how political power and the accumulation of wealth were historically intertwined and how capitalism represented a structure of social classes functionally and culturally linked to the system of production. (See Marx's own works or, for example, Robert C. Tucker's, 1972, edited work on Marxism, *The Marx-Engels Reader.*)

People's social status, wages and salaries, experiences and opportunities in life, everyday lifestyles, ideology, and inherited and acquired values had a common reference point—namely, where they fit into the class-based division of labor, whether beggar (lumpen proletariat), laborer (proletariat), bureaucrat, professional or merchant (petit bourgeoisie), or an owner of production (bourgeoisie or capitalist). Marx did not and could not predict all that transpired after his life, and many do not accept the validity of some of his most basic arguments. Nonetheless, Marx had probably the most profound influence on social science of any modern intellectual throughout the world. When one thinks of political economy as an approach to understanding the contemporary world, Marx is the most frequently cited authority. In Chapter 2, we look more fully at the political economic foundations of communication and information.

CHAPTER 2

A Political Economy
of Communications

MONEY, ETHICS, AND THE
INFORMATION SUPERHIGHWAY

Commercial media, like other private industries, are forever in search of a larger market share. Compelled by the logic of ratings and dependence on commercial sponsors, the mass media have little interest in pursuing unpopular ideas, higher educational standards, or content that discords with consumption values. The idea of allocating a broadcast spectrum on behalf of public interests and backed by government enforcement on behalf of those objectives has never had much official support in the United States and much less so in the past 15 years. Conservative economists (e.g., Milton Friedman) have pushed back Keynesian liberalism in arguing that the private market is the best guarantor of personal choice and freedom and that the business of business is to pursue profits for stockholders, not the social well-being of citizens.

Yet, the knowledge explosion that was promised by a private-sector-driven "information revolution" has yet to lift American society above a series of social crises. Information technology does not appear to be a panacea for poverty, gang violence, growing prison populations, unemployment, drug addiction, family deterioration, unaffordable housing and medical care, racism and ethnic ten-

sions, wars, urban decay, educational disintegration, and psychological alienation. Although digital communications have developed in some remarkably
forward-looking ways, such as in space exploration, medical science, and
community-based and personal information sharing networks, their more common usage has likely solidified the concentration of transnational elite wealth
and power at the expense of the majority. U.S.-based transnational corporations
are by far the largest users of digital communications, and their CEOs are the
most enthusiastic backers of the Clinton administration's program for an "information superhighway."

If the past is prologue, one might recall that when the *Titanic* was receiving
transmissions about icebergs in its vicinity, the ship's radio operators were too
busy sending private messages for the wealthiest passengers onboard to respond
to the warnings (*Mercury News* Wire Service, 1993). The broadest use of
information technology is in the service of private banking and other financial
institutions, which transmit about 80% of all international data traffic. These
organizations are less interested in elevating the knowledge of citizens than in
building empires of prodigious financial investment. Computerized networks
provide instantaneous links between offices and currency exchanges around the
world (more than $1 trillion a day), and for pennies per telecommunications use
charge, financial trades worth millions or even billions of dollars can be finessed
in a matter of seconds. Transborder data access is one core technology in the
employ of the new class of global economic actors who are consolidating market
domination on a world scale.

Television has demonstrated perhaps the most dazzling public displays of
new communication gadgetry and techniques, yet to many media critics, ranging
from the political left to the right, the general quality of American television has
been in decline. "Commercial broadcasting," writes Edward Herman (1990),
"in fact, offers a model case of 'market failure' in both theory and practice."
Although he recognizes a positive potential of television "externalities"
(spinoffs not factored into the original cost of the product), the downside he sees
includes "the exploitation of sex and violence to build audiences . . . [as] a likely
(and observable) consequence of commercialization." Herman sees the industry
trend increasingly toward ratings-driven escapist entertainment and away from
public service programming (p. 63). With far more entrants in the broadcast
market during the last 20 years, the pressure to preserve or create audience share
has intensified, resulting in increasingly sensational and tabloid-style programming, catering to the lowest common denominator of taste, intelligence, and
interest.

The Public Broadcasting System (PBS), established by the federal government as part of the Corporation for Public Broadcasting in 1967, was supposed
to give Americans the option of more intelligent and non-commercial program-

ming. On the whole, PBS offers greater dimensionality and more educational programming than other broadcast and cable TV, but with few exceptions, public television has shown a lack of spunk and permitted increasing commercial control and compromises in its intellectual independence. Moreover, many audiences are poorly served, especially peoples of Third World ancestry, leftist political organizations and thinkers, radical feminists, gays and lesbians, trade unionists, and working-class citizens. It had been argued that the low level of government support, about 16% of its budget (Britain's BBC and Canada's CBC, respectively, receive 36 and 30 times as much government support per capita), increases its dependence on private corporate capital and lack of intellectual independence, although government financing certainly never ensures a hands-off policy in broadcasting. The Carnegie Commission on Educational Television, which wrote the original plan for public television, proposed a tax on television sets and fees for commercial uses of the medium, but this approach was rejected in subsequent congressional legislation.

The real-life struggles of ordinary working people, generally ignored or ridiculed by commercial television, as in the Archie Bunker and Homer Simpson stereotypes (like the *Life of Reilly* and *Honeymooners* working-class buffoon images before them), is given only slightly better treatment on PBS. A study done by the City University of New York in 1988-89 found that less than 0.5% of prime-time television addressed the lives and concerns of workers as workers, of which more than two-thirds focused on British workers, and that a mere 20 minutes per month dealt with American workers (cited in Cohen & Solomon, 1992, p. 19). PBS does a better job of catering to elites, not well served by network television, with "high brow" entertainment (e.g., *Masterpiece Theater, Great Performances*).

Conservatives have long complained that PBS is a bastion of liberalism and beyond mainstream political culture. The weekly program *Frontline* occasionally does stories critical of U.S. government support for right-wing regimes and of big business corruptions, but more often it deals with national issues non-threatening to establishment interests. Conservatives seem to locate bias only when liberal points of view are presented but ignore the strident right-wing programs that PBS has aired, including those produced by the Reverend Sun Myung Moon's political organization, CAUSA. The bias of television is found not only in the slant in programs but more essentially in its sponsorship, and conservatives find it easier to secure sponsors and underwriters than do producers, writers, and directors critical of the political economic system. One regular basher of the so-called liberal bias in media is former Vice President Dan Quayle, who himself owns nearly $500 million of stock in Central Newspapers, Inc., and is one of its board members (Cohen & Solomon, 1992, p. 19; 1993, pp. 5, 7).

The conservative lineup on PBS includes *Firing Line, McLaughlin Group,* and *One on One,* all hosted by editors of the right-wing *National Review* magazine. A conservative, Morton Kondracke, hosts a program on foreign affairs. An African American Republican hosts a program targeted to blacks, *Tony Brown's Journal.* On the economy, PBS airs *Nightly Business Report,* representing Wall Street and corporate interests. Conservative business views are also reflected in *Adam Smith's Money World* and *Wall Street Week.* No regular program on PBS is hosted by an advocate of the left (Cohen & Solomon, 1992, p. 19).

Media critics in the left monthly *Extra!* published by Fairness and Accuracy in Reporting (FAIR) have documented a heavily white, male, elite, official, conservative bias in the guest appearances on the nightly *MacNeil/Lehrer News Hour* (and found similar results in a separate study on the ABC nightly news/ interview program *Nightline*). FAIR argues that *MacNeil/Lehrer* (and other PBS programs) defer to their sponsors' sensibilities, and in the case of the nightly news show, these include such giant corporations as American Telephone and Telegraph (AT&T) and Pepsico (whose $12 million sponsorship amounted to half of the program's budget; Cohen & Solomon, 1992, p. 19). In 1995, a subsidiary of the private cable conglomerate Tele-Communication Inc. (TCI),

Liberty Media, got two thirds ownership of *MacNeil/Lehrer* (Tolan, 1996, p. A11). They who pay the piper call the tune.

As for network news, with rare exceptions, labor is featured only during strikes, with emphasis usually given to the disruption and inconvenience issues, whereas millionaire anchors are socializing in the playgrounds of corporate CEOs. When airline flight attendants walk out, television news typically interviews only people stranded in airports, and when supermarket employees strike for minimal control over their hours, stations refuse to tell their stories. The devastating impact of strikes on striking workers themselves and their families is ignored. And although working-class crime gets a lot of attention, there is little trashing of big corporations, the affluent, the poor quality of most commercial goods, or the health effects of industrial pollution. Television is itself, for the most part, a junk food cafeteria, its formats and story lines monotonously reduplicated and driven by market surveys and supervised by advertisers. Television represents the best opportunity for big-ticket item sales, for delivering audience shares to advertisers, and for prolonging commodity shelf lives until the next marketing trend comes along. Pressured by commercial competition in recent years and the struggle to maintain ratings, American television has less self-consciously employed more tabloid formats and content.

American television also has a strong ethnocentric aspect. Among industrial countries, the United States has the lowest percentage of imported television; what little of it is aired, mainly on PBS, is mostly British. Of the British imports, the longest-running series, *Masterpiece Theater,* generally caters to conservative tastes. One might also read into PBS wilderness programs, such as *Nature,* which emphasize biologically determined animal aggression and survival-of-the-fittest themes, as essentially a conservative view of ecology and, implicitly, of human relations. The small percentage of left-oriented programming is mainly produced by stations in New York (WNET), Boston (WGBH), and San Francisco (KQED), three cities that are typically at the forefront of liberal and left political causes and movements. Yet, despite the right-to-centrist bent of most PBS programming, many outspoken conservatives insist that public television is permeated by a "dangerous" left-wing bias.

Right-wing journalists such as Pat Buchanan have attacked both the National Endowment for the Arts for its $5,000 grant and PBS for their roles in airing an episode of the Emmy Award-winning series *P.O.V.* (Point of View). The controversial program *Tongues Untied* was a documentary about the impact of AIDS on gay African Americans and included explicit sexual language. The decision to carry the program was taken by each affiliate independently, and some 100 stations that usually carry the series chose not to air the particular program. Nonetheless, some Senate Republicans reacted to PBS's rare act of non-conformity by threatening to end federal funding for the network (Editorial, 1992, p. A22).

Robert Knight, director of the Cultural Studies Project at the conservative Family Research Council and who spoke at Republican–dominated House hearings on PBS funding in 1995, has argued that the network "favors alternative lifestyles and is usually a mirror of the liberal democratic agenda" (Hartigan, 1992, pp. 1, 14).

PANOPTICAL AND
TOTALITARIAN POTENTIAL

The early 19th-century English utilitarian philosopher and economist Jeremy Bentham once described his vision of an optimum security community, an asylum, prison, school, or factory as a circular design, whereby a centrally placed surveillance telescope, called a *panopticon,* could allow its observer to view the behavior of every resident without permitting visual intercourse by those observed. This idea represented that century's ultimate conception of state power and recommended to centralized authority a secular "architecture of control" over citizens—a blueprint of "seeing without being seen" (Robins & Webster, 1988, p. 57). Through much of the 1800s, France used optical telegraphs mounted on hills across the countryside for military and police reconnaissance. Centrist control of the Bentham sort later became the model for prisons and concentration camps. In the Orwellian version, the panopticon is the all-seeing eye of "Big Brother" that compels conformity and obedience to the state through endless observation of the masses (see also Foucault, 1979).

The panopticon can also be seen as a metaphor for contemporary state systems, which carry out various sorts of institutionalized surveillance and information gathering for the purpose of command, intimidation, and control, as well as for developing elaborate disinformation and jamming techniques against adversaries. With the tools now available for locating and scrutinizing personal behavior, one can only wonder what Hitler's gestapo could have done with similar instruments. The tradition of state spying certainly did not end with the fall of the Third Reich. Indeed, it has been argued that, in the contemporary United States, one lives not so much in an "information society" as in a "surveillance society" (Gandy, 1989).

As one writer on the subject sees it, "Modern surveillance technology is an integrated system of hardware and software that includes devices for sensing, measuring, storing, processing, and exchanging information and intelligence about the environment" (Gandy, 1989, p. 62). This is occurring at several levels simultaneously: in the areas of reading, listening, and recording of written and vocal communication; camera and scanner observation of personal activity and transactions and of the physical environment; even down to the scrutiny of one's biological and genetic essence. Unknown to most people in this age of tele-entertainment, two thirds of all launched communication satellites (more than 200 in orbit from 18 nations or communication agencies in 1990) have had military functions, and 40% of these systems have been used for photographic reconnaissance (Graham & Marvin, 1996, p. 21; Mulgan, 1991, pp. 36, 74). In a society penetrated by technologies of private and surreptitious surveillance, information can be processed, stored, retrieved, and integrated into profiles of citizens and resources and later used for the purpose of social control, political manipulation, economic exploitation, or repression.

Political surveillance in France, developed by Joseph Fouche during the reign of Napoleon, has a direct lineage to U.S. Attorney General Charles Joseph Bonaparte, who organized the Bureau of Investigation. J. Edgar Hoover, who later directed the renamed Federal Bureau of Investigation (FBI) for nearly 50 years, kept secret dossiers on congressional members and their staffs, political and community leaders, political organizations, journalists, academics, writers, actors, directors, labor union organizers, thousands of other public figures, perhaps millions of ordinary citizens, and even U.S. presidents as part of an anti-communist frenzy going back to the 1917 Bolshevik Revolution. Internal surveillance was but a small part of the worldwide spy operations conducted by the Central Intelligence Agency (CIA) and other U.S. intelligence branches. The existence of Soviet socialism provided grist for anti-communist crusaders in the United States and was often used as an excuse to push the political agenda to the right or to mobilize prejudice against immigrants, non-Europeans, and Jews.

During this period, the FBI kept extensive dossiers on noted American writers including Ernest Hemingway, John Steinbeck, John Dos Passos, William Faulkner, Edna St. Vincent Millay, Thomas Wolfe, Archibald MacLeish, and Carl Sandburg ("Documents Show," 1987). Suspicion of foreign-born artists and writers during the heyday of the McCarthy era led to passage of the McCarran–Walter Immigration Act (1952), which forced immigrants to carry registration cards and threatened both them and naturalized citizens with the possibility of deportation. The act was also intended to keep communists and "fellow travelers" from entering the United States, denying visas to such celebrated writers as Gabriel Garcia Márquez, Carlos Fuentes, Graham Greene, the playwright Dario Fo, and even the politician Pierre Trudeau before he became prime minister of Canada.

Under its "counterintelligence program" (COINTELPRO), Hoover's FBI also conducted hundreds of illegal burglaries against or sabotaged legal liberal, leftist, antiwar, and black militant institutions and organizations; photographed demonstration participants; bugged telephones; opened first-class mail; enlisted thousands of informants (including Boy Scouts);

harassed and slandered critical journalists; stole private files and correspondence; and planted false information and instigated violent personal
feuds among leaders or otherwise defamed them (Neier, 1981). At the
same time, the FBI often assisted right-wing extremist groups such as the
Minutemen and the John Birch Society. They also took part in violent and
terrorist acts against innocent blacks after infiltrating the Ku Klux Klan.
One major target of the FBI dating back to the 1950s was Martin Luther
King, Jr., falsely accusing him of harboring communists among the leadership of his Southern Christian Leadership Conference organization and
being under communist control. In fall 1964, the FBI sent King an
anonymous threatening message intended to harass and intimidate him
and or induce his suicide (Katznelson & Kesselman, 1987, pp. 204-205;
Parenti, 1977, pp. 157-162).

The entire post-World War II political climate in the United States has
suffered from a cultivated paranoia about communism that has largely
served the interests of what President Dwight Eisenhower called the
"military-industrial complex" and the careers of "cold warriors" like
President Richard Nixon, President Ronald Reagan, and Senator Joseph
McCarthy. The U.S. national security state system of electronic surveillance is far more sophisticated than that of any other country and makes
the espionage efforts of the former Soviet KGB look primitive by comparison. With its surveillance infrastructure of satellites; high-resolution infrared and night photography; miniature cameras; microscale listening devices; bugging capability for telephone, telegraph, radio, and computer
transmissions; plus a nationwide informer network, the technological
foundation for modern totalitarianism is certainly upon us. The political
will to impose a full-blown system of authoritarian control through the
capture of state power is probably better organized at present than in any
period since the 1930s.

Following the death of Hoover and the fall of Nixon, COINTELPRO
operations came to light, and the FBI was forced to close down the
program, along with its secret "Red squad" operations in local police
forces set up to harass left and civil rights groups. Reagan reversed a
reformist political trend in favor of more intensified federal surveillance,
set up in the name of protecting the national security state, and pardoned
the only two FBI officials convicted in the COINTELPRO operations (Glick,
1989, p. 30). His (and George Bush's) involvement in the secret and illegal
sale of arms to Khomeini's Iran and the funneling of money and weapons
in support of the brutal *Contra* counterrevolution in Nicaragua has been
well detailed in official and media accounts.

What did not emerge during the congressional Iran-*Contra* hearings
was the role of the Federal Emergency Management Agency (FEMA), a

government agency originally set up to organize relief during natural disasters, in working with Oliver North to draft a martial law plan that would create specified concentration camps for protesters in case of a U.S. invasion, for example, of Nicaragua. Meanwhile, the federal Immigration and Naturalization Service (INS) in 1986 completed a plan, "Alien Terrorists and Undesirables: A Contingency Plan," that without public knowledge would put Palestinians in the United States in detention camps for "national security reasons" (Demac, 1988, pp. 82-83).

At home, Reagan also tightened the rules of state secrecy by rewriting an executive order on making available government documents under the Freedom of Information Act (originally passed in 1966, substantially amended in 1974) in an attempt to undermine the intent of the legislation. Much of what was illegal in COINTELPRO became legal under Reagan's Executive Order No. 12333 (December 4, 1981), which approved "counterintelligence activities . . . within the United States" by the FBI, the CIA, and the military branches "to support local law enforcement" with covert electronic and mail surveillance and break-ins when approved by the U.S. attorney general (Glick, 1989, p. 31). In 1983, Reagan issued an executive order, National Security Decision Directive 84, that imposed lifetime non-disclosure and pre-publication censorship on some four million federal and contractor employees, including cabinet officials, with access to sensitive government documents to prevent them from revealing classified or embarrassing information. The directive also imposed polygraph tests for those accused of leaking information not yet classified.

Reagan also tightened the ban on travel to Cuba; widened the denial of visas to leftist foreign playwrights, academics, novelists, philosophers, and political figures; and extended the FBI surveillance and harassment of grassroots political and religious organizations. At the same time, the Reagan government regularly used the FBI and CIA to pressure and harass leading newspapers and magazines, some of which were threatened by CIA Director William Casey with prosecution for espionage (Katz, 1987, pp. viii, 28, 48). Reagan had every reason to be close to the FBI. After World War II, Reagan became president of the Screen Actors Guild while working as an active informant for the FBI and assisted them as a "friendly witness" in the House Un-American Activities Committee hearings on communists in the movie industry. His political career took off as a result of his cooperation.

State intelligence-gathering agencies in the United States, such as the FBI, the Defense Intelligence Agency (DIA), military intelligence, local and state police, and the CIA, together with other federal, state, and local government departments (INS, U.S. Treasury Department, State Department Passport Office, Internal Revenue Service [IRS]) and private agencies, have spread a vast information-gathering network over the country and overseas. One congressional research unit, the Office of Technology Assessment (OTA), found by the mid-1980s some 3.5 billion separate government records kept on U.S. citizens, including computerized surveillance of seven million workers. Some 55% of federal files were estimated to be erroneous (Mulgan, 1991, p. 72; Winner, 1993, p. 287). It would be reasonable to ask whether this was not more the behavior of a police state than a democracy. The Republican–controlled Congress ended funding for the OTA in 1995.

In the hands of well-financed, right-wing, fanatical but highly disciplined organizations with connections in high places (e.g., the American Security Council [ASC]), computers, data banks, mailing lists, and up-to-date communication equipment become powerful vehicles to be used against their manifold liberal and left-wing enemies. The ASC has been described as an "umbrella group for Latin American death squad leaders, Hitler collaborators, followers of the Rev. Sun Myung Moon, rightist dictatorships, and anti-Semitic activists, some of whom are connected to the quasi-Nazi Liberty Lobby" (Bellant, 1991, p. 65). The ASC, whose roots go back to the 1930s' pro-Hitler America First Committee and other racist, anti-Semitic, and pro-big-business organizations of the time, was founded in 1955 by the then chairman of Sears Roebuck, General Robert E. Wood, also founder of the America First Committee and the right-wing magazine *Human Events* (Bellant, 1991, pp. 30-33).

The greatest danger of anti-democratic movements and surveillance technologies converging into a totalitarian state would likely start with a "national emergency"—a war or an internal rebellion that would be used to invoke "emergency powers." And if the suspension of civil liberties or martial law were to occur, it could be justified as legal, invoking Article I, Section 9, paragraph 2 of the Constitution, which reads: "The Privilege of the Writ of Habeas Corpus shall not be suspended, unless when in Cases of Rebellion or Invasion the public Safety may require it." *Writ of habeas corpus* refers to the right of people not to be held without formal and legal charges by a court of law. And although the words "Rebellion or Invasion" certainly limit the applications of this provision, one historical scholar has noted that "the habeas corpus writ itself does not safeguard personal liberty in the face of the enactment of laws providing for imprisonment or detention which are themselves violations of such liberty" (Aptheker, 1976, p. 79).

Indeed, at several junctures in U.S. history, federal legislation has been passed to incarcerate or banish political dissidents opposed to particular laws or policies, including acts of war, of the federal government—and in some cases to imprison purely on grounds of race. Alien and Sedition Acts (1798) were passed during the Federalist John Adams's administration that attempted to destroy Republican (Jeffersonian) opposition. A Sedition Act made it a crime to "write, print, utter, or publish . . . any false, scandalous and malicious writings against the government of the United States, or either House of Congress . . . or the President with intent to defame . . . or to bring them into contempt or disrepute" (Greenberg, 1989, p. 70). The editors and publishers of the 14 leading Republican newspapers were indicted and convicted under these laws, but this action failed to prevent the leading Republican, Thomas Jefferson, from defeating Adams in the 1800 election (Katznelson & Kesselman, 1987, pp. 259-260).

With the establishment of corporate mass media in the latter half of the 19th century, the press came to be an ally of the state in repressing the political rights of working people. Congress' passage of an Espionage Act (1917) and Sedition Act (1918) shortly after U.S. entry into World War I was directed at union organizers, as well as at those speaking or writing against U.S. involvement in that war. These acts effectively shut down the socialist press, banned marxist political organizations, blocked more than 100 publications from using the mail, and led to the arrest of thousands of dissidents, labor leaders, and antiwar activists. A special target was a radical labor organization, the Industrial Workers of the World (IWW), because their militant, though nonviolent, organizing methods and opposition to the war made them a threat to the traditional privileges of big business.

After the Bolshevik Revolution and Lenin's formation of a worker state, the New York dailies, adopting the official anti-communism of the Wilson administration, served as the main purveyors of a "Red scare." Running such headlines as "Nationwide Search for Reds Begins" (*New York Tribune,* June 4, 1919), "Russian Reds Are Busy Here" (*New York Times,* June 8, 1919), and "200 Reds Taken in Chicago: Wholesale Plot Hatched to Overthrow U.S. Government" (*New York Times,* January 2, 1920), the press fostered a public climate that bracketed Bolshevism with union demands of American workers (e.g., right to organize, length of workday, wages, regulation of child and female labor). In 1919 and 1920, Attorney General A. Mitchell Palmer took advantage of the political mood to organize a mass roundup of citizens and more than 4,000 immigrants, ordering their deportation, including at least 249 to the Soviet Union. Overseeing the roundup, the head of the Bureau of Investigation (predecessor of the FBI), William J. Flynn, declared, "This is the breaking of the backbone of radicalism in America" (Boyer & Morais, 1955, p. 210). By smashing the union movement, the Red scare made more profits for business than could have been accomplished by work speedups or new labor-saving technology (Boyer & Morais, 1955, p. 210; Zinn, 1980, p. 366).

Another domestic "enemy" was found at the outset of World War II. In February 1942, in the hysteria of war, President Franklin D. Roosevelt signed Executive Order 9066, which forced 110,000 Japanese American men, women, and children on the west coast, three fourths of them born in the United States, to be incarcerated in internment camps for the duration of the war, solely on the basis of their ancestry. They lost their land, their livelihoods, and in many cases their childhoods. A young *nisei* (second-generation Japanese American) was forced to give up his land in Cupertino, California, for a pittance, real estate that is now occupied as the headquarters of the multi-billion-dollar firm Apple Computer. The U.S. Supreme Court ruled in 1944 that 9066 was constitutional on grounds of military necessity. Not one Japanese American was ever convicted of disloyalty (Zinn, 1980, p. 407).

The First Amendment to the Constitution reads: "Congress shall make no law respecting an establishment of religion, or prohibiting the free exercise thereof; or abridging the freedom of speech, or of the press; or the right of the people peaceably to assemble, and to petition the Government for a redress of grievances." In 1901, a Sedition Act was passed to prevent Filipinos from making "scurrilous libels" against American colonization of their country (Paterson, Clifford, & Hagan, 1991, p. 251). Not only does the Espionage Act of 1917 remain in law, but the Kennedy administration even attempted, unsuccessfully, to have it applied against American journalists abroad who were critical of the Diem regime in Vietnam that the United States had installed in power (Zinn, 1980, pp. 357-358). In 1940, the Smith Act was passed, which outlawed the teaching or advocacy of the violent overthrow of the government and under which law 11 Communist Party leaders were convicted (10 sentenced to 5 years each) in 1951 (*Dennis v. United States*) for promoting the theories of Karl Marx (Wasserman, 1988, p. 182). None of the various sedition acts has ever been repealed by Congress, and not until 1964 was the Sedition Act of 1798 ruled unconstitutional.

Were A. Mitchell Palmer the U.S. attorney general today, he would not have to engage in messy raids to try to ensnare illegal immigrants and other "undesirables." During the Bureau of Investigation's sweep of January 2, 1920, Palmer's agents broke into union meeting halls, private offices, and homes, rounding up thousands of individuals, often without arrest warrants, with the idea of capturing armed revolutionary communists. His tactics, in fact, yielded three pistols and no explosives (Barnet, 1990, p. 171). In the 1990s, a political climate not dissimilar to the 1920s, he would have telephone, electronic listening, and camera devices; satellite surveillance; a wide array of computerized record keeping, governmental and commercial; and digital imaging technologies to search and destroy any and every labor or political leader he so wanted.

Moreover, the opposition-party-controlled Congress that exposed Nixon's and Reagan's crimes against the Constitution in the 1980s was no longer in power after the 1994 national election. Openings to the far-right elements in the American political spectrum were begun with the selection of Newt Gingrich as majority leader of the House Republicans and Trent Lott as floor leader in the Senate.

THE INFORMATION
PROFILING OF SOCIETY

The proliferation of information-gathering techniques has been a fact of life in late industrial society. It is a rare individual who is not regularly asked to submit personal data to complete strangers by credit card, in writing, by telephone, or by mail. People are expected to function on day-to-day faith that information rendered about themselves will not be turned against them. Schools, hospitals, workplaces, government agencies, financial institutions, retail stores, mail-order outlets, and scores of other enterprises collect, record, process, and retrieve information; construct commercial, legal, or political profiles; share them with other institutions without authorization of the respondent; and flood households with junk mail, not to mention scanning computer and cordless telephone conversations. And yet relatively few people actively contest these panoptical social practices that have established at least the technical basis for a Big Brother society—along with thousands of Little Brother spymasters.

Federal, state, and local governmental agencies, commercial interests, and various organizations keep data profiles on virtually every citizen in the United States. Moreover, many of these agencies use computer matching of personal files to either verify already existing information or seek out additional information about people. One study noted the "frequent subversion" of collected and stored information by the IRS "to political ends unrelated to its primary function of tax collection." The IRS, the study found, had an "undercover capacity" greater than all other federal civilian agencies combined (Gandy, 1993, p. 57). In late 1996, the press reported recorded conversations of Nixon and his aides discussing using the IRS to audit Jewish contributors to the Democratic Party. Nixon told his chief of staff, H. R. Haldeman, "Go after 'em [the Jews] like a son of a bitch" (Associated Press, 1996, p. A26).

By the mid-1980s, the OTA reported that multiple federal agencies were already using a wide variety of electronic surveillance technologies (see Table 2.1) "to monitor the behavior of individuals, including individual movements, actions, communication, emotions, and/or various combinations thereof, as well as the movement of property or objects" (Office of Technology Assessment [OTA],

TABLE 2.1. Categories of Surveillance Technology

Electronic Eavesdropping Technology (audio surveillance)
- radiating devices and receivers (e.g., miniaturized transmitters)
- nonradiating devices (e.g., wired surveillance, including telephone taps and concealed microphones)
- tape recorders

Optical/Imaging Technology (visual surveillance)
- photographic techniques
- television (closed-circuit and cable)
- night vision devices (use image intensifier to view objects under low light)
- satellite based

Computers and Related Technologies (data surveillance)
- microcomputers—decentralization of machines and distributed processing
- computer networks
- software (e.g., expert systems)
- pattern recognition systems

Sensor Technology
- magnetic sensors
- seismic sensors
- infrared sensors
- electromagnetic sensors

Other Devices and Technologies
- citizens band radios
- vehicle location systems
- machine-readable magnetic strips
- polygraph
- voice stress analyzer
- voice recognition
- laser reception
- cellular radio

SOURCE: Office of Technology Assessment (1985, p. 13).

1985, p. 4). These included closed-circuit and cable TV, night vision systems, miniature transmitters, electronic beepers and sensors, telephone taps, recorders and pen registers, computer usage monitoring, electronic mail (e-mail) monitoring or interception, cellular radio interception, pattern recognition systems, and satellite interception. The means are available for e-mail to be intercepted at the terminal of the sender or receiver, on the files of the computer where it is stored, while being communicated or while being printed into hard copy.

With only 25% of federal agencies responding to the OTA survey, it was found that 288 million records were kept on 114 million people. The survey found that almost all of these forms of surveillance are not protected by legislation (OTA, 1985, pp. 4, 9). Increasing levels of reported violence in the

United States has given rise to demands for more police involvement, tougher sentencing, faster executions, and more prison space, even after more than a decade of "law and order" politics and the death penalty seemed to show no deterrent effect. It is difficult to know whether the infrastructure of government and police surveillance would shrink if the level of crimes of violence and theft were to subside or would remain in place to be used against crimes of thought. Could it be that the network of public and private institutions involved in "crime fighting" has taken on a life of its own, a domestic parallel to the global military-industrial complex?

Other implications of the generalized surveillance of society affect the means and quality of people's communication. The widespread use of answering machines, voice mail, fax machines, e-mail, beepers, car phones, and other implements may have created certain efficiencies to time management but also reduced the opportunities for direct, informal, person-to-person, and spontaneous conversation. Mediating technologies make it more difficult to communicate intentions and easier to distort what the intentions may be. The wide array of surveillance equipment and the network of people employed or encouraged to monitor and report on neighbors, friends, clients, or associations change the climate of free communication in society and further limit what can and cannot be communicated.

In 1986, for example, Mead Data Central, a major electronic publisher, was told by the Air Force, the CIA, and the FBI that its clients would be monitored and in some cases denied access to unclassified information. Library administrators at Columbia University, the New York Public Library, and many other college and public libraries in the area were asked in 1987 to cooperate with the FBI's Library Awareness Program by keeping records and reporting on foreign students using their facilities or reading radical authors (Katz, 1987, pp. 41, 44). If people are made to feel intimidated about their private reading habits, those with unconventional views might be utterly silenced on e-mail or even cellular phone, media that can easily be monitored. A student living in a community that was largely in favor of school prayer might be hesitant to communicate over e-mail certain beliefs that were more consistent with the First Amendment's separation of church and state. Surveillance, combined with the financial means needed to use media at the network level, infuses communication technology with a powerful instrumental and political bias (value-oriented inclination or tendency).

One function of any government is to exercise coercive means to safeguard the stability of the state. Among the many applications for state power maintenance, communication technologies in recent decades have provided unprecedented opportunities to enlarge the scope of public monitoring and policing. Operation Chaos, launched by the CIA and FBI in 1967 as a response to President Lyndon Johnson's paranoid belief that the antiwar movement was being funded and influenced by foreign powers, became in the Nixon years part of a sweep tactic against Americans protesting the foreign and civil rights policies of the government. The CIA, illegally involved in domestic spying at the time, kept dossiers on 7,200 citizens and computerized files on 300,000 persons and groups and regularly opened mail, wiretapped telephones, bugged homes, and burglarized offices of dissident organizations (LaFeber, Polenberg, & Woloch, 1986, p. 518).

Nixon and his national security advisor and secretary of state, Henry Kissinger, routinely kept illegal secret recordings and wiretaps on journalists and government officials. John Mitchell, attorney general at the time, the country's highest-ranking law official, ran a secret fund of up to $700,000 for the purpose of carrying out a campaign of dirty tricks against the Democratic Party—including planting false news stories in the press, forging letters, and pilfering party campaign documents (Zinn, 1980, p. 531). The "plumbers" sent by the Nixon administration to break into the psychiatrist's office of Daniel Ellsberg, the former Pentagon official who leaked "the Pentagon papers," as a way of trying to discredit him, ultimately brought down the president. This arrogant, political abuse of power would be followed by the administration's even more outrageous act of lawbreaking, the attempted but botched effort at burglarizing and bugging the Democratic National Committee's party headquarters at the Watergate building in Washington, D.C., and Nixon's subsequent attempts to cover it up.

The sanctity and privacy of communication, whether personal or political, are threatened by the potential totalitarian reach of a power structure dedicated to knowing everything about everyone. Not only the actual use of electronic monitoring of citizens but also the threat to do so undermine the foundations of a democratic society because both may achieve the same ends—the creation of a submissive and conformist public and the isolation of dissidents or agents of social change. Not until the Watergate scandal and the related police state-type activities of the Nixon administration directed against ordinary citizens, the press, and even government officials came to light in the 1970s was the long-suspected underworld of government spying even partially revealed. Supreme Court Justice Louis Brandeis in 1928 warned of a day that "government, without removing papers from secret drawers, can reproduce them in court, and by which it will be enabled to expose to a jury the most intimate occurrences of the home" (cited in OTA, 1985, p. 12).

SEGMENTING CONSCIOUSNESS: LIFESTYLES, CONSUMPTION, AND THE POLITICS OF SOUND BITES

The commercial foundation of mass media and communication systems in the United States is built upon values and messages of acquisition and consumption as the chief ends in life. Eat and drink more frequently, redo your wardrobe, redecorate your home, drive and travel more, buy a new model car or truck, burn more fuel, use more electricity, use more household utensils, indulge children with more toys, indulge adults with more toys, stuff your refrigerator with more food, get a larger refrigerator, build a bigger house, borrow more money, and secure more stocks and bonds to generate more income to do all of the foregoing. Americans are using up more than 30% of the world's available natural resources. In short, the rule is consume more and reflect less about the consequences. The commercial cultivation of hedonistic and narcissistic lifestyles based on the use of non-renewable resources and non-perishable waste is on a collision course with what a growing section of the scientific community sees as the final chapter in the earth's ecology—the death of the planet's finite and closed environment, much of it already irretrievably lost.

Since 1945, the earth has lost 11% of its vegetated surface—an area larger than China and India combined. Certain forest types will vanish in a few years, and most tropical rain forests and many plant and animal species in about 100 years. Brazil has lost more than a quarter of its rain forests to export mining and cattle interests and privatized 50 million hectares of the Amazon region between 1970 and 1990. Major water shortages have occurred in about 80 countries. Fisheries are collapsing, and rivers and lakes are filling with toxic waste; enormous levels of water, earth, and air pollution have deteriorated the quality of life for all living species. Stratospheric ozone depletion, deadly ultraviolet radiation, and acid precipitation are destroying humans, other species, forests, and crops.

We are losing untold numbers of medicinal plants and other staples of the life-sustaining natural environment. Climate alteration is occurring partly because of fossil fuel burning, deforestation, and increased levels of human-generated gases, including carbon dioxide. At the current rate of resource depletion, we can expect greater and more frequent natural calamities in the years ahead (Union of Concerned Scientists, 1993). The United States is leading this ecological disaster, and an American in his or her lifetime will destroy twice as much of the environment as a Russian, 13 times a Brazilian, 35 times an Indian, 140 times a Kenyan, and 280 times a Haitian, Rwandan, or Nepalese (Barnet & Cavanagh, 1994, pp. 177-178). In the past 20 years, U.S. per capita consumption rose 45% while the Index of Social Health indicators fell by 51% (Robin, 1994, p. B7).

Corporatism, the concept of hierarchically organized activity based on functionally specialized roles, identities, and rewards and chartered or controlled by the state, impels a high level of conformist behavior both inside and outside the workplace. In late 20th-century capitalism, the corporate-driven level of consumption is the ultimate measure of "growth," which in turn represents the "health" of the economy. Market-oriented ideologues tolerate little interference with growth demands, and as a result, almost every branch of human behavior has been industrialized, whether it be the rules of production or the habits of consumption. The marketplace has come to govern who gets a livable wage, access to health care, education, and affordable housing; who gets police and legal protection; the management of incarceration; the content and definition of food; how people spend recreation time; the material of mass culture; the standards of beauty; the agenda and significance of daily news events; and other institutions of everyday life.

By the end of the 1980s, annual corporate spending on advertising, promotion, and packaging of goods and services amounted to about $620 billion, with Americans on average receiving 3,000 advertising messages daily (Barnet & Cavanagh, 1994, pp. 171-172). All of this and much more were absorbed by consumers—initial costs plus "hidden" ones, such as debt, injury, pollution, illness, or even death. Manufacturers are not required to disassemble what they have assembled, and the result is growing allocations of land for solid wastes and toxic chemical combinations. Corporate capitalism has overbuilt the physical environment, and it is obvious to many that culture and the political process are also manufactured and packaged just like product advertisements or the images on MTV. Chapters 3 and 4 discuss historically how communication technology, media, and commercial and industrial interests came to find a common home and purpose under the aegis of the modern corporate entity.

PART **II**

THE SOCIAL-HISTORICAL PROLOGUE

To have a fuller sense of how the evolution of communication and information technology has reached its current state, it is important to have a social-historical understanding of material progress. Conventional histories suggest that technology itself logically gives rise to offshoots of newer and more innovative technologies, only awaiting discovery, and that "great inventors" (and other "great men") are the conduits in the remaking of society. In such an interpretation, the real-life clashes between different social groups within the world economic system, divided by nation, class, income, ethnicity, gender, skin color, elite competition, and other identities whose conflicting needs propel change, are irrelevant to the history of technology.

Chapters 3 and 4 describe those conflicting social contexts in which technological innovation takes place as a logical outgrowth and resolution of, not the machine itself, but of the social tension. The dominant forces within society seek advantage through technological, political, and economic control. If "*necessity is the mother of invention*," as Benjamin Franklin said, the critical question becomes, What or who determines what's necessary? A political economic framework that discusses the organization of influence and decision-making power in society provides the clearest answer to this question. Using such an approach, we begin to see a pattern that links the social purposes behind the telegraph to subsequent communication developments up to the contemporary forms of digital technology.

CHAPTER 3

Historical Perspectives on Communication

Technologies do not spring forth as sudden unforeseen events, nor is their coming entirely predictable. Readings of social history, which we explore in this and the following chapter, give us a better understanding of the origin of ideas and the purposeful uses to which creativity has been put. History also helps place technology not only in the perspective of its time and lineage but also in the context of the larger social forces that give rise to its material existence, meaning, and necessity. The digital technologies that currently are so central to goods and services production and to everyday social practices are linked to earlier information and communication technologies that were developed and applied to general and particular needs of the time and that were used to shape the physical and cultural spaces we now inhabit.

As the United States unfolded as a nation state, commercial, industrial, political, cultural, technological, and other spheres of activity began to form into larger institutional structures. These structures promoted increased trade, more complex political formations (e.g., political parties), religious movements, territorial identities (e.g., the "South," Texas, the "West Coast"), the assimilation of different ethnic groups into "Americanness," a public school system and land grant colleges for popular education, the founding and expansion of limited liability stock corporations, the expansion of police functions, the creation of national armies and navies, and other changes in support of growth and nationhood. The ways and instruments of communicating and the value of information itself changed within this larger historical pattern. Although the life patterns that were forming were diverse, unmistakable features were also central to the American experience. The rapid development of large-scale commercial and industrial enterprise and of long-distance transportation and communication networks was clearly among them.

Nature provided great abundance to the native inhabitants and early settlers of the new territories. The European Americans brought with them a social-cultural ethos shaped by a material consciousness in what were then the most advanced areas of capitalist development. Scientific interest and experimentation were driven beyond the study of basic natural properties (e.g., chemistry, physics, mechanics) to the pursuit of their material and economic value. As Antonio Gramsci wrote, science and technology became of interest to the modern world as "socially and historically organised for production" and for "the development and further necessities of development of the forces of production" (quoted in Hoare & Smith, 1971, pp. 465-466).

The science and technology of the early American republic are linked to the present era through a complex constellation of people and events, but a social-historical reading of industrialization provides the most compelling explanation of these linkages. The energies dedicated to transforming the United States from a set of divided rural communities into an integrated industrial and urbanized nation were initially driven as much by ideals of a shared social community as by the lure of market conquest. But the capitalist industrial concept of development would prevail. Under its own momentum and internal logic, nothing could stand in its way of its "creative destruction"—not the forests, or the minerals, or the agrarian dream of Jefferson, or the Native American settlers of the soil.

The logic of great power enlisted skill, courage, and ruthlessness in defense and expansion of economic dominion. The age of the "robber barons" was but one phase before the captains of American industry accepted the necessity of government partnership and mediation in the marketplace. Many bloody struggles would ensue before workers would achieve the right to unions, the abolition of child labor, and the 8-hour day. And the great infatuation with the scientific education of the Enlightenment was transmuted into undertakings in applied science—technology—a new alchemy in the service of organized power, vision, and wealth.

Changes in the methods, instruments, and scale of production would be inconceivable without concurrent changes in the means of communication. Long-distance communication by post, Pony Express, wire, and later wireless was impelled by these larger economic changes. Rural communities that had long survived without telegraph or telephone could hardly go on without them as farming techniques were "revolutionized" for surplus production and geographically dispersed markets. The "industrial revolution," representing both a material and a cultural transformation, would permanently alter the relationships of people to nature, of people to their tools, and of rulers to the ruled. The expansion of capitalist trade that gave rise to the growth of port cities in Europe would absorb the United States into the world economy. Not only shipping but also railroads and telecommunications would serve as the conduits for urbanization and the formation of new centers of business, politics, knowledge, and civilization.

THE REINVENTION OF COMMUNICATION

Capitalism and modern state systems, propelled by the disintegration of feudalism, the English civil war, and the rise of mercantilism, scientific thought, secular social values, and philosophy and a spectacular renaissance in literature, music, the visual and mechanical arts, and architecture, precipitated the radical transformation of human communities. *Community* and *communication* come from the same Latin root, *communis,* meaning that which is held in common. But communities, long defined by common association and the use of common physical space, were breaking down under the currents of world capitalism and its associated culture of individualism that would transform traditional identities with place. Mercantile and industrial capitalism required mobility of goods, of people, and of information and the remaking of past social relationships.

In the early 19th century, vested notions of community would give way to mobility associated with opportunity. The Enclosure Acts of the English Parliament forced landless villagers out of the countryside, creating pauperism as well as the new working class in the rising industrial and trade cities. In the United States, the poor farmers of the east rapidly settled the lands west of the Appalachians, increasing the population of those areas from 5,000 to 8 million between 1770 and 1840 (or from less than 1% of the national population to nearly 50%). Abraham Lincoln's family moved from Pennsylvania to Kentucky, a few years later to Indiana, and again to Illinois when the future president was still only 21. These migration patterns were typical of the period (Huberman, 1970, pp. 99-100).

The transportation technologies of the European Americans, their wagons, draft animals, roads, canals, rivercraft, and later the steamboats and railroads constituted a continual threat to Native American community patterns and often even to their survival as a people. Indeed, as white migration rapidly expanded beyond the Appalachians, the number of Native Americans east of the Mississippi shrank from 120,000 in 1820 to fewer than 30,000 by 1844 (Zinn, 1980, p. 124). Private property and land grabbing could not coexist with the tribal and communal land-holding practices of the Native Americans. Settlers in the wilderness transformed their own ways of community, adapting themselves to their limited means and often learning habits of dress and farming from those friendly Native Americans who were not immediately driven out. Eventually, as the white population expanded and transportation and communication improved, the western territories became tightly linked to the trade and other developments of the country as a whole.

THE RISE OF INDUSTRIAL CAPITALISM

The 19th century is seen as the era of the industrial revolution, the rise of manufacturing, mechanical engine design, steam and electric power, the factory system, banking and industrial monopolies, the beginnings of the industrial proletariat and trade unionism, colonialism and imperialism, and other expressions of capitalist "modernity." Building on the collected scientific and technical information of the past, the 19th-century "inventor" fed the growing need for capital goods and household products with innovative designs and technological breakthroughs that would widen the opportunities for market expansion, industrial segmentation and specialization, and private fortunes. In service to the inventors were the growing numbers of skilled craftspeople and mechanical workers, many of them immigrants, who transformed concepts and designs into practical and functional implements, machines, and merchandise.

In that century, artisanship itself was becoming transfigured or disembodied of its artisanal content; mastery and authority in production gradually transferred to machine and to corporate management. Along with the transformation of manufacturing processes and the fuller exploitation of natural resources, 19th-century industrial capitalism further segregated finance and management from the sphere of direct production and converted production, in turn, into a more complex division of labor. According to Marx,

> Modern Industry was crippled in its complete development, so long as its characteristic instrument of production, the machine, owed its existence to personal strength and personal skill, and depended on the muscular development, the keenness of sight, and the cunning of hand, with which the detail workmen in manufactures, and the manual labourers in handicrafts, wielded their dwarfish implements. (cited in de la Haye, 1980, p. 147)

With the unfolding of the structure of industrial capitalism in the West, it was clear that the very scale and complexity of production it was bringing forward required reliable forms of command, coordination, and communication. This was already evident in the military sphere. As private enterprise enlarged its own territorial claims through colonial acquisition, it, too, would need to command the use of communication technology. The more communication resources that industry had at its disposal, the weaker the voice of its workers. The development of communication technology changed the rules in the articulation of class power.

Telegraphy Discovers Samuel Morse et al.

As the 18th century had unlocked the secret of mechanical production, the 19th would be the century of communication (Bernal, 1971a, p. 544). Included in the industrial revolution's pantheon of heroes is Samuel Finley Breese Morse (1791-1872), a Charlestown, Massachusetts-born portrait painter of renown, photographer and famed inventor, who is eponymously associated with the remarkably enduring binary system of telegraphic coding (more properly belonging to his collaborator, Alfred Vail) and the wired electrical instrument for long-distance messaging. In the genealogy of technology, Morse "invented" the telegraph. It would be more appropriate to say that telegraphy invented Samuel Morse as a figure in historical memory.

Many forms of telegraphy, in fact, preceded Morse's first primitive electrical transmitting device of 1832. His innovative talents established a reputation for ·him, though, through some prescient adaptations of earlier and then current work in the field, including a coding system, still in use. Binary message coding dates back, however, at least as early as the two-letter concept developed by Francis Bacon (1561-1626) in 1604 and the later work of Gottfried W. von Leibniz (1646-1716), although long-distance coded signaling goes back at least as early as pre-Christian era Greece: The fall of Troy (ca. 1193) was "announced" with a "firegram" relayed across mountain peaks over a distance of some 300 miles (Frederick, 1993, p. 25).

Morse's device prevailed technically, in part, because of the elegance of his single-wire transmission system and its associated software. The binary code of dots and dashes proved to be a "viable solution to the problem of compressing information so as to reduce the use of bandwidth: cheap skilled labour at the switches minimized the amount of information that had to be transmitted by converting words into dots and dashes and paring language down to its barest essentials" (Mulgan, 1991, p. 101). This division of labor, incidentally, excluded women as telegraphers in the early decades because of gender-based social restrictions—and also because some believed that women were not hardy enough to manipulate the switching key!

Morse lived in an age of utilitarian ethics—namely, the unfettered opportunity for capital gain as the measure of social progress—and Morse himself was a vocal expansionist, anti-immigrant, and defender of the institution of slavery. Ultimately, his place in history was made possible precisely because of the rising fortunes of commercial enterprise and finance and expansionist military interests that were in search of new and better means of defending their respective and growing empires. Such power groups gave utility to telegraphy and respectability to Morse and his contemporary telegraphers. The growing national market of goods and services needed a fast and reliable method of information exchange. Carrier pigeons that London banker Nathan Rothschild used to get news of the battle of Waterloo and make his own kill in the English financial markets signaled to capital interests and the earliest electrical engineers that communication technologies of speed were highly useful and profitable means to overcome the constraints of distance and competition.

Electric current, already known by the 18th century, was in itself a kind of telegraphic medium in its being able to signal over distances between sender and receiver, although its vocabulary was extremely, and sometimes painfully, limited. One of Morse's mentors was a professor of physics, Joseph Henry, who helped him understand the properties of electricity and demonstrated to his curious understudy the capacity of a small voltaic cell to electromagnetically "pull" and evoke the sound of a bell (Winston, 1986, p. 301). Morse would go on to successfully gain a patent (a legal monopoly of ownership and marketing privileges) in the United States for the commercial exploitation of his communicating device. The patent was a means developed under early capitalist Venetian law for promoting the interests of the individual, industry, and state. It was first granted in North America in 1641 and eventually became encharter in the U.S. Constitution (Art. 1, Sec. 8) "to promote the progress of science and useful arts, by securing for limited times to authors and inventors the exclusive right to their respective writings and discoveries" (Slack, 1984, chap. 8).

Invention properly understood is a collective enterprise that "adheres to certain fairly definite patterns of impersonal causation" (A. E. Kahn, cited in Slack, 1984, p. 107). The single inventor concept and its reification (a concept made tangible) in the Western patent system is, however, more than a particular reading or misreading of history: It is an ideological construct. Its ideological function is to help in mythologizing the "rugged individualist" in history and in attributing material progress, not to the sweat and collective intelligence of working people and their accumulated knowledge or to the collaborative process of invention, but rather to the genius of individuals in the employ of capital, thus creating legitimacy for moneyed elites. This history becomes the official story through which civic consciousness is impressed on youth, and capitalists are able to memorialize themselves. As Slack (1984) argues, the inventor concept disguises "the identification of the class nature of the capital-labor relationship" and permits inventors or their capitalist backers "a means of acquiring surplus product [profit] and, in conjunction with the subjugation of labor, the means whereby class domination is mobilized" (p. 108).

Henry and Morse were by no means alone in seeking inventor status in telegraphy. Before Morse wired his celebrated "What hath God wrought?"

message in 1844, various other innovators were developing electrical telegraphic devices in Europe and in Russia, all of whom were building on the functional concepts of non-electrical telegraphs already in existence. Danish physicist Hans Christian Oersted in 1820 had found, accidentally, that electric current could move a compass needle. A decade later, English physicist Michael Faraday explained the interactive relationship between magnetism and electric conduction, giving rise to the science of electromagnetism (Bernal, 1971a, p. 608). According to Brian Winston's colorful history of the period, Edward Davy, expanding on earlier work in Britain, had developed an electrical signaling device primarily for regulating traffic along the early British single-rail lines then in use and for facilitating the anticipated development of faster and safer locomotives.

British competitors for wealth and glory in this field included William Fothergill Cooke and Charles Wheatstone, who developed and patented in 1837 a kind of telegraphic instrument based on a galvanometer (an instrument for detecting and measuring small electric current) before eventually falling out with one another over ownership rights. A Russian, Baron Pawel Schilling, used a battery-operated galvanometer to establish his inventor credential, although Tsar Nicholas, fearful of such a device falling into the hands of the public, forbade its development and dissemination. Morse, too, had to contend with competitive patent claims to the telegraph in the United States, including those of Cooke and Wheatstone. Morse got his patent in the United States but was denied in Britain (Winston, 1986, chap. 6).

All technologies are laden with bias, both intended and inadvertent, that tend to largely benefit empowered interests such as CEOs, bankers, and large investors. The telegraph (and later the telephone and other communication technologies) reinforced the 19th-century shift to large-scale production, *scale effect,* that led to industrial and financial monopolies in the late 1800s. Telegraphy also helped companies manage their resources through timely information access, *control effect,* as well as collect information about existing or potential markets for their goods or services, *intelligence effect.* Finally, it made facts and know-how in themselves into valued and often strategic goods, the *commodification of information effect.*

From the beginning, telegraphy took on many state security and social control functions. During labor strikes, such as those on the railroads, which badly shook the U.S. economy in the 19th century, the telegraph was used to call up federal troops for quick dispatch to areas of the country where the local militia proved inadequate. On one such occasion, during the 1877 Depression that withered industrial production and sank already low wages ($1.75 a day for brakemen, based on a 12-hour-day work schedule), railroad owners sent false messages over the telegraph that workers in St. Louis and Baltimore had returned to their jobs and that spontaneous railway strike activity was breaking up in some 10 other cities. This proved to be a valuable piece of disinformation at a particular moment when, though unusual for the times, even the American press seemed reluctant to take the owners' side in the conflict. In other countries, too, the means of communication was proving to be extremely useful in maintaining political and social order. Government control of the telegraph, as well as the press, served to put down the Chartist movement in Britain in the 1840s and worker uprisings following the 1851 coup d'etat in France (Harvey, 1990, pp. 234-235; Zinn, 1980, pp. 240-246).

In France, electric telegraphy was the outgrowth of Claude Chappe's "optical telegraph," originally in 1791 a system of telescopes and semaphore signaling devices set up across the country's peaks and mountain ranges at 5- to 10-km intervals, by 1850 covering 5,000 km with 556 stations (Headrick, 1991, p. 11). The French used the optical telegraph in 1794 to transmit news of Napoleon's military victories against the Austrians, and within a decade similar systems were produced in Russia, Sweden, and Denmark (Inose & Pierce, 1984, pp. 10-11). The security value of long-distance communications was so obvious to a succession of French governments that police and military control of the telegraph was not relinquished until the 1870s. In early 17th-century Venice, too, even Galileo had to persuade the navy of the military utility of his telescope before the device could be accepted for its broader uses in observing "the heavenly universe," those applications eventually imposing on the scientist a visit to the ecclesiastical court of inquisition.

In Britain, all inland telegraphy was nationalized under the British Telegraph Act in 1868 and put under the control of the post office, which provided relatively inexpensive and efficient service (telephone service was fully nation-alized by 1912). In Germany, both the telegraph and later the telephone were under the post office and integrated as part of a call-in message service from outlying village telephones. Sweden used a mixed government-private form of telegraph ownership, with the railroads having a major share, and its telephone density initially far exceeded that of the United States. Only in the United States was the government averse to the spirit of public obligation in the development of telecommunications.

The most extensive use of telegraphy occurred within the logic of defending the imperial, commercial, diplomatic, and military interests of the British empire. By the time of the electrical telegraph, Britain was already the foremost imperial and colonial power in the world, spanning all of the oceans and most of the established international land routes. Within two decades of the terrestrial telegraph, the British were actively laying a skeleton of global submarine cable, which one historian of technology explained as "an essential part of the new imperialism" (Headrick, 1981, p. 163). Submarine cable proved to be critical to defense during times of war and extremely profit inducing in times of peace.

Telegraphy would strengthen the "long arm of imperialism" established earlier by the British navy and commercial fleet. Against their rivals, sabotage, censorship, and disinformation in the use of submarine cables were all considered by the British to be fair game (Headrick, 1991).

Early attempts at submarine telegraphy were technological fiascos. The initial cable crossing of the English Channel in 1850 ended the same day by way of a fisherman's anchor, and the first Red Sea cable in 1859 provided an £800,000 feeding frenzy for terero boreworms. At the same time, success with the extensive terrestrial telegraph in India helped put down an Indian anticolonial uprising in 1857 and invited further penetration of imperial communications. A Suez (Canal) to Suakin (Sudan) cable was instrumental in the British invasion of Egypt in 1882, and a British–Cape of Good Hope line (1899-1901) helped defeat the Boers, which led to dominion over South Africa (Headrick, 1981, pp. 158-159, 162-163).

Similarly, submarine and land telegraphy (or radio telegraphy) for logistical information and propaganda or its denial to the enemy by cutting the lines of communication was important in the U.S. Civil War (1861-1865), the Franco–Prussian War (1870-1871), the U.S. invasion of the Spanish colonies in the Philippines (1898) and Cuba (1898), the Russo–Japanese War (1904-05), and the two world wars, among many other armed conflicts. Before the U.S. entry into World War I, American newspaper editorials, abetted by the pro-British Wilson administration and bank lending of J. P. Morgan, were heavily influenced by the British government-censored cable dispatches of Reuters, then the world's preeminent news agency. Reuters did not hesitate to invent German atrocities, as Britain censored firsthand accounts of American reporters in central Europe and hid the facts of British military misdeeds. Meanwhile, as trench warfare slaughtered hundreds of thousands on both sides, military-censored British correspondents dutifully reported back home, "All is quiet on western front." Control of global information up to that time had enabled Britain to rule the waves and, whenever necessary, to waive the rules.

By the early 20th century, the United States and Canada were the only countries with wholly private forms of telegraphy. In 1844, Congress had provided a $30,000 appropriation to help Morse establish the first long-distance wire, linking Washington, D.C., to Baltimore, where, coincidentally, the Democratic Party (which controlled the House of Representatives and, therefore, appropriations) was holding its presidential nominating convention. After this first successful test of electric political communication but a subsequent unprofitable 6-month experiment of national post office control of telegraphy, the U.S. government came to regard the device economically as a white elephant.

Congress withdrew support and refused to purchase the patent rights to the technology (to Morse's disappointment), a precedent that, unlike its European counterparts, abolished the idea of public ownership of telecommunications for the next century, except for a brief government takeover during World War I. A public ownership debate resurfaced in the U.S. Senate in the 1960s concerning the disposition of commercial communication satellites, but that resolution, too, ended in privatization (Winston, 1986, p. 302). By 1870, the western European states had all nationalized telegraphy, and by 1912, telephone as well (Britain being the last private-sector holdout in both cases). The United States had not even begun to regulate telegraph until the Mann–Elkins Act of 1910.

Morse and his cohorts had employed the first practical applications of electricity, and this would give rise to the electrical engineering profession. In the narrow sense, they had "discovered" a useful artifact of technology: telegraphy. But in the larger picture, the propelling force behind all of this technological change and the adaptation of science to serve practical needs was a vigorous capitalism, continuously in search of new or larger means of expanding productive output and wealth. In the U.S. context, the real, applied meaning of the telegraph, and hence what gave it its "life" (and status to its developers), lay, not in its clever design, but in its *utility*. Whose utility (and demand) could bring the telegraph into general use?

The earliest widespread application of the telegraph was as a signaling device for the single-track railroads. By 1852, 13,000 miles of railroad and 23,000 miles of telegraph were in operation (Beniger, 1986, p. 17). (By 1900, there would be 193,000 miles of track, more than in all of Europe, controlled largely by the J. P. Morgan empire and the financial house of Kuhn, Loeb & Co.) The railroads, enjoying the benefits of extensive government-conferred land grants and "rights of way" (which also subsidized telegraph expansion), were beginning to pass the canals and rivers as the principal conduits in the flow of goods and services in a growing U.S. economy and in other industrial capitalist countries as well. On the eve of the Civil War, some 36,000 miles of transcontinental telegraph pole line and 60,000 miles of wire were carrying 6 million messages per year, but in no other country was the "telegraph so thoroughly dominated by business interests and business use" (Du Boff, 1989, p. 21).

Besides the railroads, the other major early clients of the telegraph were stock brokers, who resold wire services to the New York and Philadelphia stock exchanges, newspapers, bankers, gold and other commodity markets, and gambling syndicates. (In the 1890s, the telegraph wired professional baseball games to urban saloons and pool halls, which stimulated copious betting and odds-making and the beginning of underworld involvement in the sport.) Trade interests were encouraged by telegraphy to start up 11 commodity exchanges in major U.S. cities between 1845 and 1871. At the same time, the telegraph helped Wall Street consolidate its grip on the U.S. securities markets by 1880, which enabled its exchanges to establish prices for all other cities (Tarr, Finholt, &

Goodman, 1987, pp. 41-42). And presaging real-time, 24-hour global financial and commodities markets, the telegraph also established the basis for national and international time zones.

By 1866, telegraphy in the United States had become a virtual monopoly of Western Union, the country's first industrial monopoly and its largest corporation. In the same year, the Atlantic cable between the United States and England (an earlier one was unsuccessful) made it possible for the London financial district to communicate instantaneously with Wall Street. The well-known historian of science J. D. Bernal wrote: "The actual impetus that set a host of inventors working at the same time (e.g. Morse, Wheatstone, etc.) was not any general need of social communication but the actual money value of news of the prices of goods or stocks and of events that might affect them. News means money, and the electrical telegraph provided the means to convey news rapidly" (1971a, p. 548).

America's westward sweep brought on expansionist wars with Native Americans and an equally aggressive imperialist war with Mexico (1846-1848), resulting in the seizure of half of that country—namely, all of its territory north of the Rio Grande (presently California, Nevada, Arizona, Utah, New Mexico, parts of Texas not already annexed, Colorado, Oklahoma, Kansas, and Wyoming). In its first century, the United States had become a bicoastal nation state with an immense national-scale market. The momentum of capital expansion was aided in 1868 by the passage of the 14th Amendment, which denied the states the right to "deprive any *person* [italics added] of life, liberty, or property, without due process of law," thereby freeing up a new low-wage labor contingent. The constitutional amendment did almost nothing to assist newly liberated African Americans, whom it was supposed to serve. In fact, it was consistently interpreted by the courts to protect the rights of corporations, deemed "persons" under the law (1886) and thereby protected from state regulatory legislation (e.g., minimum wage, occupational hazards, child and female labor protection, ceilings on public utility rates) as a matter of "due process" (Hunt & Sherman, 1990, pp. 100-101).

The rise of the factory system after the Civil War led to increasing "economies of scale" (reduced per-unit production costs by way of mass production), the concentration of Wall Street investment capital, giant anti-competitive combinations and corporate mergers, and the growing importance of transportation and communication in linking these industrial chains. Hitherto disparate economic subcenters could be linked in real time, enabling industries and the commodities markets to use wired communication so as to gain advantage over rivals in terms of reduced inventory costs, efficient freight delivery scheduling, price differentials in raw materials, marketing data, advertising, legal advice, fewer middleman costs, and availability of other strategic information, including intelligence on worker organizations. Telegraphed messages could also be encrypted for commercial security or deceptively sent as disinformation to sabotage competitive firms.

Before the Civil War, the telegraph (later the telephone) had already greatly helped consolidate the trade in grain and other commodities in regional urban trading centers such as New York, Chicago, St. Louis, Philadelphia, Boston, Buffalo, Kansas City, and Milwaukee, and later in Toledo, Omaha, Minneapolis, Duluth, and New Orleans (Beniger, 1986, p. 249; Du Boff, 1984, p. 571). In the latter half of the 19th century, Western Union (created in 1856 following the mergers of Mississippi Valley Printing Telegraph and New York Telegraph) completed its dominant position in national telegraphy with the buyout of the American and Pacific Telegraph Company (owned by the robber baron and stock manipulator Jay Gould) and United States Telegraph, making it a bicoastal monopoly by 1881. Gould would turn the tables by securing controlling interest of Western Union and the national telegraph system in 1881, running the company with ruthless corruption. By 1880, after a monopolistic horse trade with Bell to separate their areas of operation, Western Union held 92% of telegraph messages and 89% of total telegraph revenue in the United States (Brock, 1981, p. 85).

The telegraph was to be vital in tying together a national market of stocks and commodities and, with the laying of the transatlantic cable in 1866, in international commerce as well. In an age of rogue capitalists, swindlers, wholesale bribery, and stock manipulation, the telegraph's intelligence-gathering capability was part of its instrumental appeal. It also helped centralize management of resources in increasingly dispersed monopolistic corporations. With high-speed information at their disposal, the larger industries in the United States could maintain headquarters in one city, operate manufacturing in another, get access to raw materials in other regions, and make use of legal, marketing, distribution, and other services in other parts of the country or even abroad. By the end of the century, telegraphy greatly facilitated the trend toward regional and industrial monopolies, the standardization of goods and prices, and the dominance of Wall Street financial houses.

Communication and privileged access to it by large-scale users (with lower rate charges) would continue to be a critical element in the elimination or forced merger of many small- and medium-scale enterprises and their substitution with brand name products and monopoly capitalism (domination of each major industry by a single corporation). The telegraphing of accurate or inaccurate information could also precipitate banking panics and contribute to economic depression, much like the way computerized, large-batch "program trading" of securities by big institutions "fueled" the New York stock market crash of 1987 (Associated Press, 1988, p. 27). When Morse introduced the telegraph, U.S. Postmaster General Cave Johnson, who favored government ownership, was concerned enough to warn that "an instrument so powerful for good or evil" should not "with safety be left in the hands of private individuals uncontrolled by law" (quoted in Winston, 1986, p. 302). He was remarkably prescient.

The telegraph also made possible urban-centered mass communications in the United States, abetting the newspaper reporting of national daily events and largely displacing, or at least appearing to displace, the weight of editorial opinion. News had become a product—and a valuable commodity. During the war with Mexico (1846-48), the major eastern newspapers seized the opportunity to speed up sensational and highly partisan coverage of events, no longer relying on postal service, although few miles of telegraph line existed at the outset of the hostilities. By the war's end, however, the business manager of the *New York Herald,* Frederic Hudson, could exclaim: "With the brilliant conflicts on the Rio Grande the Telegraphic Era of the Press began. What a commencement! What a revolution!" (Thompson, 1947, p. 220). The jingo press had found their magic carpet.

Urban centers in general monopolized the flow of commercial information even before the Civil War, New York City in 1851 being the terminus of 11 telegraph lines; big-city newspapers only added to the sense that large metropolitan areas were where the action was. To further consolidate their position, newspaper publishers quickly seized upon telegraphy to organize themselves into a press consortium in 1848. Born of a newspool covering the U.S. war in Mexico, the publishers later called their cartel the Associated Press (AP), which sold news to subscribers—and formed the basis of a national mass media. Newspapers would never again simply be a medium for opinion, for better and for worse, and with the aid of the telegraph, daily "news stories" would increasingly be turned into dramatic, attention-grabbing spectacles or mere razzmatazz for selling advertising. And those papers that could not afford AP dispatches "starved for distant news or used less reliable services founded after the AP" (Tuchman, 1978, p. 19). Together with the European powers, the United States was beginning its ascent as both an industrial and communications empire, and before the end of the century, virtually the entire world would be wired by the Atlantic nation states.

By 1866-67, the same year the U.S.–U.K. oceanic cable was opened (under the direction of Lord Kelvin, born William Thomson), AP established a bureau in London. By 1873, the national press agencies of the United States, the United Kingdom, France, and Germany had set up agreements that essentially formed a global news cartel. In 1875, AP agreed to not distribute news in Europe or South America in exchange for its having a monopoly over the distribution of foreign news within the United States (Frederix, 1959, p. 226), an arrangement that lasted 40 years. Domestically, the AP worked out another "sweetheart deal" with Western Union, in which the press agency would not report negatively on the telegraph monopoly as long as the latter refused to serve competing news agencies, both benefiting from the bargain through the news and information grip that it conferred (Du Boff, 1984, pp. 581-582). In Britain, the Reuters press agency and one of the country's telegraph companies had worked out a similar compact.

Although the telegraph had a host of irregular uses, as diverse as the "sweet nothings" wired between separated lovers, parsimonious death announcements, and a novel long-distance form of competition between chess aficionados, its most profitable applications were in the restructuring of the U.S. economy. Electric long-distance communication was assimilated into the mainstream because of its broad utility to a nascent but expanding American industrial capitalism. The personal familiarity of buyers and sellers in the early part of the 19th century was transformed by century's end into more abstract monopoly capitalist exchanges between direct producers of goods and services on the one hand and far-removed middlemen and consumers on the other. Telegraphy became a boon to big business by drastically reducing long-distance information and transaction costs and inducing other kinds of savings and by speeding up interactive communications and the pace of capital turnover and investment (Tarr et al., 1987, p. 69).

The industrialization of communications had other social impacts. Along with the two other major infrastructural developments of the 19th century—the railroad and the telephone—the telegraph was critical to the process of nationalizing markets, in standardizing and stabilizing commodity quality, inventory, and price in different parts of the country (to the advantage of large-scale manufacturers), in formalizing a national language of business, in facilitating cross-regional financing and stock ownership, and in establishing service norms and contributing many other information functions, including control (and blacklisting) in labor organizations.

By the late 19th and early 20th centuries, telegraphy, joined by telephony, took on other functions in the service of property. These included the spread of urban fire and police call boxes that gave local governments more explicit control in limiting damage to industrial and commercial enterprises and alerting authorities to potential mass actions by workers (the 1863 anti-draft riots in New York City and various pro-strike demonstrations) or burglary in the neighborhoods of the new bourgeoisie (Tarr et al., 1987).

The logic of capitalism in the early 19th century required a more refined system of long-distance communications to help overlay the key nodes of information switching and exchange on an existing infrastructure of economic and technological activity. It was not popular social needs that impelled the imagination of telegraphy, but rather the material necessities of expanding private interests giddy with newfound powers to communicate and transport their goods, money, media, and services. In the capitalists' perennial search for talent and new ways to raise the threshold of wealth accumulation and technological and information control, telegraphy and Samuel Morse were among their great discoveries.

Telephony Discovers Alexander Graham Bell et al.

The telephone, like the telegraph, was not the inspiration of a single individual, but rather a composite of ideas, rooted in industrial capitalism, whose time had come. The emergence of the modern corporation and the boom in office buildings after the Civil War made necessary such technologies as typewriters, adding machines, and elevators, all of which were conceptually feasible much earlier, from 20 to 250 years earlier, in fact. But corporate-scale capitalism, with its emphasis on large industrial and service centers, conceived their actual materialization in the 1870s (Winston, 1995, pp. 68-69). Integrated with the organizational applications of these new office technologies, electric communications would help satisfy growing industrial demands on time, space, and labor.

Alexander Graham Bell, son and grandson of well-known elocutionists of their time, migrated to the United States from Scotland and Canada after having studied with the German electrical scientist, Hermann von Helmholtz. Bell, truly a remarkable and creative mind, had spent his first years in Boston as a teacher of the mute and deaf while working to develop a hearing device and pursuing an interest in the electrical transmission of sound. Urged on by the president of the school where he taught and the inspiring figure of a fellow Scottish emigrant and millionaire steel magnate Andrew Carnegie, Bell raced against time and equally skilled competitors to patent a "talking machine."

It was 1875, "the heyday in America of laissez-faire venture capitalism" without the constraints of regulation or anti-trust laws (Brooks, 1976, p. 42). A period of unbridled faith in machinery, it was also the era of the robber barons. People like Jay Gould, Jim Fisk, Daniel Drew, Cornelius Vanderbilt, E. H. Harriman, J. P. Morgan, and John D. Rockefeller corrupted the institutions of democracy and the republic to build an industrial aristocracy of concentrated wealth and came to be indicted by the writers of their time, like Edward Bellamy (*Looking Backward*), as destroyers of civic values. The transcontinental railroads, telegraph, and telephone had transformed the regional and local cultures of the country into a unified economy dominated by urbanism, large-scale manufacturing, national finance, stock and commodities exchanges, plutocracy, interlocking directorates of power, and a growing industrial proletariat, 22,000 of whom lost their lives or were injured working on the rail lines in 1889 alone (Zinn, 1980, p. 250).

The development of telephony would repeat and magnify the experience of telegraphy. Both technologies had close historical connections in terms of their technical aspects, personalities, and political economic dimensions. By the time of the telephone, however, the market value and private control of long-distance communication was uncontested. In the United States, Western Union had become a virtual singular entity in telegraphy, the U.S. economy by the 1870s shifting from a relatively competitive to a more monopolistic form. Western Union and American Telephone and Telegraph (AT&T) would be among the leaders in the rise of a free-swinging American industrialism and in the consolidation of national markets.

For a time, Western Union would challenge the Bell interests in the control of telephone service, hiring such creative technical minds as Elisha Gray (who would later be put in charge of the AT&T manufacturing arm, Western Electric), Amos Dolbear, and Thomas A. Edison. The brilliant Edison quickly provided his new employer with the carbon transmitter, which was acknowledged as superior in its voice quality to Bell's instrument, thereby giving Western Union a short-lived edge on its competitor. For their part, the Bell interests fought patent and monopoly wars for some 40 years, capturing the lucrative markets of the New York brokers who had been customers of Western Union's Gold and Stock Telegraph Company (Brock, 1981, p. 93). The two competing monopolies, together with the remaining smaller companies in the field, were locked into wasteful legal and resource battles. AT&T was famous for using the strategy of purchasing patents of auxiliary equipment to block their use by competitors, which severely held back the growth of telephony in the United States, especially residential access, until well into the 20th century.

AT&T's monopoly pricing structure in the 1890s made telephone service inaccessible for most Americans. Residential telephone in Washington, D.C., cost between $36 and $96 per year (roughly equivalent in current prices to $2,500 and $6,500), compared with $16 in Stockholm, $18 in Paris, $36 in Berlin, and $100 in London (where the post office held back telephone development to protect its investments in telegraph). In New York City, residential rates ranged from $125 (two-party line) to $180 (direct line), Boston at $50 (two-party), and Philadelphia and Chicago between the Boston and New York rates. In terms of long-distance charges, those in Europe were much lower at the time (e.g., a Boston to Washington, D.C., stretch cost $4; the same distance in England, $1.30; and in France, $0.70).

Telephone density (telephones per 100 population) as a measure of access suggested a very slow start in the American system (e.g., New York City with less than 1%), compared with the largely state-owned networks in Europe. Only 266,431 telephones, mainly for business, were in service in the United States by 1893 in a population of some 65 million, reaching 2.3 million 10 years later (in a population of 80 million). Rates would come down after the turn of the century, but by 1920, less than 20% of American homes had access, only 40% at the time of Pearl Harbor (about 15 phones per 100 population). By World War II, more American families owned automobiles than telephones.

Yet one more indicator of the weak public commitment of AT&T was that, by the 1920s, telephone operators serving business were twice the number as those serving residences. Alexander Graham Bell himself acknowledged that he developed the telephone "as a means of communication between bankers, merchants, manufacturers, wholesale and retail dealers, dock companies, water companies, police offices, fire stations, newspaper offices, hospitals and public buildings, and for use in railway offices, in mines and [diving] operations" (quoted in Winston, 1986, pp. 329-330). Nary a word about families (Danielian, 1939, p. 101; Stehman, 1925, pp. 45-48).

Bell and his backers had focused the direction of investment in the early years almost exclusively for the large profitable urban areas of the Northeast. The first Bell regional operating company was the New England Telephone Company (1878), with its original telephone exchange in Hartford. After Massachusetts began to impose tighter restrictions on holding companies and trusts in the 1890s, Bell Telephone moved all of its assets to the more free-wheeling New York City, which is still its international headquarters. The company went on to try to control the emerging radio and film divisions in the expanding communication industries. Its main customer base has always been large businesses, and the rhetoric of "universal service" would not become close to a reality for U.S. households until after World War II.

Constantly in search of market domination in both equipment and service provision, the predatory management of the Bell Telephone Companies (put under the parent corporation AT&T in 1885) eventually bought out all of its major competitors and, for a time (1909-13), even established controlling (30%) interest in Western Union. The Sherman Anti-Trust Act of Congress (1890) was a weakly defined and enforced attempt "to protect trade and commerce against unlawful restraints and monopolies" in the corporate economy, but it did create a public climate that focused more attention on AT&T's attempts to monopolize telecommunications. The U.S. Supreme Court was unenthusiastic about the act, and both Democratic and Republican administrations did little until reformist "progressive era" administrations (Roosevelt, Taft, and Wilson) began to initiate legal action against some of the more scandalous behavior of the industrial monopolies (e.g., see Carman, Syrett, & Wishy, 1967; Ginger, 1975; Josephson, 1962; Wasserman, 1972; and Zinn, 1980, for colorful accounts of this period).

AT&T aspired for the telegraph monopoly in the interest of blocking potential competition and using Western Union's vast resources to develop nationally wired telephony. The tension between them was ultimately resolved through the intervention of the U.S. Interstate Commerce Commission and the U.S. attorney general, which resulted in the "Kingsbury Commitment" of December 19, 1913. This agreement severed Western Union from AT&T, separated service rights in telegraph and telephone (retaining for AT&T the rights to the private line telegraph market), and forced the telephone company to open its long-distance lines to interconnection by smaller telephone operating companies, to limit further acquisitions, and to accept other forms of government regulation. By this time, the telephone was out of the hands of its original tinkerers, craftspeople, and creative designers, who were now gone or comfortably removed from everyday decision making in the complex, national enterprises.

Following the expiration of the master Bell patents in 1893 and 1894, many small companies sprang up around the country (more than 3,000 by 1902), handling largely the rural traffic in which Bell had shown little interest. One important legal battle in 1880 pitted the high-powered American Bell Telephone Company (as it was then called) against a small upstart based in rural Eberly's Mills, Pennsylvania, the People's Telephone Company, organized by a local mechanic, Daniel Drawbaugh. People's Telephone was seeking to sell its telephones, closely resembling the Bell design but claiming to have antedated it, in New York City, the most lucrative market of them all and Bell's headquarters. Testimony in the trial filled more than 6,000 pages, and it was not until the case went to the U.S. Supreme Court in 1888 that a tight 4 to 3 decision was rendered for Bell (Brooks, 1976, pp. 78-81).

According to one study, New York Bell Telephone allegedly later blocked People's Telephone from starting service in New York City by virtue of the former's stock control of the monopoly Empire City Subway Company's underground system, through which telephone lines were required to be run. Bell denied any anti-competitive behavior in the case. People's Telephone was part of a small-scale telephone operators consortium, the National Association of Independent Telephone Exchanges, which would challenge Bell for service rights in many other metropolitan locations (Stehman, 1925, pp. 56-57, 80-81). Defeated by Bell, the rural mechanic and tinkerer Daniel Drawbaugh, whose name is largely forgotten and in any case never had the same ring to it, died nearly penniless in 1911 (Goulden, 1970, p. 46).

The most formidable claims to telephone rights would come from the telegraph monopoly Western Union, a company that earlier had turned down an offer to buy out the Bell patents for $100,000 and that, ironically, would be later taken over by Bell. Opportunities for monopoly profits also drove many less serious claims. One of the more frivolous was that of Antonio Meucci, a brewer and candlemaker by training, who displayed for the courts his non-electrical "telephone" made of a tight wire connecting two tin cans—the "lovers' telegraph" that had already been in circulation for more than two centuries. Another was based on a claim that the would-be inventor had heard the croak of a bullfrog transmitted over a telegraph line. The obvious commercial utility of the telephone had not been missed by Europeans either. In French and English newspapers, it was reported that telephony could be achieved by placing plates of silver and zinc in the mouth, "the one above, the other below the tongue" and attached to a telephone line, "words issuing from the mouth so prepared are conveyed by the wire" (Brooks, 1976, pp. 36, 77).

Popularly, the name of Alexander Graham Bell is most closely associated with telephony in the United States, not because he was the inventor of the telephone (as with the telegraph, many creative individuals were involved in its collective development), but because his company was the most successful at buying out and legally contesting the competing patents of his time. In many instances, this happened as a result of manipulation of patent law, by which Bell and his associates bought out auxiliary patents (e.g., the Berliner microphone), without which a fully developed telephone system could not work. By doing so, he and his financial backers prevented rivals from competing. The Bell interests also manipulated the patent system by bringing a number of infringement suits for no other reason than to stall potential competition in the courts and thereby winning through financial attrition.

The device that Bell's main financial backer, Gardiner Greene Hubbard, brought to the U.S. patent office on February 14, 1876, just 2 hours before their principal rival, Elisha Gray, registered his own claim to the telephone with the same office, was in fact not a telephone but an incomplete design for one. For many years, the dispute about its authenticity would lead to a litany of litigation and questions about the rightful ownership of the title to this extraordinarily lucrative piece of technology. But Bell's Patent #174,465 was awarded even though it never actually claimed to transmit speech but only to advance telegraphy "and other uses . . . such as the simultaneous transmission of musical notes, differing in loudness as well as in pitch, and the telegraphic transmissions of noises or sounds of any kind" (quoted in Winston, 1986, pp. 320-321). This patent served not only as the legal basis of Bell's infringement suits but also as a dowry. Hubbard had demanded it, initially granting Bell only 10 of 5,000 shares issued, before permitting his daughter, Mabel (with 1,497 shares), to marry the "inventor of the telephone"—effectively making the marriage proposal one between a fiancee and financier (Schement & Curtis, 1995, p. 30).

Hubbard, unable to raise sufficient capital for telephone diffusion or keep financial pace with the wave of litigation over the Bell patents, soon lost his emerging empire to the big money in Boston, which would form AT&T. Under the aggressive management of Theodore Vail, cousin of the man who was Samuel Morse's assistant, the Bell/AT&T fortunes grew. Vail left the company in 1887 after pushing hard for expansion, patent protection, and business reorganization but would return 20 years later (Fischer, 1992, p. 38). In the meantime, Western Electric, the company of Bell's former competitor, Elisha Gray, and of Western Union, was taken over and became the exclusive equipment supplier to Bell. (This would be an issue of considerable irritation to the smaller U.S. telephone operators for the rest of the century—until the 1984 breakup of AT&T forced the corporation to give up regional telephone service but retain its long-distance services and equipment sales, the latter two segments by far the most profitable slices of their business.)

Alexander Graham Bell's fascination with communicating devices originated with his interest in the physiology of human speech, a tradition handed down by his father and grandfather in his native Scotland. After achieving his 1876 patent, his involvement with telephone development was actually short-lived, pulling out of the company that bore his name following disagreements with its new president, a Boston brahmin, William H. Forbes, in 1880. Hubbard had been forced out the previous year. Bell sold his shares in the company and went on, comfortably endowed, to involvements with other experimental technological enterprises, including Columbia Records in 1887, but nothing of the technical sophistication or importance of the telephone. He also returned to the teaching of the deaf.

Bell had played a significant role in the development of communication technology and the infrastructure of a national corporate economy—but had the Bell company patents not been awarded, the development of long-distance voice communications surely would have been attributed in his own time to another "inventor." Research and development of technology was moving in toto to corporate laboratories, where applied science could be more closely supervised by business management. As part of AT&T, Bell Laboratories would become one model of such a hub for commercial research. Scientists, engineers, and inventors who followed Bell were likewise involved in projects of much larger orders of power than mere personal glory, but their enshrinement in the history of technology helps keep alive public faith in the wisdom of America's political and economic institutions and its mystical cult of the individual.

Regulated Monopoly Communications

In 1909, with a $30 million check, AT&T bought controlling interest in Jay Gould's Western Union, which it continued to hold until the forced separation of services by the Kingsbury Commitment of 1913. AT&T itself came under the suzerainty of the foremost tycoon of the era, John Pierpont Morgan, who at various times had controlled U.S. banking and finance, steel, iron, and shipping. In 1905, he bought $150 million in AT&T's convertible bonds and gained controlling interest, afterward bringing back Vail as president of the corporation (Brock, 1981, pp. 150-151). In 1892, Morgan had also featured in the merger of the Edison General Electric Company and its rival, the Thompson-Houston Company, and subsequently forced out the Edison General Electric president and the founders of Thompson-Houston, while putting two of his own partners on the board of the new company, General Electric. By 1912, the House of Morgan controlled not only huge banking and insurance companies but also nonfinancial corporations, including AT&T, United States Steel, General Electric, International Harvester, and Western Union (Kotz, 1978, pp. 31-32, 36).

Under the Kingsbury Commitment, AT&T had traded government regulation for its effective monopoly control of the market in long-distance telephony and local services. Vail had said in the 1907 AT&T Annual Report that he saw no problem with government control, "provided it is independent, intelligent, considerate, thorough and just, recognizing, as does the Interstate Commerce Commission . . . that capital is entitled to its fair return and good management or enterprise to its reward" (quoted in Brock, 1981, p. 158). Vail must have found the government's control "independent" and "just," given that AT&T took over 500,000 independent company telephones in the next 3 years (Bolter, McConnaughey, & Kelsey, 1990, p. 75). AT&T had neglected technical improvements, popular access (especially in the Western states), and affordable service to this point but had won the battle for command of the market.

With AT&T firmly in control of long-distance and international communications after the Kingsbury Commitment, the fortunes of Western Union began to slide by the mid-1920s. Air mail made its appearance at this point, and together with telephone eroded the demand for telegraphy. By 1914, the United States already had some 70% of the world's telephones (Pool, 1983a, p. 6), although some smaller European countries (e.g., Sweden) maintained higher levels of access. Toll telephone had 59.6% of all long-distance revenues by 1926 and 79.1% by 1949, with telegraphy slipping to 2.9% and postal telegraph to 12.7% by 1949, and all telegraphy to 1.3% by 1983 (Bolter et al., 1990, p. 77). Before its breakup in 1984, AT&T was, by some measures, the largest corporation in the United States and certainly the largest communications entity in the world (second in the 1990s to Japan's NTT).

Internal telephone expansion in the United States eventually became more comprehensive than almost anywhere else, but not because of private monopoly control. AT&T was interested primarily in equipment manufacture, the larger, more profitable urban and business exchanges, and in controlling patents and interconnections. It made little effort to bring telephone to small towns and rural areas, and the main reason why telephone developed on a broad geographic basis was because of the initiatives of smaller local service companies, many of them little more than village cooperatives (Brock, 1981, pp. 106, 123-124). Even in the large metropolitan areas, the Bell interests were primarily interested in business lines: In New Jersey and New York in 1891, there were 7,322 commercial customers, compared with 1,442 residential telephones. Telephone service for homes, at one tenth a worker's monthly wages in Los Angeles (higher in Northeastern cities), "was a stepchild in the system" (Fischer, 1992, p. 39).

Despite the promise of "universal service" in the Kingsbury Commitment, access to telephone was long in waiting for most Americans. In exchange for accepting nominal government regulation, AT&T was able to ward off antitrust laws, with protections broadened under the Willis–Graham Act of 1921 for its monopoly position, including exemptions from important sections of the Sherman Antitrust Act. Regulation protected AT&T from political control and market competition at home, although in 1924 it was pressured to sell off its huge global manufacturing subsidiary, the International Western Electric Company, to another U.S.–based company, International Telephone and Telegraph (ITT), which became the most powerful overseas telephone and telegraph entity and eventually the biggest telecommunications conglomerate outside its home country. Domestic regulation included guarantees of AT&T interconnection with smaller telephone companies and thereby gave them a technical stake in the monopoly's long-distance trunk lines, which precluded any further serious competition in those markets (Brock, 1981, pp. 12, 156). By 1930, 80% of telephones in the United States would be AT&T's, with virtually all of the rest tied by interconnection to the Bell lines (Fischer, 1992, p. 51).

TAYLORISM GOES GLOBAL

By the early 20th century, U.S. capitalism was well into its monopolistic phase, the "trust busting" of Republicans and Democrats having accomplished little in deterring the agglomeration of capital into ever larger units. The small workshop and the family-owned enterprise would fade from the nucleus of the U.S. economy. Having a steady labor supply, corporations began to shift their emphasis from control of production to the disciplining of the labor force, widely accepting the approach taken in F. W. Taylor's time-motion studies (see Chapter 1). Taylorism also encouraged the professionalization of management, the basis of the scientific study of the workplace (Howard, 1986, pp. 18-19).

A few years after Taylor's celebrated work, Henry Ford opened his earliest plants based on assembly-line mass production techniques that reduced the costs of manufacturing automobiles while further fragmenting the division of labor in the workplace. A new modernist era was in the making that would ultimately touch every aspect of social and personal existence, continually standardizing mass-oriented products and lifestyles, and bringing with it the logic of planned obsolescence. Before the end of the century, the industrial model would be extended throughout the world, and communication technologies would assume a central function in the conquest of time and space. With long-distance communications, the segmentation of work and corporate operations would be dispersed on a national and eventually international division of labor and production.

The last quarter of the 20th century saw an intensification of "Fordism" and "Taylorism" in industrial markets, extending the principles of mass production and scientific management to larger and, at the same time, more specialized groups of workers and consumers. A wide range of technological development within the capitalist world system, consolidated after World War II, enabled global corporations increasingly to escape national regulation and supervision and transcend the limits of national boundaries. A constant search for new markets and lower "factor costs," particularly labor, along with competition among the leading capitalist countries, propelled industrial transfers to the Third World. As the champion and center of the rehabilitation of industrial capitalism in the postwar world economy, the United States took the lead in such transfers.

By this time, factory and even a majority of office and service workers had become little more than appendages of the machine, the automated workplace. This meant that little training and skill was involved in most areas of production, especially in routinized and assembly-type work, which made low-wage Third World labor very attractive. In many industries, but particularly those in Third World countries, female labor was brought in to perform the kinds of tasks that, from management's perspective, required small motor skills, discipline, patience, willingness to work long hours under stressful conditions, low propensity toward rebellion and unionization, and a large pool of hungry workers.

Poverty that would ensure an oversupply of workers made the Third World locations very appealing as sites of transnational production, unless strong unions were present to bargain the conditions of employment (as in South Korea or the Philippines). Developments in transportation (e.g., jet aircraft, containerized shipping) and communications (e.g., satellite, oceanic cable, microwave, computers, fax) were other prerequisites for this transnational phase of capitalist development. AT&T, RCA, Western Union, and ITT were among the major U.S.-based corporations (joined by western European and Japanese corporations) with well-established global telecommunications pipelines that facilitated the movement of other TNCs to offshore locations.

Industrializing Sound and Image

The executive class in the United States that turned wired communications into national-scale monopolies presented the rest of the world with a political economic model of how telecommunications could profitably be run. In most of Europe and in Japan and their colonies in the Third World, state-owned telecommunications systems was the prevailing type until the 1980s. The post-World War II breakthroughs in new communication technologies, especially satellite, fiber optics, signal compression, digital electronics, and computers, would stretch, and in many cases overwhelm, the capability of the public sector to handle the technical and regulatory demands of private corporate users.

Before this stage could be reached, long-distance communication had to prove its value to corporate hierarchy, and that it did. In the 19th century, vertical integration (higher- and lower-end industrial units under common ownership) was speeded up by electric communications, tying together credit, collections, orders, traffic and shipping, advertising, and accounting (Beniger, 1986, p. 256), as was the corporate form of nonindustrial firms (e.g., law, finance, trade). When Alexander Graham Bell uttered his celebrated first telephonic words to his subordinate, "Mr. Watson, come here, I want you," he demonstrated the technology's utilitarian value as a command structure over distance.

Over time, the telephone became widely used for personal communication (to the chagrin of the sternly business-minded), but its more lucrative value was in its structuring and standardizing the use of space in long-distance business transactions, much as the clock had done in the workplace. Business telephone uses permitted major transactional cities like New York, London, and Paris remote control over factories located elsewhere, even while the concentrated northeastern business establishment in the United States prevented diffusion of telephone to Chicago for 12 years and transcontinental connections for 29 years. With wired communications, those colossal monuments to corporate egos, the skyscrapers, could function without messengers having to clog elevators. And, together with the automobile, the spread of telephony helped segregate the lives of the rich and famous to the suburbs, far from the maddening urban proletariat (Inose & Pierce, 1984, p. 14).

The sound of telephone and radio and the image of photography and (silent) film were disembodied forms of communication, characteristic of the abstract and depersonalized relations of production and consumption in industrial capitalism. The convergence of advertising, public relations, and broadcast media would foster larger utilitarian relationships and identification between products and the commoditized citizen, the modern "consumer." Consumption replaces rational thought and discourse as the raison d'être and proof of existence, standing Descartes on his head: *Consumo, ergo sum* (I consume, therefore I am).

Electric, electronic, and synthetic communications (telegraph, telephone, film, and broadcasting) helped normalize the forced separation of working-class family members as demanded by an expanding industrialism, Americans being by far the most migratory of all industrialized peoples. Having little practical choice but to accept the conditions of relocation from city to city or from rural to urban settlement, the telephone, the movie theater, the radio, and later television provided pleasant escapes from the drudgery, isolation, and alienation of blue-collar and clerical livelihoods—but only to the limited extent that they were affordable. And it was many years—in the case of telephone, about 75 years—before the majority of working people had the means for regular communication and media access. The full social cost of commercial media diversions is difficult to measure, but their general level of banality and substitution for aesthetics and learning has left craters of cultural debris, what Kennedy–era FCC Commissioner Newton Minow called "a vast wasteland."

In the 1890s, newspapers, like industrial corporations, were big businesses and the prototype "mass media." Publishers and editors of the time were often straightforward about the business mission of their product. A South Dakota editor commented in 1891 that "a newspaper office, in country village as in city, is a business established by which editors and printers must make a living," and the editor of the *Boston Journal* said in 1894, "One thing should be clearly understood, and that is that this property has been bought for business purposes, to be run on business principles. . . . You cannot put that too strongly" (cited in Baldasty, 1993, pp. 98-99). A. J. Liebling wrote, "Freedom of the press is guaranteed only to those who own one" (cited in Lee & Solomon, 1990, p. 75).

Broadcast media would take their cues from the press, avoiding strong political party identification and analytic reporting and, instead, pushing short, entertaining items, advertising, and variety show styles to news and features. It was important to patronize, not offend, the reader and, most of all, the advertisers, from whom the bulk of revenues was generated. Good news was no news. Regular beats (courts, police and fire stations, city hall, the hospital) were the most reliable sources for dramatic, often gory copy and relied on overworked reporters padding, sensationalizing, and frequently fabricating stories for sales appeal and their own job security (Baldasty, 1993).

In the early 1960s, one scholar observed that the mass media serve "not as a link between persons with something to say and an audience with a cause to listen, but rather as a marketing device, with the needs of marketing rather than the creative impulses of authors or the needs of listeners as the determinant of content" (Dan Lacy, cited in Schiller, 1969, p. 20). From the beginning of the American republic, the federal government had little economic power, abdicating decisions about development largely to the private marketplace. The very corporations directing the public imagination toward the life of consumption became the owners and interlocking directors of the media organized for that purpose. Gannett, the largest newspaper chain, with some 90 newspapers, is essentially an interlocking directorate of Fortune 500 corporations (as of the mid-1980s)—Merrill Lynch, McGraw–Hill, Philips Petroleum, Kellogg, New York Telephone (now Nynex), McDonnell Douglas, Kerr McGee, Sohio, Eastern Airlines, 20th Century Fox—while the *New York Times* interlocked with Merck, Morgan Guaranty Trust, Bristol Myers, Charter Oil, Johns Manville,

American Express, Bethlehem Steel, IBM, Scott Paper, Sun Oil, and First Boston Corporation (Bagdikian, 1987, p. 12).

Although public resistance occurred at each step in the process of corporate consolidation of mass media, and alternative media were always there, the main battles for the airwaves and the printing presses were among the media giants themselves, sometimes on global fronts. World War I was a turning point in the United States assuming a leading role as an industrial, financial, military, and diplomatic world power, which gave rise internally to a new wave of business expansion. In 1920, in the context of a world carved up for colonial exploitation by western Europe, the United States, and Japan, "there was minimal regard . . . for the radio spectrum as a resource useful to general development or as a helpful instrument in the conservation and maximization of other resources. To the contrary, in the United States, radio quickly became an adjunct to the mass production way of life" (Schiller, 1969, p. 21).

COMMUNICATION IN WAR AND PEACE

It is often claimed that communications is the ointment of peaceful relations and that its withdrawal incites or prolongs hostility. On that assumption, it has become a cliché that the Battle of New Orleans during the War of 1812 between the governments of Britain and the fledgling United States would not have occurred had the telegraph already existed. The argument is made that peace accords had been signed before the battle took place but that the military forces contesting New Orleans had not received word of it. The claim may be technically reasonable, but the problem is that its logic is *only* technical and quite misleading in the real world of global political and economic power struggles, which the War of 1812 was certainly one example.

Communication technology does not prevent or terminate wars and is not used to that end unless protagonists so wish it. In 1812, Britain was the world's most powerful imperial, naval, and economic power, and the aggressively anti-Anglo initiatives of its rival, Napoleonic France, provided the former with the excuse to cut off U.S. commerce with the European continent. Britain's naval blockade was also an attempt to recover for itself some degree of control over the former Atlantic colonies and try to contain further expansionism of the young American republic into such areas as Florida, Canada, and the Indian territories (documentation of the causes of the War of 1812 is found in Paterson, 1989, pp. 133-179). Were telegraphy available at the time, it would just as likely have speeded up military aggression and increased casualties.

It is likely that telegraphic communications on the eve of World War I contributed to the breakdown of diplomacy by speeding up and forcing the pace of decision making, inducing ultimata in an already hostile and suspicious environment and eventually resulting in electronically transmitted declarations of war (Headrick, 1991, pp. 139-140). Native American tribes that destroyed telegraph poles in the late 19th century, as the Cheyenne and Sioux did in the western states, understood that the lines were circuits of military information dedicated to their annihilation. Charles Mazade, a French historian, wrote in 1875 that telegraphic exchanges exacerbated diplomatic tensions between France and Prussia in 1870 by leaving out important details and chronological context while curtailing the time needed for calmer deliberations (Headrick, 1991, p. 74). The result was a war that only intensified nationalism, international rivalry, and political cynicism.

Another unfortunate reality is that the nature of war has become increasingly brutal with each advancement in technology, including communications. Long-range artillery, automatic and semiautomatic weapons, airborne strike power, grenades, missiles, fragmentation bombs, chemical, biological and nuclear ordnance, and the like have all added to the likelihood of involving civilian populations in warfare and the possibilities of indiscriminate killing, in comparison with 18th-century field armies facing off against one another. A telegraph in the hands of either belligerent in 1812 would have provided opportunities for faster military mobilization and better logistics and simply raised the level of violence. Indeed, as Daniel Headrick chronicles it, aggression in Europe in 1914 following the assassination of the Austrian archduke by a Serbian nationalist, was accelerated by the use of telegraphed ultimata among the subsequent belligerents; and when Austria declared war on Serbia, it was the first time that such a message was delivered by telegram (Headrick, 1991, p. 139). Aware of its strategic importance, the U.S. government took over all telecommunications in World War I.

As another scholar notes, the events leading to the onset of the "Great War," a time when modern communications had greatly contributed to global integration,

> the men in power lost their bearings in the hectic rush paced by flurries of telegrams, telephone conversations, memos, and press releases. . . . Newspapers fed popular anger, swift military mobilizations were set in motion, thus contributing to the frenzy of diplomatic activity that broke down simply because enough decisions could not be made fast enough in enough locations to bring the warlike stresses under collective control. Global war was the result. (Kern, 1983, pp. 260-261, cited in Harvey, 1990, p. 278)

It is naive to overlook the instrumental and repressive potential of techno-logical development, especially in the hands of powerful interests, and see only its benign or benevolent possibilities.

Electronic communications did not prevent, and could not have pre-vented, World War II. U.S. radar in the Pacific was able to detect the coming of the invading Japanese fleet to Pearl Harbor, but this did not alter what happened next, because political calculation, not technology, was in control. Distracted by the European conflict, the U.S. government and its military simply were not prepared to believe that Hawaii would be attacked and refused to accept technical evidence to the contrary. Had the United States acted more prudently, its casualties on that "day of infamy" might have been reduced, but not the devastation of the full-scale war that followed. The war was fought over the larger issues of access to resources in Asia and competing imperialisms, not because of any momentary planning lapses, logistical advantage, or desire to test military weapons and tactics.

Communication technology did not prevent the use of atomic weap-ons on Hiroshima and Nagasaki on August 6 and 9, 1945, even though the United States and Japan were in discussion about ending the war prior to the attacks. Literature was dropped over those two cities warning of imminent bombing, but the nature of the bomb was not disclosed. The suffering of the Japanese civilian population as a result of the attack is not well understood in the United States, and the climate of McCarthyism, national chauvinism, and the Cold War that followed the cessation of hostilities stifled any possibility of broadly educating Americans about atomic energy or the imperial origins of the war itself. If the use of atomic weapons had any positive outcome, it was in sparking public fear and outcry against its inhumanity and future use, by the 1960s reaching the idiom of popular culture as portrayed in such films as *Dr. Strangelove* and *Seven Days in May.*

Clearly, the advent of radio, radar, oceanic cable, computers, lasers, and satellites has not inaugurated an era of peace and social tranquility in the world. There have been more wars in the age of modern commu-nication technology than in any previous period of human history. Simply put, communication technologies in the hands of those who seek power advantage will serve aggressive ends; in the hands of those seeking egalitarian objectives, such technologies would have other utilities. The same telegraph line can transmit Valentine sentiments or coercive terms of unconditional surrender, depending on the social and political context in which it is employed.

During the U.S. Civil War, the telegraphic technology of the Union forces proved to be invaluable to its supply lines and other logistical needs (making a fortune for Armour, Gould, and other war profiteers), just as the railroad was to troop movement—opportunities that were denied to the Confederacy. There was even a book titled *Lincoln in the Telegraph Office,* about the long periods the president spent in the War Department Telegraph Office, receiving and coordinating battle plans with his generals (Oslin, 1992, p. 120). The Union forces, with their Military Telegraph Corps and more than 15,000 miles of new line, dominated the tactical information flow in the war, and "in Grant's final campaign, wires radiated from his headquarters to every salient point, enabling him to manipulate every movement of this troops in concert with others" (Oslin, 1992, p. 127). In the aftermath of that conflagration, the east-west rail and telegraphic lines provided a distinct advantage over the largely destroyed southern lines, which helped overwhelm the southern economy and solidify the northern Atlantic to Pacific vector in U.S. economic expansion. Military telegraphy in the Civil War would be largely transferred to the postbellum monopoly, Western Union.

Another advantage to the Union forces that telegraph expedited was in the ideological war, aided by near real-time reporting from the front. The press was obstructed from reporting the dark side of the conflict and the extent of northern resistance. In fact, desertion rates were extremely high on both sides, and anti-draft riots, often racist and proslavery in tone, broke out throughout the northern cities, including New York, where free blacks were lynched. Anti-war tensions were severe enough that the government shut down for a time the *New York Journal of Commerce,* the *New York World,* and the *Chicago Times* (Dennis, 1991, pp. 8-10). Almost nothing was reported about blacks who defended the Union cause, including the 186,000 army enlistees in the "United States Colored Troops" (Franklin & Moss, 1988, pp. 195-196).

During the U.S. intervention in Vietnam, television news did little to alter the public perception of the war presented by the Departments of Defense and State (Hallin, 1989). The television media showed that the United States was not *winning* the war, but without critically examining the official rationale for the intervention. Far from being hostile to television, President Lyndon Johnson promoted the use of the medium in southeast Asia as, what he and his adviser, Walt Rostow, considered, an antidote to communism. Johnson ordered the United States Information Service to help bring television to South Vietnam, and the U.S. Agency for International Development assisted importers in getting outdated black-and-white sets to sell in the country so that Vietnamese government propaganda and American programs like *Gunsmoke* could be broadcast (Gibson, 1988, pp. 286-288). Television brought pacification but no peace to Vietnam and helped prop up a military dictatorship and its U.S.

sponsor that rained violence and destruction on the land, "more deaths and misery . . . by Allied firepower than Communist terrorism" (Robert Chandler, cited in Gibson, 1988, p. 288).

The American media were far more critical of Soviet intervention in Afghanistan. In that situation, the Muslim fundamentalist forces fighting the Soviet Union and the government it backed in Kabul were treated by the American media as national liberators even though many were well known to be involved in international drug smuggling and represented political values similar to those the United States opposed in Iran under the Ayatollah Khomeini. Battles and personalities were cited in American television news, but the issues were left obscure and secondary, compared with the spin put on the "Soviet Union's Vietnam." When Gorbachev came to power and pulled Soviet troops out of Afghanistan, the story died on the spot in the American media, despite the fact that fighting thereafter intensified. The media, says Noam Chomsky, pay "tribute to the soundness of our self-correcting institutions, which they carefully protect from public scrutiny" and in the process organize a system of thought control to shield American business from the threat of democratic discourse (Chomsky, 1989, p. 20).

Radio Daze and Happy Talk Media

As the U.S. Civil War was so central in the diffusion of wired telegraphy, U.S. participation in World War I would bring about a great expansion of radio and the major corporations involved in its development. The Russo–Japanese War of 1904-1905 had already demonstrated the importance of radio in naval combat, as well as signaled the emerging trade rivalries of German and British telecommunications equipment companies (Telefunken and Marconi, respectively). World War I brought a huge U.S. military demand for radio vacuum tubes, and General Electric (GE), already involved in the production of electric light bulbs (and other electrical equipment), was at the outset the most favored supplier. Tubes for transmitters and receivers were needed as part of the land, sea, and airborne strategies in the war, the Signal Corps on one occasion alone placing an order for 80,000 tubes (Barnouw, 1975, p. 18).

Electric communications was increasingly becoming central to control, command, and intelligence in military (and business) operations. Developments in radio represented another important stage of modern warfare and in the evolution of the military-industrial complex. The use of "wireless" in point-to-multipoint communication maximized the strategic value of the technology but also offered more opportunities for interception. Hence, the U.S. military's reluctance to allow public access to the airwaves.

In the United States, the origins of radio go back to experimental wireless radio telegraphy (1895), public radio broadcasting (1906), and radio telephony (1915), which were developed initially for ship-to-shore and other commercial communication, before the advent of commercial radio broadcasting (1920). Some mistakenly believed that radio would close the era of oceanic cables, just as others later believed that the advent of television would spell the end of radio. However, radio presented some new problems for classified transmissions. By 1912, radio had already become a toy of amateur hobbyists, but in that year the federal government, through a Radio Act, began to license transmitters, using a national security state argument that hackers might interfere with sensitive U.S. Navy transmissions at sea.

After the United States entered World War I, all radio equipment—commercial, amateur, and armed services (except the Army's)—were put under the control of the Navy, a military monopoly it sought to extend after the armistice. With peacetime and a conservative, pro-business political tide, however, the profitable big corporate sector was not to be denied. A long-term collaboration of government, the military, and industry in the communications and electronics fields was foretold by the Department of War's appointment of Gerard Swope, an executive at AT&T's Western Electric, who, immediately following the war, became president of General Electric. Those military-industrial connections remained intact up to and beyond the U.S. presidential election of a former public relations prop for GE, Ronald Reagan (Nash, 1989, p. 278).

Strong industrial opposition to the idea of government control, backed by the Department of Commerce, led to the formation of a private "national" radio monopoly. Headed by General Electric, with Director of Naval Communication William H. G. Bullard on its board of directors, the idea was to take over the patent rights and assets of American Marconi (a British radio equipment subsidiary). GE then formed a radio patent pool with Westinghouse, AT&T, and the United Fruit Company (a major user of ship-to-shore communications for its plantations throughout Latin America and the Caribbean and holder of minor patents), which together created the Radio Corporation of America (RCA) in 1919, GE holding the largest share. GE, together with RCA and Westinghouse, established NBC radio broadcasting in 1926. Under a cross-licensing agreement, radio receivers and parts would be allocated to GE and Westinghouse, marketed under RCA trademarks; transmitters and telephone-related service and equipment would be retained by AT&T and its subsidiary Western Electric; transoceanic, marine communications, and limited wireless telephony would be RCA's; and government purchases would be open to any of the partners. Their biggest competitors were hundreds of household "amateurs," some turning entrepreneurial, who, through individual ingenuity, put together crystal receiver sets and even transmitters, not unlike hackers of subsequent generations (Barnouw, 1975, pp. 22, 37-39). The RCA "combine" lasted until 1930, when it was broken up under antitrust legislation intended to open the radio industry to greater competition (Rossi, 1985). Western Electric, AT&T's manufacturing subsidiary, was one recipient of the many highly profitable wartime contracts from the U.S. Army and Navy. The government postponed all patent claims for the duration of the European conflict to avoid problems of military procurement. GE, in the meantime, was encouraged by the Wilson government to avoid further trade with American Marconi and to buy up its patents in order to reduce British influence in what was considered a highly strategic communication resource, especially given Britain's already overwhelming control of international submarine cable infrastructure. By the mid-1920s, RCA radio telegraphy would be a major intercontinental competitor to British cable interests. Once again, it was the political economic motivations behind war and territorial advantage that had served as the "mother of invention."

RCA, Westinghouse, and General Electric came to dominate the fields of radio and electrical appliances, but AT&T, with its control of telephone lines and radio transmitters, did not easily give up the fight for control of radio. Following its success with KDKA radio in Pittsburgh in 1920, Westinghouse, and later RCA, set up broadcasting stations in major cities across the country. AT&T in 1922 started its own "toll broadcasting" operations with WEAF in New York City, leasing space to any commercial or other interest wishing to use its "phone booth." WEAF was the first of AT&T's "long lines" group of 13 radio stations, stretching from New York and Boston to Minneapolis–St. Paul and Davenport, Iowa, that could be picked up by smaller stations across the country, with license fees ranging from $500 to $3,000 (Barnouw, 1966, p. 176). Industry had created a high-powered electronic pitchman.

AT&T also began making its telephone lines available to smaller telephone companies for radio networking, in competition with RCA and other radio telephony and radio broadcasters, but held back its biggest transmitters for its own stations (Kellner, 1990, pp. 29-30). RCA, GE, and Westinghouse found radio transmission over telegraph lines to be a technically unsatisfactory alternative. Under growing pressure from its partners in RCA and the U.S. commerce secretary, Herbert Hoover, AT&T settled this conflict in 1926 with the sale of WEAF to RCA (later becoming WRCA and WNBC) and by pulling out of the partnership. This act separated their areas of operation, taking the telephone company out of toll broadcasting, and opened AT&T wire services to the Radio Corporation on the condition that the latter relinquish its transmission arrangements with the telegraph companies. Wired radio and wireless telephony thereby would become an exclusive domain of the Telephone Company (Danielian, 1939, pp. 126-128). On yet another front, threatened by American Marconi's move into radio telephony, which would have made it a competitor for long-distance and transatlantic telephony, AT&T's Western Electric deceptively gained control of Lee de Forest's crucial Audion vacuum tube patent and subsequently withheld it from use by its rival.

Despite the competing and highly contested property claims on radio (and later television) in the national patent office and in the courts, such ownership assertions in reality, if not in law, were, for the most part, fatuous. Independent "inventions" were actually built on the foundations of diverse technological precedents, constituting a shared compendium of knowledge (sometimes illicitly acquired), private and governmental sponsorship, public subsidies of all sorts (including schooling), *and* individual brilliance. The idea of transmission of pictures by wire goes back to 1842, and early successes in long-distance phototelegraphy by Abbe Casselli occurred in 1862. In its practical meaning, an invention is but a legally recognized conceptual innovation established within a setting of private ownership and market incentives, what has come to be known as "intellectual property rights."

Engineers at Bell Laboratories (established by AT&T in 1925) would be part of a race to patent and market radio and later television and other video offshoots (e.g., the videophone) and to enable AT&T to dominate the entire communications industry. But radio relied on a wide range of talents from several countries, building upon the research that went into telegraphy and telephony and the work of Heinrich Hertz (Germany). Hertz's development of wave theory would educate Guglielmo Marconi (Italy), the person most often credited with the earliest transmission of wireless telegraphy (for which, in 1909, he shared the Nobel Prize in physics). The popular science writer Carl Sagan attributes the development of broadcasting, microwave transmission, and communication satellites mainly to the theoretical breakthroughs in wave theory by the mid-19th-century Scottish scientist James Maxwell (Sagan, 1995, pp. 10, 12).

Amos Dolbear, who had previously developed a telephone device, had also patented a radiotelegraph in 1882. The Russian, Alexander Stepanovitch Popoff, transmitted Morse code in 1895, and the Soviet Union would later honor him as the inventor of radio. Edison obtained a radiotelegraph patent in 1891, selling it to Marconi in 1903. Besides Hertz, Marconi drew on the work of Frenchman Edouard Branly, who produced the "coherer" wave detector to produce a primitive but successful radio transmitter over short distances (Oslin, 1992, p. 274). In actuality, no single individual invented the radio, but many innovative minds made important contributions to its development.

Marconi's ideas, not taken altogether seriously in his early adulthood in Italy, had received a better reception on his arrival in Britain in 1896, where the queen, the media, commercial interests, the Royal Navy, and the sporting world gave notice to and made various uses of radio. (Marconi's very first sale of radio was to the British War Office during the Boer War.) In 1899, Marconi (1874-1937) and his British company executives sailed for the United States and found similar enthusiasm, especially from the U.S. Navy. With mainly British capital, the American Marconi subsidiary was established in 1902. During this heyday of monopoly capitalism, corporations knew how to exploit their technological advantages, and American Marconi controlled most ship-to-shore transmission and much of the domestic commercial and international press traffic. In exchange for leasing Marconi equipment, for example, commercial shore radio stations initially agreed to transmit exclusively, except in emergencies, with company-equipped ships. Marconi was not to hold a privileged position in radio for very long, however (Headrick, 1991, pp. 118, 124, 126).

Nationalism and nationalist competition, presaging the bloodbath of World War I, led to the refusal of British Marconi Wireless and German Telefunken to allow intercommunication between rival radio systems onboard ships at sea, contributing to the *Titanic* disaster in 1912 (Headrick, 1991, pp. 130-131). The U.S. Navy, becoming anxious about what it perceived as British technological hegemony at a time when President Theodore Roosevelt was espousing U.S. claims to global power, preferred to use its own shore stations. It also supported the radio research of homegrown developers, including Lee de Forest and his Wireless Telegraph Company and a few years later the De Forest Radio Telephone Company (Barnouw, 1966, pp. 23-25). De Forest radio transmitters were installed onboard the Great White Fleet that Roosevelt sent across the Pacific in 1907 as a symbolic demonstration of America's global military superiority.

Marconi's patent rights in the United States would be continually challenged by the Navy and in the courts and, under government pressure, would later be taken over by a GE-led consortium and their RCA radio monopoly subsidiary. According to accounts of the *Titanic* disaster, Marconi pressed his company radio operators at the time to suppress the news and even the list of survivors in order to cash in on the inevitable rise in sales and stock prices for his equipment (Lewis, 1991, p. 106). After World War I, Marconi returned to Italy to serve under the fascist government that took control in 1922, a fact that hagiographers of his life usually do not mention. When Marconi married for the second time in 1927, the dictator Benito Mussolini was his best man. Mussolini appointed him to high posts, including president of the National Council of Research in 1928 and president of the Royal Academy in 1930 (Oslin, 1992, pp. 294, n9).

With the creation of RCA, it did not take long for its profitable and powerful utilities to be recognized and applied. In 1922, the first commercial programming began, with sponsorship going for $100 for the first 10 minutes (Beniger, 1986, p. 362). By early 1923, more than 500 broadcasting stations were already operating, 69 of the licensees being newspapers with cross-media ownership. Other licensees included 29 department stores, promoting jingles for new goods in stock; 12 religious organizations, now involved in the mass merchandizing of the Bible; some city governments; along with car dealers, theaters, and banks.

As the eminent historian of broadcasting Erik Barnouw noted, the first two presidential candidates to use the airwaves in their campaigns, Warren G. Harding (Republican) and James M. Cox (Democrat), were themselves newspaper publishers. On Harding's death in 1923, his successor, Calvin Coolidge, became the first "radio president." The convergence of mass media and electoral politics was established (Barnouw, 1966, p. 4).

Once in place, broadcasting was a commercial and public relations phenomenon. The National Broadcasting Company's (NBC) parent company, RCA, had already mass produced and sold 1.25 million radio sets by 1924, 4 years after the first licensed radio station, Westinghouse's KDKA in Pittsburgh, went on the air. By the late 1920s, commercialization of radio was firmly imprinted by the two dominant networks, NBC and the Columbia Broadcasting System (CBS), created in 1927, together controlling almost 90% of transmitting power in the country. William Paley, president of CBS, hired the World War I public relations expert Edward Bernays to improve its advertising operations, as one of Bernays' contemporaries in the same field, Ivy Lee, became a dedicated fascist and went to work for the Nazi government in Germany. (Bernays later chose to serve the U.S. mission in the Cold War, directing the propaganda campaign to destroy the political reputation of Guatemala's reformist president, Jacobo Arbenz, and helping the CIA overthrow his government in 1954.) (Chomsky, 1989, p. 29; Lewis, 1991, p. 183).

Advertising on radio, despite some initial government opposition to it, brought windfall revenues to stations in the 1920s and 1930s, a revenue flow disrupted only briefly during the Depression and by the rise of television after the war. Apart from creating new advertising opportunities and bringing new forms of popular and consumer culture to millions of people, network and local radio would help extend the frontiers of commercialism in other ways. Access to radio information and culture, together with the automobile and the growing reach of telephone, meant that the urban nouveau riche could choose alternative living spaces, removed from the grimy congestion of the cities, in nicely manicured suburbs. Radio also helped solidify the grip of monopoly or monopolistically competitive industries, benefiting national-scale advertisers that formed the foundation of major broadcast networks, while giving programming advantage to networks over financially less secure independent stations.

Another function of radio was to introduce new ways of indoctrinating people politically. While the Hitler dictatorship in the 1930s was cultivating the use of radio for Nazi propaganda of German "race, blood, and nation" and organizing collective listening, and the dictator Mussolini was doing the same for Italy, the United States had its own fascist broadcaster in the person of Charles Edward Coughlin, a Catholic priest based in Royal Oak, Michigan. Coughlin preached over WJR in Detroit and dozens of other stations (including WOR, New York) about Jewish financial conspiracies, the evil of communism (often linking the two), the world leadership of Hitler and Mussolini, and an assortment of "nativist" and populist causes to millions of Depression–era weekly listeners. Coughlin's messages, backed by Bishop Michael James Gallagher of Detroit, also had a lot of resonance with the right-wing owner of WJR, George Richards, who controlled at one point 34 other stations. Richards despised Franklin D. Roosevelt, whom he depicted as a "communist" and a "Jew lover." The FCC had no problem renewing Richards' radio broadcasting license annually, and the ideologically conservative Du Pont Company awarded Richards its national public service award in 1945 (Cirino, 1972, pp. 78-79), the year that marked the beginning of the Cold War and the McCarthy–era political witch-hunts.

Coughlin's and Richards's counterparts at the Ministry of Propaganda in Nazi Germany, created in 1933 by Joseph Goebbels, attempted to enforce "a peculiar staidness and solemnity" in radio broadcasting, to instill discipline in listeners, and to purge the medium of its "Jewish spirit" (satire) and "negroid" musical influences (rhythm and jazz). Radio was regarded by Goebbels as a propaganda apparatus for controlling and manipulating the passions of German citizens and "educating" them "into the National Socialist [Nazi] way of thinking and feeling"—that is, the totalitarian vision of the Führer, Adolf Hitler. Goebbels defined broadcasting as an "instrument for forming a political will" and serving as "a disseminator of culture." Anyone not proven to be completely loyal to the Nazi program was purged from radio and other media organs. To quarantine "political pollution," German law also made it a crime to engage in "illegal listening" to foreign radio broadcasts (Sington & Weidenfeld, 1942, pp. 272-277).

In a very different political context, President Franklin D. Roosevelt also made use of the airwaves in those years with his periodic "fireside chats," which many say demonstrated his mastery in building support for the policies of the government's New Deal programs and in general creating trust with listening audiences during the Depression and the war years (Barnouw, 1968, pp. 7-9). Roosevelt would also become the first president to appear on television. When Ronald Reagan became president, he would attempt to capture Roosevelt's style, though clearly not his political message, with weekly radio addresses during the 1980s. Unlike Nazi Germany's crude manipulation of radio for statist ideology and Britain's more clever government uses of the medium, especially during Winston Churchill's wartime prime ministership, American uses of broadcast propaganda were more typically handled by the private sector.

The unregulated and wildly spontaneous spread of broadcast radio and the contest for control between public and private, large-scale and small-scale interests, eventually brought the federal government into the fold. The use of the spectrum by small-time operators had particularly concerned RCA ownership, who were relieved once the government stepped in. In 1927, Congress passed a Radio Act that created a Federal Radio Commission to license and periodically renew licenses of broadcasters and ostensibly to prevent monopoly "directly or indirectly, through the control of the manufacture or sale of radio apparatus, through exclusive traffic arrangements, or by any other means or to have been using unfair methods of competition." That year, the second major radio network came into existence, the Columbia Phonograph Broadcasting System, later Columbia Broadcasting System (CBS). Also at this time, the government established the standard of "public interest, convenience, or necessity" as the rationale for licensing, while nonetheless favoring big-time operators. Although the commercial radio operators did not own the bandwidths *de jure,* anymore than they owned the Mississippi River, they effectively did *de facto.*

AT&T's toll broadcasting had imprinted radio programming with commercial values, and even though a number of non-commercial stations were created in the early years, the advertising interests would prevail. The licensing practices of three Republican administrations (Harding, Coolidge, and Hoover) in the 1920s favored the more powerful stations in granting licenses and in distributing the more desirable frequencies. This was a decade of Republicanism, loosely regulated capitalism, wild financial speculation, anti-reform policy attitudes, and some of the worst political scandals in the nation's history (including Teapot Dome). Experiments in not-for-profit radio broadcasting would be short-lived, overcome by an alliance of advertisers, bureaucrats, network executives, and big commercial operators.

Nonetheless, in the first years of radio, networks did provide broadcast space for "public service" and other non-commercial programming, such as the *University of Chicago Round Table* and *America's Town Meeting of the Air.* These seeming acts of public spiritedness, according to a leading broadcast historian, were driven largely by a concern on the part of the networks that the Federal Radio Commission was about to be amended to require a 25% allocation of stations to nonprofit organizations. Although many educational, municipal, labor, and religious groups supported the proposed Wagner-Hatfield amendment in the Senate, it was easily defeated through the combined influence of radio, advertising, and business interests (Barnouw, 1975, pp. 73-76). By peak year 1947, 97% of radio stations were network affiliates, and by 1979 only 10% were non-commercial (Sterling, 1979, pp. 66, 117). The vernacular of broadcasting would remain essentially commercial, not educational, intellectual, artistic, or spiritual, redirecting the momentum of educational progress and aesthetic creativity in democratic societies toward indulgent consumption, garish amusements, and gullible spectatorship.

CHAPTER 4

Mass Consumption, Mass Communications, and Politics

When broadcast radio started in the 1920s, RCA's National Broadcasting Company (NBC) packaged itself as committed to being a "public service corporation rather than a traditional for-profit corporation, which would sell only the advertising that was necessary to subsidize high-quality non-commercial fare" (McChesney, 1993, pp. 224-225). RCA's president, David Sarnoff, would tell listeners over the NBC network in 1938 that commercial radio was rooted in American democracy and its "free economic system" in which "no special laws had to be passed to bring these things about" (McChesney, 1993, p. 247). Because it spares its listeners and performers from distracting visual artifices, radio is potentially a more talkative and contemplative medium and, contrary to Marshall McLuhan, invites greater imaginative and intellectual involvement of audiences. With few exceptions, such as National Public Radio's audience-interactive *Talk of the Nation,* radio has not turned out that way. Talk radio is dominated by the likes of shameless self-promoters, such as Rush Limbaugh, who mobilize public backlash and prejudice against favorite strawmen and often screen out callers who disagree.

CULTIVATING COSMETIC
AND MATERIAL IDENTITIES

Driven by the 19th-century capitalist appropriation of land, labor, industry, science, and technology, the United States was continually being redefined, increasingly toward mass-market values. The dominance of northern industry and banking in the Civil War, the rise of the modern stock corporation, the expanded capabilities for national-scale production, transportation, distribution, and postal and wired communication were interrelated within an expansionist economy. Settlers, armies, navies, the belching smoke of the railroad locomotive, telegraph poles across the horizon, and the crusading leadership of American presidents and captains of industry overcame obstacles in the way of this expansion: the British, the Spaniards, the French, the Russians, the Mexicans, the Filipinos, the Puerto Ricans, the Hawaiian monarchy, the Native American tribes, and other peoples who occupied the lands west and south of the original colonies. The "Indian Removal" policies set a precedent for colonial attitudes and practices that would make the United States an imperial power by the turn of the century.

Once the United States had become an area of contiguous states and territories in a huge bicoastal landmass, the need arose for economic infrastructure to consolidate and exploit the country's vast resource potential. Before the development of chain supermarkets and retail outlets, there already existed in the 1870s and 1880s mass circulation magazines, newspapers and full-page advertising, and by 1889 an advertising industry trade journal. Mail-order catalogs, flourishing by the 1880s, would absorb the disposable income of some 76 million citizens at the turn of the century. Montgomery Ward was sending out a 540-page catalog of 24,000 items by 1887, and the circulation of the Sears and Roebuck catalog and its other mass mailings grew from 318,000 in 1887 to 75 million by 1927 (Beniger, 1986, p. 19).

The international advertising agency J. Walter Thompson was founded in 1864, originally as one of the early developers of market research. In 1878, James Walter Thompson took over the company that has since borne his name, and it remained an American company until its takeover by a British group, WPP, in 1986. Between the two world wars, JWT and McCann–Erickson were the preeminent transnational advertising firms, some calling them the "imperial models," or "colonial models," in the industry (Mattelart, 1991, pp. 4-5). By 1970, these two agencies controlled 56% of advertising billings in all of Latin America, concentrated in the biggest markets of Argentina, Brazil, Mexico, and Venezuela. As of 1997, the biggest ad agency, Saatchi & Saatchi (UK), operated advertising firms in 80 countries and marketed 20% of the world's broadcasting commercials.

Relatively big salaries offered by the transnational ad agencies in Third World countries attracted local talent away from domestic advertising and other industries and made them agents of foreign tobacco companies, automobile manufacturers, and other units of global capital (Barnet & Müller, 1974, pp. 143-144). Frank Shakespeare, former CBS executive and director of the U.S. Information Agency, reflected on the imperial reach of American media: "We dominated motion pictures and television for years; we still do. 'Madison Avenue' has become a worldwide cliché for referring to the technique of marketing and that's the dissemination of ideas" (cited in Barnet & Müller, 1974, p. 145).

The 1920s had ushered in a tidal wave of commercial values and "salesmanship," using newly available media to induce consumption and brand loyalty. Electrical appliances for sale were valued at $92 million in 1899, and by 1927 they were worth $1.6 billion (Carman et al., 1967, p. 528). Automobiles had opened up the American continent to relentless mobility and the culture of itinerancy and "passing by." American commercialism converted the 19th-century European Saint Nicholas from a sometimes stern disciplinarian who flogged naughty children to an ever-jovial advertising star for Christmas retailing, and Coca-Cola capitalized on Santa's colors, which matched those of the soft drink ads. Enduring values of the past would continually be challenged by the ephemeral qualities of the new and the now, while salespeople, fast-buck artists, and hucksters of every stripe were constantly on hand, at one's door or over the airwaves, to ply the need for lifestyle "change" and "self-improvement."

Advertising, started in the United States by Volney B. Palmer in 1841, was assuming its place at century's end as one of the most-read "literatures" of American society, stimulating new forms of taste, temptation, identification, and lifestyle. Modesty, frugality, and conservatism in behavior and demeanor were overwhelmed by an industrialized culture of fashion, smoking, urbanity, and mass consumption. By 1929, advertising revenues in the United States, now by far the leading manufacturing country in the world, were estimated to have reached $2 billion, reaching well beyond traditional local markets toward national-scale mass consumption targets.

The privatization of the airwaves assured manufacturers and the advertising industry that mass-marketed messages about their goods and services could be transmitted in ways that required virtually no education on the part of the receiver, which made radio and television "the prime instruments for the management of consumer demand" (Galbraith, 1967, p. 218). A Venezuelan consultant to McCann-Erickson and J. Walter Thompson and a university professor of "social communication" revealed that the most enduring finding in her research on the impact of advertising was that the poor had "lost their perception of class differences" in the belief that most consumer goods advertised on television were, in one form or another, available to all. A typical response among the poor in her research was: "I don't have a floor to wax, but I can buy the [Johnson's] wax if I want to" (quotes cited in Barnet & Müller, 1974, pp. 175-176).

A heady optimism among the well-off, spearheaded by the "return to normalcy" catch phrase of the conservative Republican Harding and Coolidge administrations in the 1920s, distracted Americans from the destruction of World War I, the growing polarization of wealth and immiseration, and the rise of European and Japanese fascism. By the end of the decade, the 200 largest corporations (0.07% of the total) had half the country's nonbanking corporate assets and 22% of national wealth of some $361 billion. The poorer 60% of American families had less than 24% of national income, whereas the richest 21.6% had 60% of the total (Huberman, 1970, pp. 244, 254, 269).

At the same time, proportions of migrant and tenant workers were growing, as farmers experienced economic depression a decade before the rest of the country. Unemployment was steadily on the rise (4,270,000 in 1921, over 5% in the early 1920s, close to 25% by 1932). Squatter communities ("Hoovervilles"), composed of people evicted from apartments and homes, were sprouting in urban garbage dumps, where the people waited for trucks to unload refuse, in hopes of finding scraps of food or something of value. (In the 1990s, thousands of people are forced into the same degraded existence in the former U.S. colony, the Philippines, where marginalized citizens live off smoldering garbage dumps like Manila's infamous "Smokey Mountain," and similar scenes, with dumpsters instead of open garbage pits, are repeated in cities across the United States.)

Following World War I and the Bolshevik Revolution, a wave of cultural chauvinism, so-called nativism, resurfaced in the United States, an outlook suspicious of foreigners, hostile toward nonwhites, and quick to associate progressive social values with communist conspiracies. The polemic often was mixed with a populist distrust of big money, the "eastern establishment," and "Wall Street." Oriental exclusion acts and tightly limited immigration of Italians, Jews, East Europeans, Russians, and other non-Anglo-Saxons and non-Teutonics, Jim Crow laws, the revival of the Ku Klux Klan in 1920 (4.5 million members by 1924), lynching of African Americans and white America Firsts were all part of that era's political culture. A silent film blockbuster, *Birth of a Nation* (1915), long regarded as a "classic," condemned African American political enfranchisement in the South, portraying abuses supposedly committed by blacks, and the Ku Klux Klan as white people's liberators.

Subcultures of radical political movements (socialism, communism, anarchism), black liberation in literature, art, intellectual, and popular culture (the Harlem Renaissance) and in the New Negro Movement (Marcus Garvey, W.E.B. Du Bois, the NAACP), together with the Depression-era minstrel talent of Woody Guthrie were also part of the scene. F. Scott Fitzgerald wrote of the age: "It was borrowed time anyway—the whole upper tenth of a nation living with the insouciance of a grand duc and casualness of chorus girls" (cited in Zinn, 1980, p. 374). And yet a gilded image of America seemed to be what would-be immigrants were more apt to receive, as in the 1930 depiction of Mike Gold (*Jews Without Money*):

> One picture had in it the tallest building I had ever seen. It was called a skyscraper. At the bottom of it walked the proud Americans. The men wore derby hats and had fine mustaches and gold watch chains. The women wore silks and satins, and had proud faces like queens. Not a single poor man or woman was there; every one was rich. (cited in Ewen & Ewen, 1982, p. 209)

Radio and early television programming did little to present the complex and bewildering fabric of American culture and society. The hardships of American farmers and farm workers; the oppression of African Americans, Native Americans, Mexican Americans, and Asian Americans; and the soup kitchens and poor farms that kept so many of the unemployed alive during the Depression were not the stuff of broadcast programmers, although a few stations ventured into real-world discussions. Network radio (and later, television) was more interested in marketing goods and services with staged entertainment and using the wired household as the "theater" of presentation. The idea was to simulate for the largely immobile masses a vicarious exposure to the world of power and glamour. Radio and television thus became the cultural center of the struggling classes, allowing the well-off and nouveau riche to entertain themselves without guilt in the urban opera houses, theaters, and symphony halls and to experience firsthand the far-off places of which others could only dream.

Postwar Tele-Visions

The postwar proliferation of television sets in the United States took corporate industry and the culture of consumerism a step farther in solidifying the links between factory and home, between work and leisure, and between global production and global consumption. Erik Barnouw, commenting on radio in the 1930s, said: "Commercials, which had been brief and diffident in NBC's first days (1926), were becoming long and unrelenting—but successful instruments of merchandising" (cited in Beniger, 1986, p. 362). With television, the visual allure of commodities became all the more important, and the medium responded with wholesomely pretty, cheerful housewives ecstatic over their new detergent, talking camels making testimonials about their favorite candy bar, and cigarette boxes tap dancing to a snappy jingle.

Television became the vehicle par excellence for bringing home the message of consumption, a pitchman that could enter without knocking, and each year its frontiers were extended. By 1950, the television industry broke down another barrier to commercial penetration by introducing programming and advertising late into the night and early morning hours and putting commodities themselves into starring roles—in programs like *The Price is Right* (Ewen, 1976, pp. 208-209). The friendly, ingratiating sounds and images flowing from the set offered household companionship and created captive audiences for its sponsors, who in the early 1990s were paying networks $200,000 per minute, prime time. By 1988, American households, at near complete television saturation and with almost 100 million viewers at prime time, were tuned in an average of 7.1 hours per day (Volti, 1992, p. 176).

Television news in the early years borrowed from the newsreel tradition developed for movie houses. NBC and CBS came up with 15-minute news programs, the *Camel News Caravan* and *Television News with Douglas Edwards,* that added pictorial effects to selected news topics. In more recent years, television news has come under the increased influence of advertising and public relations, incorporating fast-paced entertainment spiced with sound and video bites that caters to the marketing preferences of their sponsors far more than the needs of their viewers. Spaced between frequent commercial interruptions, no story can be developed in any detail or depth. Given that networks "are essentially in the business of selling a national audience to advertisers," they must generalize the product to the comprehension of the least attentive consumer in order to enlarge market share, which is the basis on which advertisers pay them (Epstein, 1974, pp. 79-80).

As profitable an enterprise as television has become, its discovery and development as a technology had lapses of decades. The movement of electron beams was explained by the English scientist J. J. Thomson in 1897 and the same year led to Ferdinand Braun's cathode ray tube, an early forerunner of the modern television screen. The Scottish scientist John Logie Baird publicly demonstrated a working television system in England in 1926, and though available in the mid-1930s, its effective diffusion did not actually begin until the late 1940s. Philo T. Farnsworth had procured a patent for an electronic television camera in 1930 (at the remarkably young age of 24), which he later sold to RCA. Television broadcasting in the United States, except for a few experimental transmissions in the 1930s, did not take off until after World War II; the British and Germans started up general broadcasting in the mid-1930s, although on a very limited basis. And although CBS developed color television in the 1940s, it did not become widely available until the 1960s.

This had been the pattern in radio as well. Reginald Fessenden, a Canadian pioneer in "continuous wave" radio voice transmission, broadcast a program with music, poetry, and a speech from Brant Rock, Massachusetts, as early as Christmas Eve 1906. The U.S. Navy would hold on to a monopoly of radio patents for years before relinquishing them. Broadcast radio from its first years was an advertising medium, and the early dominance of the set manufacturers would eventually give way to the power of station owners (Schiller, 1969, pp. 22-25). Only when the commercial consolidation of broadcast time and the definition of the major markets (the "audience") and retailing possibilities took shape, however, did the medium take on national network proportions, entering 80% of American homes by the late 1930s. Even facsimile machines, so widespread in the 1990s, were available for home use in the 1930s, as was FM transmission in the early 1940s and UHF television in the late 1940s, but all suffered suppressed development at the hands of government and industry to protect the monopoly interests of emerging communications corporations (Kellner, 1990, p. 40). Marketeers and political stakeholders, not technology, prevailed.

The technological determinist notion that technology gives rise to historical and social change ignores and obscures its political economic aspects and is easily refuted by the history of broadcasting. Television technology and its program content were shaped by a mix of visions and values, but ultimately the self-empowering motivations of the captains of industrial capitalism in America, such as David Sarnoff, formerly an assistant manager with American Marconi and the founder of RCA, were quite clearly ascendant. In 1935, Sarnoff, then president of RCA, spent the then astounding sum of $1 million (about $15 million in 1995 dollars) in demonstrating the use of television programming. In 1947 alone, with a dominant position in broadcasting and consumer electronics (including $40 million in television product sales), RCA made $300 million (Galambos & Pratt, 1988, p. 173).

As with the telephone, where private corporate patent conflicts and often inept government involvement held back the diffusion of that technology, broadcasting technology was similarly restrained by the monopolistic inclinations of its patent owners and the lack of a clear public vision by federal regulators. As one historian of technology has stated:

> The development of television was slow not because its principles were not grasped at an early date (Campbell Swinton's proposals [for cathode ray tube scanning of images] on essentially the same lines as are now used were made in 1911) and not because of the technical difficulties of scanning or of broad-band transmission. It lagged essentially because the big electrical firms, even the new firms that had grown up with radio, were too intent on immediate profits to indulge in expensive development. It was left to enthusiastic amateurs like Baird (1888-1946), using primitive equipment, to make the decisive advances and convince the commercial world that there was money in it. (Bernal, 1971b, pp. 780-781)

Television got its start in ways much like its electronic forerunners. Again, it is hardly the case that a clever individual with an eye for markets simply "invented" a new technology. The concurrent technological development of television in several countries is discussed in considerable detail by Brian Winston, who shows how the industrial competition among the Western capitalist economies and Russia (later the Soviet Union) established the logic of television development, even as the specific technical and social design of the technology varied in each case. Winston argues that the Russian claims to inventing television and radio are no more absurd than similar claims by the British, the Germans, or the Americans, for indeed several Russians—Boris L'vovich Rozing, Aleksandr Stepanovich Popov, Vladimir Zworykin, and others—made important technical contributions to the development of broadcasting. Although different state and industrial interests were at hand, Soviet, western European, and North American science shared a stage of technological evolution and a pool of knowledge from which they generated technically similar communication artifacts.

Zworykin, a former student of Rozing, would leave Russia following the Bolshevik Revolution and eventually join RCA. He is credited with creating the first workable electronic television camera (*iconoscope*) while electrical engineers in Nazi Germany stayed and advanced their technical knowledge with the help of purchased patents from the Western democracies. Meanwhile, in the United States, Philo Farnsworth managed to beat off challengers, including Zworykin, to several key patents. Zworykin and RCA had to lease his patent to start up their first regular broadcasting out of New York City in 1939—to a few thousand flickering 5-inch screens that could not be viewed in bright light.

From a political economy perspective, these contributions to technological change occur within a larger societal context. Brian Winston sees the need for a "supervening necessity" within capitalist markets in order for "inventions" to become marketable commodities. In the case of American television, the excess productive capacity of the electronics industries, with a labor force raised to more than 300,000 in the radio industry alone, just before the end of the World War II, created that necessity (Winston, 1986, p. 52). A major increase also occurred in civilian consumption, 25% growth from 1939 to 1945 (Melman, 1985, p. 15), which meant that a large market for new goods, postponed by the war and the last years of the Great Depression, would be opened to manufacturers. The continuation of large-scale military spending after the war, anticipated by 1944, would make television a beneficiary of the Cold War, defense research, and the postwar military-industrial complex.

The aftermath of World War II marks a major turning point, not only in media penetration of American homes, but more essentially in the reconstruction of the U.S. economy, social institutions, and the family. Industrial expansion in the United States was phenomenal. Built on an enormous wartime production schedule that had directly changed the outcome for Europe and Asia, American prosperity continued into the early 1970s. By that time, major global and domestic economic shocks (e.g., the fallout of an estimated $120 billion in Vietnam War expenditures from 1960 to 1973, the rise in oil prices, the loss of technological leadership, increasing trade imbalances, the declining value of the

dollar, rising unemployment) began to revise downward the material expectations of most Americans. A sense of gloom over the weakening economy and
real income, the decline in educational commitment by the federal government,
lowered job expectations, family instability, the rise in violent crime, the
deterioration of the cities, racial tensions, the depletion of the physical environment and the ozone layer, unending military involvements, and other anxieties
have since become part of the public mood.

The first 25 years (1945-70), however, brought the fantasies of Hollywood and
Madison Avenue image factories, what one writer called an "infantilizing of culture"
(Dorfman, 1987), into almost every home in the United States. The television
industry built on the values carried over from radio and movies in reinforcing
cultural landmarks in the American mind—frontier builders as the Lone Ranger and
Tonto, African Americans as Amos 'n' Andy, modernity as smoking "Luckies." Even
if the media images of society did not correspond to the lives of most people, it
was difficult to argue with the apparently successful formula for material success
that the United States seemed to offer in this period. The inability to sustain the
basis of that achievement, built on an intact industrial economy at the end of
World War II and the rehabilitation of Europe and Japan as trade and security
partners, was not understood by most Americans, who were absorbed at varying
levels in the biggest wave of material consumption in human history.

Successful commercialization of radio established the pattern that television
would follow. The household served as the perfect unit of consumption, a
one-way channel of information from commercial broadcaster, with its advertising inserts packaged in seductively entertaining appeals of sound and image
to individual viewers. More than 11 million soldiers in uniform would be
returning at war's end, starting millions of new households and the baby boom,
which together with pent-up demand for material goods during the war years
would lead to a bonanza in the production and sale of consumer goods as diverse
as refrigerators and ballpoint pens. Although punctuated by recessions, severe
inequalities, Cold War repression, and callous indifference to the poor, the
country experienced a 25-year heady confidence about its material and technological standards, and television helped Americans ignore the downside of this
experience.

The three national broadcasting networks that would emerge—NBC (owned
by RCA), CBS, and ABC (a federally mandated spinoff of NBC in 1943)—were
all outgrowths of early radio monopolies that, by the mid-1930s, controlled 97%
of nighttime programming across the country (McChesney, 1994, p. 29). Programming served the interest both to sell radio and television sets (which RCA
produced and whose manufacturing rivals CBS actively courted for sponsorship) and to sell the advertising of sponsored programs (Lewis, 1991, p. 183).
In this sense, the medium was the message *and* the messenger.

Government, under the Federal Communications Commission (FCC), did little to supervise the cultural value of television other than to grant and renew station licenses, which was done almost automatically, to encourage a modicum of public affairs programming and to oversee matters of political fairness and public morality. Indeed, the FCC, even with a flurry of reformist thinking in the late 1960s and early 1970s, largely represented a "revolving door" of common interests with the industries they were purportedly overseeing: 21 of the 33 commissioners who left from 1945 to 1970 went to work as managers, lawyers, or engineers for the same communication corporations they were previously regulating (Kellner, 1990, p. 130).

For industry and advertising, television has been a kind of electronic genie, fulfilling fantasies of sales and profit. In the 1950s and 1960s, television was an elixir that pacified concerns about jobs and justice and gave the nation news of America's righteousness in a world of danger and demons, which CBS's Walter Cronkite ended each nightly broadcast with his signature sign-off, "And that's the way it is." CBS all along was a frequently willing participant in self-censorship of information that contradicted the State Department's foreign policy line or the real situation during the war in Korea. CBS reporters Charles Collingsworth, Edward R. Murrow, and Griffin Bancroft and programs such as *The Facts We Face* (hosted by Cronkite), *World Briefing,* and *Diplomatic Pouch* staged live "interviews" with State Department spokespersons that were supposed to be spontaneous but that, in fact, were rehearsed with government-approved scripted questions that set up propagandistic responses. During this period, hundreds of television and newspaper journalists, mostly with CBS, *Time* magazine, and the *New York Times,* worked hand in glove with the CIA, regularly briefing one another, with correspondents and stringers often running intelligence errands for the spy agency—all the while presenting to the public an image of an untarnished and independent press (Bernhard, 1993, pp. 301-307).

Hollywood and Madison Avenue Imprints

Television in the United States had its takeoff in the late 1940s, at the time the antitrust legislation was being brought against the Hollywood studios to break their grip on movie exhibition (e.g., *United States v. Paramount Pictures*). While Hollywood and the emerging television industry were undergoing the purges and blacklisting of the McCarthy era, a few in the movie industry began to size up television as a new opportunity because the over-the-air medium was already eating into the big screen's box office receipts. The Hollywood talent agency MCA, led by Lew Wasserman, started the crossover by recruiting directors, actors, producers, and writers for network shows and was aided in its legal battles by the president of the Screen Actors Guild and a client of

Wasserman's, Ronald Reagan. With close ties to the Mafia, MCA later bought out Universal Studios and became a full-fledged production company, the biggest in television. Reagan was made host of an MCA television drama series (Gitlin, 1985, pp. 145-146; Moldea, 1987).

Many of the new television network executives also came from the ranks of talent agencies. Television, with its constantly shifting trends, shows, and faces, was a natural for Wasserman and headhunters like him. As television moved away from live shows, the movie studios moved in to take up the broadcast space, and most of the filmed serials and miniseries came out of lots in southern California—America's dream machine. Hollywood's integration with television also led to its surge in world markets, as it had done previously in the early 1920s, the early 1930s, and again in the immediate postwar years. The ubiquitous presence of Hollywood movies and canned television shows would lead to visible protest in many countries against U.S. "cultural imperialism."

Compared with European films, which often took on more serious social topics that could elevate the viewer, Hollywood productions more typically were concerned with generating and responding to mass appeal and the bottom line. They were also driven by huge production costs, mostly multi-million-dollar budgets by the 1960s (and averaging over $18 million per picture by the late 1980s). Nowhere did studios, producers, directors, actors, writers, other talent, and financial backers make the kind of money that was turned over by Hollywood. Just as quickly turned over were the vast majority of program series that either never made it to pilot, to airtime, or beyond one season. With sponsors directly developing their own television programs in the 1950s, the advertising industry played a big part in the management of America's cultural tastes.

The debris of American television is no different from the country's other vast wastelands. With few exceptions (e.g., *Roots, I'll Fly Away*), it is hard to recall over the past 50 years serialized television programming that made serious contributions to American social outlooks. Some have cited the work of Norman Lear (*All in the Family, Mary Hartman, Mary Hartman,* and a few of his other productions) and James Brooks and Allan Burns (*Mary Tyler Moore Show, Lou Grant*) in providing some food for critical thought, but as one scholar notes, no one, not even the sometimes controversial Lear, dared to skirt the limits of tapioca liberalism, individualism, and sentimentalism. Serious political conflict as story line material was, and remains, taboo. Dramatized collective social action as a response to injustice is virtually unknown in the television universe (Kellner, 1990, pp. 58-60). Few advertisers would ever invest in social reality, and just as unlikely would the three major networks, with 86% of national audience advertising (Parenti, 1992, p. 184).

As with newspapers, expanding dependence on advertising required commercial television to avoid messages that might distress segments of the mass audience and the advertisers themselves—such as corruption in big business, the health risks associated with many industrial products, and the real conditions of America's underclasses. Although several cities had educational television stations from the 1950s onward, non-commercial television was not a widespread option for most viewers until the advent of the Public Broadcasting Service (PBS), which began national programming in 1970. Not until 1964 were set manufacturers even required to produce UHF reception capacity, long after its availability and at a time when most educational stations were already transmitting over those higher frequencies. The conditioning of American audiences to think of television as a purely "entertainment" medium has limited interest in noncommercial alternatives to mainly better-educated viewers.

The advertising industry ("Madison Avenue") and its patrons in the corporate world are the main source of broadcasting revenues. Even in so-called public television, advertisers have a big hand both in program sponsorship ("underwriting" and direct advertising) and system legitimation, the biggest underwriters being those corporations with the worst public image problems or record with regulators (e.g., oil, the defense contractors, AT&T). A legitimation function is performed in public and commercial broadcasting through, for example, the Advertising Council, which was started in 1941 for wartime propaganda. This group places public affairs advertisements in print and broadcast media that often whitewash the images of corrupt and polluting businesses.

Critics say that the Advertising Council, which by the mid-1980s was second in advertising expenditures (behind Procter & Gamble), trivializes serious issues (e.g., crime: lock your car doors; traffic congestion and high-speed auto design: wear seat belts; energy conservation: turn down your thermostat; the environment: don't litter and listen to Smokey the Bear). The Advertising Council emphasizes personal but not corporate responsibility for these and other social problems. Millions spent by the Advertising Council in opposition to government regulation paved the way for the Reagan program of deregulation in the 1980s (Domhoff, 1983, p. 105; Parenti, 1993, pp. 77-79).

Modern advertising has its roots in the fields of public relations and World War I propaganda. In the early part of the 20th century, public relations virtuoso Ivy Lee worked at converting the public's image of John D. Rockefeller from ruthless and greedy pursuit of power to great philanthropist. Lee's successor, Edward Bernays, a nephew of Sigmund Freud and member of Woodrow Wilson's wartime propaganda instrument the Creel Commission, viewed propaganda as little more than the harmless, indeed essential, need for public persuasion in a democratic society. The "engineering of consent," he offered, "is the very essence of the democratic process, the freedom to persuade and suggest."

Fortune magazine endorsed Bernays' formulation, adding that "it is as impossible to imagine a genuine democracy without the science of persuasion [propaganda] as it is to think of a totalitarian state without coercion. . . . The daily tonnage output of propaganda and publicity . . . [is] an important force in American life" (Bernays and *Fortune,* cited in Carey, 1987, p. 39). In 1928, during the rise of fascism, Bernays wrote: "If we understand the mechanism and motives of the group mind, is it now possible to control and regiment the masses according to our will without their knowing it. . . . [T]heory and practice have combined with sufficient success to permit us to know that in certain cases we can effect some change in public opinion . . . by operating a certain mechanism" (cited in Ewen, 1976, pp. 83-84).

A reading of the *London Gazette* and the *Salem* (Massachusetts) *Gazette* following the battle of Lexington and Concord in 1775 would make it obvious that observance of this doctrine was there from the start. That is why the media and politics in democratic societies, despite occasional tensions, are interdependent. Political spectacles put politicians in the spotlight but also mean billions of dollars to print media and, especially, television. So-called negative advertising in politics reflects not only the candidate and the candidate's political consultants but also the demands of the commercial television medium, which, as the glut of talk shows reveals, thrives on attention-grabbing controversy, the more contrived the better.

For all their complaints about the media (particularly conservatives who bemoan their supposedly "liberal" bias), politicians are among the most avid and reliable clients of broadcasting and the press and, through their consultants, among the most skillful manipulators of mediated language and symbols. The structure and imperatives of the commercial media, with its emphasis on images, stereotypes, and sound-bite junk-food journalism, is the perfect vehicle for mobilizing citizens to participate in electoral exercises and, in many ways, their actual disenfranchisement. As the celebrated writer Gore Vidal once observed, "Persuading the people to vote against their own best interests has been the awesome genius of the American political elite from the beginning" (cited in Schiller, 1973, p. 2). When Soviet athletes in the 1980s began to festoon their athletic apparel with the corporate logos of American sports and consumer industries, how long could it have taken before socialist ideology and idealistic Soviet youth would be swallowed up by the imperial appetites of the West?

Apart from helping subdue Soviet state-bureaucratic socialism (a bizarre expression of socialism to begin with), the advertising industry has had other subversive callings. Although the networks gained more control over programming from the advertisers in the 1960s, the advertisers could still pull their money from programs of which they disapproved. And they regularly did, more so with the rise of the religious right in the 1980s and its attempts to mobilize against "indecency." Television's biggest packaged goods advertiser on television, Procter & Gamble (P&G), has regularly pulled sponsorship from dozens of television programs. For many years, P&G had a policy that stated:

There will be no material that may give offense either directly or by inference to any commercial organization of any sort. There will be no material on any of our programs which could in any way further the concept of business as cold, ruthless and lacking all sentimental or spiritual motivation. . . . Members of the armed forces must not be cast as villains. If there is any attack on American customs, it must be rebutted completely on the same show. (cited in Parenti, 1992, p. 186)

In 1990, a peace organization, Neighbor to Neighbor, got the popular television actor Ed Asner ("Lou Grant") to do a public affairs television spot that criticized Folgers, a coffee subsidiary of P&G, for buying coffee beans from El Salvador, then under a brutal military dictatorship. Asner asked viewers to boycott Folgers. P&G threatened to pull its sponsorship from any station airing the advertisement. WHDH in Boston ran the ad and lost $1 million in advertising revenues. *Lou Grant,* a program about a social activist newspaper editor, was cut in 1982, though high in the ratings, because of advertiser cancellation (Kimberly-Clark) in retaliation for Asner's support for the anti-regime forces in El Salvador (Parenti, 1992, pp. 187-189).

Televising the Military-Industrial Complex

While academics debate the symbolic representations of soap opera characters, advertising firms and their patrons laugh their way to the bank. Under the founding legislation of the Communications Act of 1934 that established the FCC and reaffirmed broadcasting as serving "the public interest, convenience and necessity," private radio and television stations and networks have long had a virtually rent-free residence on the "public" airwaves. Television stations in 1993 paid minimum license renewal fees ($5,000 to $18,000, depending on market size) while NBC and ABC alone had combined profits that year of $70 million. AM radio paid $2,500 and FM $2,300 in fees and a token additional amount for licenses ($250 to $900).

Whatever modesty of manner within American culture in the 1920s and 1930s resisted the inculcation of consumerism, it changed with the mass marketing by urban-based commercial television in the 1940s and 1950s. Urbanism and suburbanism (e.g., the Levittowns) provided ideal demographic concentrations of workers and middle-class consumers, with the television set a fixture in 90% of homes by 1960. The broadcast media would indoctrinate audiences to the pleasures of department store "sales" and easy credit buying. Later, cable TV and the Internet would "narrowcast" the advertising message for more customized standards of taste and consumption.

Plays, symphonies, children's educational programming, and news documentaries that were part of early American network television and demonstrated a degree of social responsibility on the part of stations have largely been dropped through the years and left to pay channels and public broadcasting. In the 1950s, television presented stereotyped visions of the past (cowboys) and present (*Father Knows Best, I Remember Mama*) that "would lead its audience to an uncompromising adulation for the market economy and the universals that it projected" (Ewen, 1976, p. 210). Network television has since turned to more tabloid formats of programming and advertising. And despite the momentum of liberal and conservative criticism about the poor quality of television content (including its advertisements) and its excessively violent and sexually degrading story lines and images, the industry seems almost immune to demands that it serve its original mandate. With virtually free access to the airwaves, television stations have become a veritable license to print money.

More than simply an instrument to fill leisure time for viewers or to capture consumers for advertisers, television also has important political and ideological functions in the service of corporate and state interests. In the 1950s, both television and film were bludgeoned by McCarthyite repression. The broadcasting historian Erik Barnouw tells of the devastating impact of a 1950 document, *Red Channels: The Report of Communist Influence in Radio and Television,* put out by a right-wing, anti-communist group of three former FBI agents who published the newsletter *Counterattack,* starting in 1947.

The "Red Channels" blacklist of 151 broadcasting talents, mostly writers, directors, and performers, was essentially a who's who of people who, over the years, "had opposed Franco, Hitler and Mussolini, tried to help war refugees, combated race discrimination, campaigned against poll taxes and other voting barriers, opposed censorship, criticized the House committee on un-American activities, hoped for peace, and favored efforts toward better U.S.-Soviet relations" (Barnouw, 1968, pp. 265-267). It was, effectively, an attempt by right-wing extremists to isolate and censure liberals and progressives, to strictly limit the effective practice of free speech, and to unfetter the most hysterical forms of Cold War propaganda. Among the early listees were, then or since, such well-known figures as Arthur Miller, Langston Hughes, William L. Shirer, Orson Welles, Lena Horne, Zero Mostel, Gypsy Rose Lee, Henry Morgan, Judy Holliday, Lee J. Cobb, Aaron Copland, John Garfield, Lillian Hellman, Dashiell Hammett, Oscar Brand, Abe Burrows, Pete Seeger, Joseph Losey, Ruth Gordon, Burl Ives, Leonard Bernstein, Edward G. Robinson, and many others. More extensive blacklists would later appear.

Pressure was brought to bear on advertisers, producers, and broadcast executives to prevent the listees from working in the industry. The television host and newspaper columnist Ed Sullivan was a big supporter of the blacklist, as was most of the broadcasting and film industry. Many of those blacklisted faced the further humiliation and dread of being forced to testify about their assumed past associations with the Communist Party before the House Un-American Activities Committee (HUAC), a modern-day court of inquisition. Films and television programs that evoked social concerns were highly suspect, and the entertainment industry, through outright censorship or intimidation, turned to light musicals, fantasy, and anti-communist television dramas about ferreting out "Reds" (*I Led Three Lives, Crusader, China Smith,* and *Foreign Intrigue*).

Philip Loeb, an actor who played the character Jake on the popular weekly family-oriented television program *The Goldbergs,* was forced off the show by the blacklist. Depressed by the loss of work, he soon thereafter committed suicide. The African American singer and actor Paul Robeson (Eugene O'Neill's *All God's Chillun Got Wings*), who did not hide his support for the Communist Party in the United States, was forced to leave the country. Before the purge tired out, HUAC had blacklisted or "greylisted" some 2,000 artists in the film and broadcasting world (Parenti, 1992, p. 41).

On the surface of it, McCarthyism and the hostility toward the Soviet Union seemed to be an irrational reversal of the wartime collaboration that proved so victorious. Why would the U.S. government suddenly turn against its erstwhile ally, the Soviet Union, which had indeed carried on most of the fighting against their common enemy, Nazi Germany? In fact, conservatives and Republicans had generally not supported President Franklin D. Roosevelt's diplomatic wartime relationship with the USSR and Stalin and harshly criticized the allied Yalta agreements on ending the war and the plans for postwar political reconstruction of Europe and Asia. Both liberals and conservatives, following Roosevelt's death in April 1945, had ideas of their own on "containing" or "rolling back" the Soviet Union in the postwar world. As a U.S. Senator, Harry Truman had urged that the Germans and Soviets be encouraged to destroy one another so that the Anglo-American alliance could then enter Europe to pick up the spoils of war. But even his anti-communist credentials were suspect to the enemies of the "World Communist Conspiracy."

McCarthyism and fear of world communism extended deeply into the American political psyche in ways that suppressed controversy and induced social conformity. Schoolchildren were required to regularly take part in air raid drills, inducing nightmarish scenarios of an imminent nuclear attack, and to submit to "emergency"-justified authoritarian behavior of teachers and school officials so inclined. In some parts of the country, booklets put out by the right-wing extremist group John Birch Society were used to teach high school history and current events. Classic works, such as *Robin Hood,* were banned in Indianapolis, Indiana, school libraries; the professional baseball Cincinnati Reds were renamed the "Redlegs"; and loyalty oaths were conducted at every level, from government service to "Miss Loyalty" beauty pageants in Dearborn, Michigan. In one state, professional boxers and wrestlers were required to take loyalty oaths as a condition for entering the ring. (The best account of the McCarthy era is Caute, 1978.)

Most liberals and conservatives supported some version of the anti-communist crusade abroad, but some of the more ardent conservatives saw the opportunity for political capital at home, casually labeling members of the opposition as "soft on communism," "pinkos," or outright traitors. This was the grand power strategy of the most zealous among them, the U.S. Republican senator from Wisconsin, Joseph R. McCarthy, a man who had risen to political power by smearing the patriotic reputations of his opponents. With the U.S. decision in 1950 to enter the war in Korea, both "containment" and "rollback" advocates got the chance to argue their respective anti-communist positions before the American people, intensifying the political hysteria. The McCarthyites demanded a purge of State Department "Kremlin agents," who were blamed for the Communist Party takeover of China in 1949. Eventually challenged by a few members of the media and censured in the Senate, McCarthy and his supporters had extended the witchhunt to left political organizations, the universities, the unions, the U.S. State Department, the Department of Defense, and even to the Eisenhower Executive Office.

Although formal blacklists may no longer exist, television still has little tolerance for serious writers with critiques of the U.S. political or economic system. Some of the most interesting television is material that never gets on network air but is available for university, activist, and artistic video or public access cable distribution. One critical television production and distribution outfit is a little-known group called Paper Tiger Television in New York City, which feeds programs via satellite to cable operators around the country willing to download it and carry it on public access channels. Cable has enabled other groups like Austin Community Television (described in Kellner, 1990, chap. 5), the Tualatin Valley Community Access station in the Portland, Oregon area, and other community-run public access channels to carry alternative programming not available on over-the-air television. Typically, however, local programming of this sort lacks the financing capabilities and therefore the "look" and "star quality" of commercial broadcasting and cablecasting. For hard-core television junkies, this makes it difficult to change viewing habits.

Delivering Audiences

Commercial broadcasting, with its sponsored programs and advertising inserts, clearly has become the standard of cultural transmission with which other media (feature films and film documentaries, book publishing, public relations campaigns, newspaper and magazine formats, even classroom instruction) are forced on some level to compare themselves. During prime time, television has an average 95 million captive viewers available for mass marketing and, with 6 to 7 hours per household use per day, absorbs more of Americans' time than any other activity except work and sleep. Also nearly 1,800 newspapers have more than 62 million readers, about 10,000 radio stations (roughly half AM band, half FM) reach nearly every residence and automobile, with daily listening averaging 2.5 hours, and almost all of about 5,000 magazines (a tenth of all broadcasting stations are noncommercial) overwhelm each individual with messages about consumption. In U.S. daily newspapers alone, advertisers spent $32.3 billion in 1991, representing about 80% of press income. Television was a close second in capturing the advertising dollar (Altschull, 1995, pp. 91-92).

Advertising, programming, and broadcast monitoring (Arbitron and the national organization, Nielsen) work in tight symbiosis to garner sales and ratings. Slick advertising inserts have altered—in fact, dictated—the tempo and pacing of television scripting, requiring quick character, plot, story, and dramatic development. The success of one television program format quickly results in "cookie cutter" responses, and soon there are monotonous repetitions for every winning situation comedy (sitcom), game show, talk show, cops 'n' robbers series, and so on. Fifty cable and broadcast options, with rare exception, generally means more formula television geared to the minimum standard that the audience will accept. Sponsors concerned about mass-market returns on their advertising dollar would not have it any other way.

In 1935, the Arthur C. Nielsen Company developed an "audiometer" to monitor the listening behavior of radio audiences. Sponsors and advertising agencies took over the production of programming to ensure control over the largest markets for their products, "leaving the networks to fill less attractive time slots with more educational and cultural—but less popular—programming" (Beniger, 1986, p. 368). After many years of surveying people's television viewing habits through a combination of written and electronic self-reporting methods, Nielsen has had to acknowledge that its Taylorist techniques of consumption management have failed to accurately reflect viewers' program and advertisement interests. Hoping to perfect the mouse(human)trap, the next generation of monitoring technology is supposed to include built-in surveillance equipment that does not require household reporting, combined with self-responses, using UPC code scanning devices, on consumer purchasing behavior. This is the modern market's version of the ancient pursuit of mind reading.

News is paced and constructed for advertising and ratings just as much as other forms of programming. Talk show radio and television, following the voyeuristic leads of supermarket tabloids, put together a daily schedule of village gossip and invite just enough audience response to fill the airtime left by commercial sponsors. Consumer culture and broadcasting is now so integrated that the radio and television industry can be seen as the priesthood of the modern commercial ethos, taking confession of every basic human instinct and impulse and sanctifying their gratification as rituals of consumption (e.g., drinking, driving, eating, dressing, demolition, polishing off enemies, sexual consumma-tion). Many people will argue that there is a state religion in the West and that it is capitalism.

Delivering audiences to advertisers is certainly not the only function of the private media, however, inasmuch as media corporations are multi-billion-dollar institutions in their own right, with powerful connections and extremely well paid executives, producers, and talent. At the time of Capital Cities' takeover of ABC-TV in 1985, the CIA director and former Wall Street lawyer William Casey held $7 million in shares in the acquiring company, for him both a political and financial coup in that he had been pressuring ABC to tone down its criticism of the agency and personally profited from the merger (Mazzocco, 1994, pp. 2-3). Anything that stands to contradict the ideology of capitalism or the way it does business (including using intelligence agencies to fight its turf battles) is seen as an enemy of the state. As Herman and Chomsky (1988) have argued, "Communism as the ultimate evil has always been the specter haunting property owners, as it threatens the very root of their class position and superior status" (p. 29). Corporate media are viscerally defensive about existing property rela-tions and have every stakeholding reason to be.

Within the United States, the media isolate the left—not only communists, but virtually all socially progressive organizations—mainly by ignoring them. Their ideological ammunition is primarily aimed at leftist political movements abroad, soft-pedaling or overlooking repression under pro-U.S. right-wing regimes while employing "human rights" criteria to attack nationalist or socialist parties that restrict foreign economic domination of their countries. The stand-ards that the media purport to apply in investigating corrupt and criminal behavior in the United States or in friendly allied countries are conveniently overlooked when condemning socialist or nationalist governments, rendering the mass media ideological tools of class propaganda. Herman and Chomsky (1988) argue that were the media inclined to investigate the repression that takes place in friendly regimes such as Indonesia, Turkey, or Guatemala, "they would elicit flak from government, business, and organized right-wing flak machines, and they might be looked upon with disfavor by the corporate community (including advertisers) for indulging in such a quixotic interest and crusade" (p. 31).

That nugget of business wisdom, "time is money," is infused within the mass media every bit as much as those financial institutions that transmit buy-and-sell orders across continents in brilliant flashes of instantaneous arbitrage and new wealth. Television drama has neither the time nor the inclination to invest in the development of serious or real-life screen characters, because concision and action leading to hurried resolution is a better formula for making money. And the more turnover, the more profit.

Fast lives, like fast food, are the idiom of modernity, and it is remarkable how quickly Third World and erstwhile socialist societies, lacking basic amenities for their citizens, are eager to embrace McDonald's as a landmark of development. The evolution of new television outlets, a fourth network (Fox), possibly a fifth (Paramount or Time Warner), and cable stations have made some difference in entertainment options, but for the most part offer more of the same—more soap operas, more sitcoms, more Hollywood and television reruns, more ball games, more murder-and-mayhem news, more game shows, more advertising inserts— limited and precious broadcast spectrum but little that educates, uplifts, or inspires. Mass consumption is not about stirring the critical faculties.

REMAKING TELECOMMUNICATIONS

The tremendous expansion of the Western, especially U.S., economy in the early postwar period, a growth spurt initiated by the militarization of production in the war effort, the enlargement of government bureaucracy, and the widening power of U.S. transnational corporations (TNCs), went hand in hand with rapid innovation in communication technologies. Postwar orders for digital computers came foremost from the Pentagon, the Census Bureau, the Atomic Energy Commission (AEC), and other government agencies. It has been argued that had industry not concentrated on developing the computer when it did, the United States would have faced an early postwar "catastrophic crisis" in banking, the armed services, the stock and commodity exchanges, and other huge computational tasks. The federal government funds almost half of all research and development in the United States, and of this some 90% is directed by the Department of Defense, the National Aeronautics and Space Administration, and the AEC (Volti, 1992, p. 23).

Government-industry space and weapons projects in the 1950s and 1960s are often associated with the development of sophisticated computer technology. The means of business and government to administer their far-flung operations have become easily integrated with the management tool of electronic computing. Concentrations of current information processing, in turn, tend to occur close to business centers or historically established corporate information nodes: 90% of computer capacity in the case of Canadian corporations; 75% of all British value-added services based in London; and 30% of all U.S. international telephone calling from New York and Los Angeles. Digital information systems investment by business, escalating to $75 billion by 1987, came into place worldwide starting in the late 1970s at a point when the capitalist economies began to suffer a severe downturn and contributed to new levels of global economic integration and recovery in the 1980s (Mulgan, 1991, pp. 223-224).

One leading computer scientist, known for his irreverent ideas about science, technology, and politics, believes, however, that the notion of the computer arriving "just in time" is a fallacy of circular reasoning, a tautology that, first, ignores the monumental tasks of the period accomplished *without* computers. These included the atomic bomb (the Manhattan Project, in which early computers were used only marginally as little more than calculators), international trade, the mobilization of forces in World War II, and the Nazis' coordinated efforts in launching rockets and organizing the "final solution." Second, if the arrival of the computer was propitious, it was only so in the context of the commanding view at the time that increasingly large-scale enterprise and global power ambitions were natural and logical stages in human evolution. The computer was necessary as an instrument of control, it is so argued, only when one assumes that the centralized power design was the only possibility at the time and that technological innovation thereby obviated a breaking up of elite institutions and interests (Weizenbaum, 1976, pp. 27-30).

Like other technologies, the computer has no single national or individual progenitor. Charles Babbage, a 19th-century English mathematician, drew plans for a "universal analytic engine" that had the core design elements of a modern computer but was never built. Babbage's idea for the use of punched cards was borrowed from the French design of the Jacquard loom. Herman Hollerith, working on the problem of government census taking, perfected the punch card concept with an electronic design that formed the basis of what became the IBM corporation. Babbage's ideas were also taken up by the American mathematician John von Neumann and the British mathematician Alan Turing, both of whom contributed to the World War II effort to develop allied computers for code breaking, the Manhattan Project and to measure bomb and missile trajectories, a context for technological breakthroughs that can be identified as the "supervening necessity" (Winston, 1986). No one can be said to have "invented" the computer, anymore than anyone invented the television, because as one author has noted, "Computer technology is team technology," requiring multiple sets of participants with specialized knowledge (Bolter, 1984, p. 35).

The UNIVAC computer, an improved version of the wartime ENIAC model, was developed for commercial purposes in the early postwar years by two engineers, J. Presper Eckert and John Mauchly (also a physicist). During the war years, Eckert and Mauchly had put together a team of 10 other engineers at the University of Pennsylvania to work on the ENIAC. And like the subsequent generation of computer specialists who developed the personal computer, they had drawn their ideas from many others working in the field, including von Neumann, Herman Goldstine, Richard Clippinger, Howard Aitken, and John Atanasoff before commercial ambitions drew them into the computer business. Computer research had been going on concurrently in England and Germany (inspired by the research of Konrad Zuse), although not with the same degree of funding, technological infrastructure, or government support. Ironically, despite his efforts during the war to help defeat the Axis and his untainted business motives after the war, Mauchly was banned from federal contracts during the early Cold War. As a former member of the American Association of Scientific Workers, which right-wingers in the government branded a communist "front organization," Mauchly was joined to the ranks of the McCarthyite blacklist, a stigma that took him 6 years to clear (Winston, 1986, p. 168).

In 1948, before the blacklisting, Eckert and Mauchly had five contracts for their UNIVAC machine: two were from the government, two were from the consumer research firm A. C. Nielsen, and one was from Prudential Life Insurance (Beniger, 1986, p. 397). One of UNIVAC's crowning symbolic achievements was in 1952 when it was used by CBS-TV News, anchored by Walter Cronkite, to project the winner of the Eisenhower–Stevenson presidential election. CBS had arranged for a very visible mockup of a futuristic-looking "UNIVAC" (and an invisible telephone feed from Philadelphia) to wow viewers into believing that they were witnessing the real thing. The UNIVAC "brain" had made an impressive national television debut (Winston, 1986, p. 169). In the 1990s, the computer mystique has been magnified in blockbuster films like *Independence Day*. In this special-effects bonanza, the hero saves planet Earth by using his Macintosh Powerbook to penetrate and plant a virus in a hostile alien computer, all accomplished within minutes and with no compatibility problems.

The further development of digital communications was spurred under the imprimatur of two of the biggest TNCs—IBM and AT&T—which accelerated their global presence on the strength of the U.S.'s huge defense industry infrastructure, the former accounting for three quarters of worldwide sales of computers in the 1950s and 1960s. Bell Labs of AT&T, meanwhile, developed the first transistor in 1947, "which eventually replaced the vacuum tube and opened the way to more reliable, smaller, and cheaper machines" (Siegel, 1986, p. 111). Semiconductor development led to the integrated circuit (a silicon-based "chip"), on which as many as millions of microscopic transistors could be implanted, and the accelerated growth of the business, military, and consumer electronics industries. The wave of innovation in electronics brought about the microprocessor, essentially an inexpensive and programmable computer on a chip, and several new generations of microcomputers. The computer has helped rationalize the worldwide operations of transnational banking, manufacturing, services, shipping, travel, tourism, media, and other industries, as well as governmental and military command, control, and information structures.

Because of the relatively easy access to this digital technology, there emerged in the 1960s and 1970s a new breed of so-called countercultural hackers, who learned how to "crash" corporate computers, circumvent automated billing systems, and delight in opportunities to electronically trespass into or cause havoc to ("liberate") business, educational, and sometimes military data systems. Much of this counterculture was California–based and innocently utopian or anarchist in outlook, although some of its more talented and organizationally minded members, such as Apple Computer founders Steve Jobs and Steve Wozniak, quickly moved from being precocious hackers to reputedly benevolent managers within the Fortune 500 class (Siegel, 1986, p. 114). In 1985, Jobs, one of the brilliant young developers of the personal computer, was forced out of managerial control of Apple, ironically because his loose organizational style was deemed ill-suited to maintain the company's growing financial position that had spun from his (and Wozniak's) technical achievements—a fate shared with the founders of Bell Telephone. The same year Jobs went, so did 1,200 other employees.

In 1984, Apple introduced the Macintosh computer through a television ad campaign. The scenario portrayed bleakly oppressed workers, one of whom is seen evading the police until, in a symbol of defiance and revolt, she smashes an immense telescreen image of Big Brother (an intended metaphor for IBM, the icon of the computer industry). In adopting the Orwellian dystopian specter, it was as if "little" Apple was rebelling against the Establishment. Having generated a critical mass of users for the Apple and Macintosh personal computers a year later, Jobs' services were no longer needed, and he was replaced by a former executive from Pepsi-Cola. Both Jobs and Orwell had served their use, though Jobs need not be pitied, as he made $217 million at Apple. Later, at his Pixar animation studio, Jobs coproduced with Disney the film *Toy Story,* providing him earnings of $1.1 billion in just the first few weeks of its release. Modern capitalism does not rely on either Horatio Alger or counterculture heroes (Fallows, 1994, p. 36; Rebello, 1995, p. 6). At the end of 1996, Jobs returned triumphantly to Apple, no longer the garage tinkerer but now the "great white hope" of a sagging and despairing corporation.

There remains, nonetheless, a certain level of progressive and anti-capitalist sentiment among some computer cognoscenti. Many generously share software and computer information and help start up bulletin boards and networks for public access. There remains some remote possibility that computer hackers, alienated by a repressive political economy, could bring the system down. This very potential, real or imagined, could activate some form of the "Big Brother" scenario that Apple had exploited in its rise up the Fortune 500 ladder. Enormous information-gathering capabilities of the present generation of surveillance equipment is available to whatever political forces may seek to use them.

The expansion of computers and communication electronics as a component of the postwar military-industrial complex ensured the major U.S. corporations continuing huge government subsidies, posited on the exaggerated threats of Communist imperialism and a Soviet nuclear attack—the ideological underpinnings of the Cold War political economy. By the 1970s, TNCs could carry on global communications by way of satellites, transoceanic cables, and computers, which were now indispensable to their operations. About half of all international telecommunications traffic is undertaken within TNCs, and 90% of all data traffic, not counting military and diplomatic flows, passes within the Organization for Economic Cooperation and Development (OECD), the main economic association of the 24 leading industrial countries.

And like the telegraph that preceded it, information concentration in major metropolitan areas, though now more global in scope, is reinforcing the urban-centeredness of major commercial activities, such as accounting, law, marketing, advertising, insurance, and consulting. The influential communication theorist Marshall McLuhan, in contrast, assumed in rather linear, technological-determinist, and perhaps utopian fashion, that electronic communication would take on the role of creating democratic opportunities in a great worldwide "global village." "Electric media," he wrote, "abolish the spatial dimension rather than enlarge it. By electricity, we everywhere resume person-to-person relations as if on the smallest village level" (McLuhan, 1964, p. 224). In reality, the media are no more democratic than the owners and executives who run them.

Although telecommunications markets have changed since the 1980s, and the telephone, television, and computer industries have many more entrants in the field, the major players continue to dominate. In the United States, AT&T retains by far the largest share of long-distance telephone service (and remains one of the world's largest telecommunications firms); and the three established television networks, joined by Fox Broadcasting, represent most of what Americans view on both over-the-air broadcasting and on cable. Starting in 1989, the Department of Defense began collaborating with a number of U.S. corporations to develop ahead of the Europeans or the Japanese a high-definition television (HDTV) standard, which would provide near-film-quality imaging in over-the-air transmission. The U.S. bloc included AT&T, Apple, and IBM (Mulgan, 1991, p. 192).

Despite worldwide competition, IBM remains the standard in mainframe computers and has 85% of personal computers, desktop, and laptop versions, based on its design, but with a growing presence of Apple and other brands. IBM ventured into telecommunications in the 1980s with joint investment in Rolm, a major digital private branch (telephone) exchange (PBX) manufacturer, Satellite Business Systems, and MCI telephone. A more competitive technological and transnational environment forced IBM and other communication corporations into new national and international alliances and the acceptance of compatibility standards, like those found in the VHS system in videocassette recorders and the Hayes standard in modems.

The Clinton administration and Republican support for international free-trade policies, as embodied in Congress' support for the "Uruguay round" of the General Agreement on Tariffs and Trade (GATT) and passage of the North

American Free Trade Agreement (NAFTA), exemplify victories for the advocates of transnational corporate mobility at the expense of the working class. Telecommunications is crucial to increasingly complex and dispersed global operations, both in terms of the need for information transfer and its ideological function in transmitting entertaining diversions and favorable cultural representations of capitalist society. And nothing is sacred in the hunt for new consumption markets: During the Persian Gulf War, U.S. corporations vied with one another to set up "patriotic" photo opportunities of American G.I.s using their products (cited in Schiller, 1993b, pp. 112-113). At the same time, however, the revolt in southern Mexico in the state of Chiapas is one overt response by the poor in challenging their government's cooptation by international capital, and similar political responses are likely in other parts of the world from both the left and the right.

Cable Babel, Home Recording, and Household Reproduction of State and Corporate Culture

Cable TV (CATV) initially opened a niche in television for clever entrepreneurs but eventually moved toward a form of monopoly power, in some ways more immutable than network TV. What started as an auxiliary service to bring broadcast and network television and radio via coaxial cable to residences out of range of over-the-air (OTA) signals in the late 1940s has led to a struggle for television program control in the 1990s. Along the way, CATV changed from being a local to a long-distance retransmitter of OTA signals and then a microwave and satellite distributor of specialized programming and an originating program source itself. Cable was limited to small markets until the 1970s, when interest in it as an alternative to the networks' lock on local television began to spread. Policy makers in the United States began to take cable seriously, some imagining that the system of delivery could be made to serve educational, public information, social, and cultural interests, as well as expanded commercial users (Streeter, 1987, pp. 177-178).

Indeed, under FCC rules (Cable Act of 1992), communities can require cable operators to carry public, educational, and government channels. Moving away from the stronger regulatory environment in the 1960s, government restrictions on cable have been chiseled away. The "must-carry" provisions requiring cable operators to include all OTA broadcast stations in their basic tier of service no longer pertains, and in 1993 the cable companies fought off efforts by network television and its affiliates to collect transmission fees for carrying OTA stations. In effect, cable companies have managed to establish nearly equal footing with network television in transmission rights, although the latter still dominate the areas of program production (with competition from independent producers) and program distribution. Rather than truly compete head-to-head, the networks have partly bought into the ownership of cable companies (e.g., NBC into the news station CNBC).

The television networks, though still the most popular fare carried on CATV, have faced some discriminatory treatment from the cable giants. Tele-Communications Inc. (TCI), the biggest of the cable systems, with a 24% market share, refused to carry CNBC unless the parent company agreed not to compete with CNN, of which TCI is a major stakeholder. TCI is one of five corporations that control over half the U.S. cable markets, the industry as a whole covering 85% of U.S. residential areas and wiring 60% of them. Moreover, these corporations own both systems and program services (e.g., TCI owning part of Black Entertainment Television, Discovery Channel, Showtime, Turner Broadcasting, and Home Sports Network; Time Warner part of BET, Turner Broadcasting, HBO, and Comedy Channel). After Time Warner's takeover of Turner Broadcasting in 1996, TCI held 9% of that conglomerate. Price restrictions imposed on cable by the FCC in 1993 after rates had skyrocketed over the past decade were lifted in the 1996 Telecommunications Act.

From the outset, the various competing entities in communications have gone head-to-head in trying to monopolize the various media. Weakly regulated monopolistic competition was prevalent from the 1930s through the 1970s. But the new means and range of communication opened by satellite, fiber optics, computers, mobile telephone, coaxial television cable, packet switching, CD-ROM, and other digital technologies and the trend toward converging or repackaging them into new formats have given rise to a "revolutionary" lexicon about the present era. The regional telephone companies in the United States offer Americans options to be wired to anyone, anytime, anywhere: reduced long-distance charges, 800-prefix telephone numbers, caller ID (and its neutralizing counterpart, ID protect), voice mailbox, call forwarding, call waiting, and other customized services). For those who can afford it, the distractions offered by most of these gee whiz gadgetry and services are, if anything, counterrevolutionary—in the sense that they divert public attention and investment from other social, political, and economic concerns and injustices, including the 7% of households that are without telephone.

When CATV is not soaking up the non-school waking time of today's "vid kids," other electronic simulations of the real world are filling in. Nintendo, a Japanese company that has nearly monopolized the video game market (85 to 90% worldwide) with its computers and pastimes like "Donkey Kong" or "Super Mario Brothers," had sold some 60 million of its machines by the early 1990s (Fallows, 1994, p. 37). For many, if not most, children and young adults in the United States, owning a Nintendo machine is required cultural currency. And for millions of other children in the United States (and many other countries), familiarity with games like the violence-charged "Power Rangers" (a transnational joint venture) and its sequels is virtually certification of preadolescent citizenship.

FAX AMERICANA

Associated with the rapid expansion of desktop computers has been the proliferation of facsimile (fax) and electronic mail (e-mail) transmission. It is almost standard in new computers to have a built-in fax and modem (a device for converting information into a form that can be sent or received over a telephone line). Fax machines are technically standardized so that one machine can easily exchange information with another. They have been particularly useful in transmitting languages based on ideographs, such as Chinese and Japanese, which are not easy to create on keyboards. Computer operating languages, however, are not standardized and require "interfacing" systems for converting one computer language into another for remote communication, making e-mail functions more complicated.

Fax has been used as a substitute for both voice telephone and mail. Although the startup costs are high, over time it is less expensive than current post and more efficient than telephone in terms of volume of information transmitted per dollar. Despite its initial cost, the fax, like the cellular telephone, has caught on and already has an installed base of several million in the United States, much of it residential. In time, the fax may well be phased out as advanced optical character readers and other reproduction technologies are more fully integrated into computer communications. There is no way of knowing, however, how long it might take for such a multifunctional home computer to become popularly accessible.

In fact, although computers are already widely dispersed, the hoped-for explosion of household use has yet to be realized. It was argued by the public relations departments of personal computer manufacturers that the new machine was going to take over an assortment of household tasks as diverse as budgeting, shopping, and controlling security systems and appliances. As of 1993, some 33 million personal computers were in American homes (none in two thirds of them), but only 17% of households have any use for them other than as extensions of school and office work. Most are using them simply as electronic typewriters (Saltus, 1993a, 1993b). What computers seem to have successfully taken over in the household is the attention of many married men, and there are reports of marriages becoming one major casualty of this extramarital affair between man and machine.

Even if the household frontier has yet to be fully exploited by the computer industry, there is no question that business, government, and other societal institutions have made the conversion to digital modes of information and record keeping. The Roman Catholic Church has made a similar leap of faith, hiring General Electric (GE) to run its global data network linking the Vatican to church terminals in every continent (Barnet & Cavanagh, 1994, p. 336). It was often assumed by some early enthusiasts of the information society that computers would make life more stimulating and raise the general intelligence level of citizens. Alvin Toffler waxed about the "revolution in the info-sphere" bringing forward "a flood of new theories, ideas, ideologies, artistic insights, technical advances, economic and political innovations" and "a new civilization" (Toffler, 1980, pp. 193-194).

Toffler's musing was more of a hallucination than empirically well thought out. Not only was there little evidence for such happy talk, but the reasoning itself was founded on the quaint notion that, in corporate America, technology, rather than organized interests, initiates historical change. Two more scholarly observers see organized transnational and political interests using technology to extend the frontiers of "Fordist" automation in the workplace into everyday life. One finds more control exercised in such areas as office technology, political management, police and military activities, communication, and commodity culture and in reorganized forms of "leisure" as "increasingly privatized and passive recreation and consumption." Information technologies are helping bring the control of wealth, they argue, into a "new phase of accumulation" (Robins & Webster, 1988). *Fax Americana* can be thought of as a dominating global information structure, but also as a broader metaphor for American-style capitalist culture that has industrialized, commodified and networked most aspects of everyday life in the United States and reproduced itself in other societies.

As with all forms of technical and artistic production, opposite and antagonistic social tendencies arise. Of its own necessity, capitalism broke the shackles of feudal society by opening up activity in politics, the arts, commerce, the academy, and communication. Its offspring, liberal democracy, offered new social uses of the objects of its production—its tools, its literature, its communicating devices, and so on. Works of art and access to information would never be confined to the palace and the nobility of an earlier era, but at the same time, as Walter Benjamin noted, their place in industrial capitalist society would be mediated by a new social class ritual—exhibition in the marketplace (Benjamin, 1969). Mass media offered at least the hope of exposure to new forms of culture, but at the same time, it was a type of culture mechanically mass produced, largely stripped of unique conception and constrained by the screens of property, wealth, and capital. In the next chapter, we look at power and wealth as context-setting realities in the dream of a true and democratic "information society."

PART III

POLITICAL ISSUES IN
THE "INFORMATION
SOCIETY"

Chapters 5 and 6 cover a broad range of social and political issues related to new communication and information technologies. We look at issues of ownership and control, government efforts at regulation and deregulation, and the uses of communication technology to manage the political process. We also consider the excited claims about "artificial intelligence" and the possibilities of "computer reasoning." Counterposed to these hierarchical forms of technological control are grassroots activists, from serious community organizers to computer anarchists, who are challenging the dominant forms of information control with ideas and tactics of their own.

In this section, we also look at how information technology has accelerated the pace of commercialism and consumption in U.S. society. Although banks and other financial institutions are able to move trillions of dollars around the globe on a daily basis, more and more people experience anxiety about their next paycheck and whether their jobs, if they are lucky enough to have them, will be there in the morning. Huge transnational industrial and financial conglomerates have the size and means to make the most use of information, and these enterprises have the most to gain in material terms from the "communication revolution." And to ensure that their assets continue to expand, private corporations have organized electronic dossiers on citizen behavior to try to profile every act of consumption. This form of electronic spying and the personal records that have long been kept by government intelligence agencies on millions of citizens have created what amounts to a surveillance society. We can run, but it's nearly impossible to hide, from the all-seeing eyes of government and commercial investigators, wondering what we are up to politically and what we last ate for breakfast.

CHAPTER 5

Modern Power Structures and the Means to Communicate

BRAHMINS OF THE "GLOBAL VILLAGE"

As Ben Bagdikian (1989, 1992) and others have noted in studies of the global media empires, very few private transnational corporations (TNCs) dominate the news, information, ideas, entertainment, and popular culture that most people in the world receive. Before U.S. corporations assumed control of the news wire, it was held largely under the aegis of British interests. British control of international cables enabled Reuters' world news service monopoly to tutor Americans about the "causes" of World War I and (after having cut all of Germany's worldwide oceanic cables) about who were the righteous defenders of Western civilization. In the early 1990s, Reuters Television was supplying television news to more than 400 broadcasters in 85 countries, with an audience of some half-billion households (Hamelink, 1994b, p. 83).

In an earlier era it was said that, in Europe, "all roads lead to Rome," but by the early 20th century, what was more consequential was that the all the world's cables led to London. By the end of World War II, it was yet more compelling that the world media were largely American (see Tunstall's 1977 study on cultural imperialism) and that a handful of film and television studios, press agencies, government radio services, book and magazine publishers, and advertising and public relations firms controlled the global distribution of mass communications and, in effect, what most people electronically heard and viewed and what they daily read. About 60% of the world's population has access to television, and of television audiences outside the United States, about half watch some American programming, mainly old serial reruns (Mander, 1991, p. 77).

Within the United States, such decisions have come down to 20 multimedia *Fortune* 500 CEOs who manage the country's cultural industries, the number of entities shrinking every few years. In the 1980s, General Electric (GE) bought out RCA (and NBC Broadcasting), Capital Cities got control of ABC Broadcasting (both later taken over by Disney), and Loews Corporation gained effective control of CBS, whereas the fourth network, Fox, was captured by Rupert Murdoch's News Corporation (see Table 5.1). Tele-Communications Inc. (TCI), the largest cable operation in the United States, was nearly overtaken by Time Warner in mid-1995 with the latter's $2.3 billion acquisition of Houston Industries (cable properties) and $2.2 billion purchase of Cablevision. By 1996, after its $7.5 billion merger with Ted Turner's cable operations (e.g., CNN, Headline News, Turner Network TV, TBS), Time Warner had 11.5 million cable subscribers to TCI's 14 million (Kramer, 1996, p. A1). Time Warner has a cash flow larger than the three major television networks combined (Downing, 1990, p. 32). Viacom moved into major new film, television, music, and publishing businesses with its $9.7 billion purchase of Paramount in 1994 and into new video and cable operations with the $8 billion acquisition the same year of Blockbuster, which in turn owns the Florida Marlins and Miami Dolphins sports franchises.

Associated Press (U.S.), United Press International (U.S.), Reuters (U.K.), and Agence France Presse (France) distribute most of the wire service news, Hollywood studios dominate the film and television imports of most countries, and a few huge publishers produce the most influential books and magazines circulated around the world. In the United States, a few film studios control about a third of the country's movie screens, although some of the most important capital ownership in these companies is Japanese (until 1995, Matsushita in MCA and Universal Studios; Sony in Columbia Films) and Italian (Pathe, S.A. in MGM/ United Artists).

Fox Broadcasting, which is 99% owned by Rupert Murdoch's News Corporation, is based in Australia. In 1994, NBC challenged the Federal Communications Commission (FCC) to enforce a regulation from the 1934 Communications Act that no foreign corporation may own more than 25% of a U.S. broadcasting company (in Australia, the limit is 15%; Carter, 1994). The chances of Congress enforcing FCC proscriptions against Murdoch, a political conservative, were diminished with the Republican electoral victory in 1994.

The Time and Warner (U.S.) merger in 1990 brought together an estimated $10 billion in annual revenues from book, magazine, film, cartoon, comics, television, recording, cable, and other operations, including a worldwide readership of some 120 million and 23 million paid subscribers to Home Box Office and Cinemax. Greater than the gross national product of most countries, this $14.1 billion merger made Time Warner the largest communications and entertainment corporation in the world and opened huge distribution channels for its music, data, video, films, and so on. And in view of GE's venture into broadcast ownership, one can only speculate the extent to which NBC's restrained coverage of nuclear power and other technologies, including military hardware, in which GE has a major stake, was influenced by the parent company. GE "designed, manufactured or supplied parts or maintenance for nearly every important weapons system" used by the United States in the Persian Gulf War, including the Patriot antimissile missiles and Tomahawk cruise missiles, the Stealth and B-52 bombers, Apache and Cobra helicopters, NAVSTAR spy satellite, and AWACS airborne radar system, each of which NBC extolled in its news commentary (Bagdikian, 1989, p. 808; Barnet & Cavanagh, 1994, p. 347; Hamelink, 1994b, p. 92; Kellner, 1990, p. 66). In 1995, rival Westinghouse, another electronics, communications, military systems, and nuclear power conglomerate, gained control over the CBS television network in a $5.4 billion deal. A year later, under new federal telecommunications law, Westinghouse announced a merger with Infinity Broadcasting in a $3.7 billion deal that brought together the country's two largest radio station networks.

As of early 1995, the other leading global media corporations in the 1990s included Bertelsmann AG (Germany), Fininvest (Italy), Hachette SA (France), News Corporation Ltd. (Australia), and Capital Cities/ABC (U.S.). Bertelsmann runs a television satellite channel and *Stern* magazine in Germany, Britain's two largest book clubs, and in the United States a number of publishing houses, including Doubleday, Bantam Books, Dell, and the Literary Guild book club, in addition to magazines (*Parents* and *Young Miss*) and the RCA and Artista record labels. Its other worldwide holdings include about 40 magazines and book clubs with 22 million members. Organized by German World War II veteran Reinhard Mohn, who spent 2 years in Kansas as a prisoner of war, Bertelsmann is made up of 375 companies in 30 countries, with 44,000 employees and $7.5 billion in 1994 sales revenues (Barnet & Cavanagh, 1994, pp. 68-69).

One of Mohn's competitors was British war veteran Robert Maxwell, a Czech Jew (born Ludvik Hoch), who fled the Nazi terror and went on to build a mass media empire. Maxwell's crown jewels, until he died in 1991 under mysterious circumstances, included Macmillan publishing (U.S.), the Mirror Group Newspapers P.L.C. (U.K.), the *Daily Mirror* (U.K.), Pergamon Press (U.K.), and for a brief time in 1991, New York City's *Daily News,* the last of which he apparently used as bait to secure loans of $238 million for money-laundering purposes from

TABLE 5.1 Top Four U.S. Media/Entertainment Corporations

GENERAL ELECTRIC

Military/Industrial/Consumer/Industrial Electronics:
GE Aircraft Engines, GE Transportation Systems; GE Power Corporation, Electrical Distribution, and Control; GE Motors and Industrial Systems; GE Plastics; GE Medical Systems; GE Appliances (e.g., GE, Hotpoint); GE Lighting

Communications:
GE Americom (satellites), GE Capital Communications Services (long-distance telephone), GE Information Services

Financial/Insurance:
GE Capital, GNA Corporation, and other insurance firms

Television:
WNBC-New York, KNBC-Los Angeles, WMAQ-Chicago, WCAU-Philadelphia, WRC-D.C., WTVJ-Miami, WNCN-Raleigh-Durham, WCMH-Columbus, WJAR-Providence, NBC Network News: *Today Show, Weekend Today, Meet the Press, Dateline NBC, Nightline, NBC Nightly News, NBC News at Sunrise*

Cable:
CNBC, Court TV (33% with Time Warner), (50% with Rainbow/Cablevision), American Movie Classics (25% with Rainbow), America's Talking (50% with Microsoft), A&E (25% with Disney and Hearst), and through NBC up to 50% of History Channel (with ABC and Hearst), Independent Film Channel (with Rainbow), Prime, News Sport, Prism (with Rainbow and Liberty Media/TCI), Romance Classics, and seven sports channels (Sports Channel Cincinnati, Chicago, Florida, New England, Pacific, Ohio, Philadelphia)

TIME WARNER

Financial/Retail/Commercial/Consumer Goods:
Capital Group Companies, Warner Bros. stores (> 100), Seagrams (14.5% ownership), Houston Industries

Television Programming/Motion Pictures:
Warner Bros. Television, Witt Thomas Productions, Warner Bros., Warner Brothers Animated

Book Publishing/Magazines:
Oxmoor House; Sunset Books; Little, Brown & Co.; Time-Life Books; Warner Books; Book-of-the-Month Club; Turner Publishing; *Time, Life, Fortune, Sports Illustrated, Vibe, People, Money, Sports Illustrated for Kids, Parenting, In Style, Asia Week, Sunset, American Lawyer;* DC Comics (50%); *Health, Hippocrates, Entertainment Weekly, Who, Southern Living, Dancyu, Martha Stewart Living, Baby Talk, President, Cooking Light, American Lawyer*

Multimedia:
CNN Interactive, Turner New Media (CD-ROMs)

Entertainment/Home Video:
Time Warner Entertainment, Six Flags, Cable franchises (11.7 million subscribers), Time-Life Video, HBO Home Video, Warner Home Video

Music:
Warner/Chappell Publishing, Atlantic Group, Time Warner Audio Books, Elektra Entertainment Group, Warner Brothers Records, Warner Music International, SubPop (40% ownership), Columbia House (50% ownership)

TABLE 5.1 *Continued*

Turner Broadcasting:
  Atlanta Braves, Atlanta Hawks, Turner Retail Group, Turner Home Entertainment, Goodwill
  Games, Programming/production (World Championship Wrestling), Hanna Barbera
  cartoons, New Line Cinema, Castle Rock Entertainment, Turner Entertainment Co. (MGM,
  RKO, pre-1950 Warner Bros. films), Turner Original Productions, Turner Pictures
Cable/Turner Cable:
  Sega (33%), E!, Bravo, Cinemax, Comedy Central (50% Viacom), HBO, HBO Direct
  Broadcast, Time Warner Entertainment, Six Flags, other cable franchises, TBS Superstation,
  Turner Classic Movies, TNT, Cartoon Network, CNN Airport Network, CNN International,
  CNN, Headline News, Sportsouth, CNN Radio, CNNfn (CNN financial network)

DISNEY/CAP CITIES

Industrial/Retail/Insurance:
  Sid R. Bass (crude petroleum and natural gas: partial ownership), 429 Disney stores,
  Childcraft Education, State Farm (partial), Berkshire Hathaway (partial)
Theme Parks/Resorts:
  Disneyland, Walt Disney World Resort, Disneyland Paris, Tokyo Disneyland, Disney
  Vacation Clubs, WCO Vacationland Resorts, Disney Institute, Celebration, Disney Cruiseline
Sports Teams:
  Mighty Ducks, California Angels (with Gene Autry)
Newspapers/Book Publishing/Magazines:
  *Fort Worth Star-Telegram, Kansas City Star, St. Louis Daily Record, Narragansett Times,
  Oakland Press and Reminder* (Pontiac, MI), *County Press* (Lapeer, MI), *Times-Leader*
  (Wilkes-Barre, PA), *Belleville News-Democrat* (IL), *Albany Democrat* (OR), *Daily Tidings*
  (Ashland, OR), *Sutton Industries* and *Penny Power* (shoppers), Hyperion Books, Chilton
  Publications (books and trade magazines), Fairchild Publishing (W., Women's Wear Daily),
  L.A. Magazine, Institutional Investor, Disney Publishing, Inc. (Family Fun and others)
Multimedia:
  Disney Interactive, Disney.com, ABC-Online
Television, Cable, and Home Video:
  Disney Channel, Disney Television (syndicated programming), Touchstone Television
  (*Ellen, Home Improvement*), A&E (with Hearst and GE), Lifetime Network (50%), ESPN
  (80%), ESPN 2 (80%), Buena Vista Television (*Home Again*), Buena Vista Home Video
Music:
  Hollywood Records, Wonderland Music, Walt Disney Records
ABC Television, Network News, and Radio:
  ABC-TV: WABC-New York, WLS-Chicago, KFSN-Fresno, KTRK-Houston,
  KPVI-Philadelphia, KGO-San Francisco, WTVD-Raleigh-Durham, WJRT-Flint, MI,
  WTVG-Toledo, KABC-Los Angeles, and part owner of 8 television stations, *Prime Time
  Live, Good Morning America* (weekdays and Sunday), *World News Tonight* (weekend
  editions), *World News Tonight with Peter Jennings, World News This Morning, World News
  Now, Nightline, This Week with David Brinkley, 20/20,* ABC Radio (12 stations, network
  with 3,400 stations)
Motion Pictures:
  Walt Disney Pictures, Touchstone Pictures, Hollywood Pictures, Miramax Film
  Corporation, Buena Vista Pictures (distribution)

(continued)

TABLE 5.1 *Continued*

WESTINGHOUSE

Industrial/Financial/Insurance:
Thermo King (mobile refrigeration), Bankers Trust, Bandywine Asset Management, FMR
Corp. (partial), Westinghouse Pension Management, WPIC Corporation

Power/Nuclear Power/Waste Disposal: (hazardous and radioactive)
Power generation (electric power plant parts), Energy Systems (40% of world nuclear power
plants use Westinghouse engineering), Resource Energy Systems, Scientific Ecology Group,
Westinghouse Remediation Services, GESCO (including four government nuclear facilities)

Communications and Information/Satellite:
Telephone, network, and wireless communications systems; security systems; Group W
Satellite Communications (satellite distributor of television programs)

Television, Radio, Cable:
KCNC-Denver, WFOR-Miami, KYW-Philadelphia, KUTV-Salt Lake City, WWJ-Detroit,
WCCO-Minneapolis, WFRV-Green Bay, WI, KCBS-Los Angeles, WCBS-New York,
WBBM-Chicago, KPIX-San Francisco, KDKA-Pittsburgh, WBZ-Boston, WJZ-Baltimore,
*Up to the Minute, CBS Morning News, CBS News Sunday Morning, 60 Minutes, CBS This
Morning, CBS Evening News with Dan Rather, Face the Nation, 48 Hours,* 21 PM and 18
AM radio stations, Country Music Television, Home Team Sports, the Nashville Network

NEWS CORP.

Television, Cable, Film
Fox Broadcasting network (some 8 TV stations and 150 affiliates in the United States),
Twentieth Century Fox, New World Communications Group (broadcast distribution) (50%),
CBS/Fox Video (part), FX (cable), Star TV (Asian satellite service), BSkyB (British
satellite service)

Newspapers, News Networks, Magazines, Publishing
New York Post, Boston Herald, New York magazine, *Seventeen, Economist* (part), and many
other magazines, *The Times, Sunday Times, Financial Times* (part) and the *Sun* (London),
Village Voice, Weekly Standard, TV Guide, HarperCollins, Scott Foresman, Viking Pub.
(part), Penguin Pub. (part), ITN (British news supplier)

Multi-Media
Delphi Internet Services

SOURCE: Miller and Biden (1996), pp. 23-28; Steinbeck (1995), passim; Bagdikian (1992), passim.

New York–based Bankers Trust corporation and other sources (Bagdikian, 1989,
pp. 810-814; 1992, pp. 240-241; Janofsky, 1994b).

On July 31, 1995, the day before the Westinghouse buyout of CBS, ABC/
Capital Cities (see Table 5.2 for data on salaries, compensation, and stock
ownership) was sold off to Disney in a record communication payoff of $19
billion. Disney now directly owns or interlocks with a series of newspapers and
television stations in most of the major U.S. market, Hollywood studios, the
ESPN sports cable network, a major share of Arts & Entertainment and Lifetime

TABLE 5.2 Salary, Compensation, and Stock Ownership of Top Capital
Cities/ABC-TV Executives, 1991

	Salary	Additional Compensation*	Value of Stock Holdings in Company**
Thomas S. Murphy, Chair	$834,349	$2,999,254	$61,641,500
Daniel B. Burke, Pres., CEO/COO	$960,742	$2,972,569	$31,285,500
John B. Sias, Exec. VP/Pres. ABC	$833,770	$2,945,884	$10,121,000
Michael Mallardi, Sr. VP/Pres. Broadcasting	$726,056	$3,406,970	$7,109,000
Ronald J. Doerfler, Sr. VP/CFO	$816,332	$2,839,142	$11,088,500

SOURCE: Mazzocco (1994, p. 108).
NOTE: *Includes incentive compensation plan and supplementary compensation plan; **Estimated value of
their stock based on listed number of shares held.

cable stations, the *Washington Post,* a block of trade and consumer publications,
the huge advertising-revenue-generating RJR Nabisco food and tobacco corporation, and the worldwide television distribution of ABC. In Italy, the leading
media magnate is Silvio Berlusconi, a billionaire media tycoon who also has
major newspaper and television holdings in Germany, Spain, Tunisia, (former)
Yugoslavia, and France.

Jean-Luc Lagardere, who is chair of Hachette, France's largest communications corporation and a conservative activist, is the world's biggest magazine
publisher (74 magazines in 10 countries), including his *Paris-Match,* and the
largest distribution network for Spanish-language publications. In 1988, he
bought out Grolier Publications (U.S.), which publishes *Encyclopedia Americana* and several book series. He also runs one of France's largest defense
corporations, Matra SA, which builds guided missiles, other weapons systems,
and military telecommunications. Legardere used his friendship with French
Prime Minister Jacques Chirac to build the Hachette empire (Bagdikian, 1989,
1992).

A fellow conservative, the Australian-turned-American Rupert Murdoch,
who owns News Corporation, Ltd., and more than 90 communications companies, started out as a young newspaper tycoon in Adelaide before moving on to
bigger media ventures in Sydney, England (*Times, Sunday Times, Financial
Times, News of the World,* and the *Sun,* the last two having daily circulations of
5 and 4 million, respectively). He also controls the *Economist,* 7% of Reuters,
Viking and Penguin book publishers, and *TV Guide,* the largest-selling magazine

in the United States. In Australia, Murdoch commands two thirds of newspaper circulation; in New Zealand, half; and in England, one third.

In the United Kingdom, his mix of elitist and sleazy tabloid newspapers and control of the all-European Sky Channel and other satellite television channels were orchestrated to elect and reelect Margaret Thatcher as prime minister. In the United States, his control of the *New York Post, Boston Herald,* and Fox (Metromedia) Broadcasting network helped build support for the Reagan administration. Murdoch bought *TV Guide* (circulation 17 million) and Walter Annenberg's other Triangle Publications in 1988 for $3 billion, one of the treasures of his unending multimedia organization (HarperCollins, *Seventeen, New York,* many religious book titles, the 20th Century-Fox movie studio, STAR satellite system in Hong Kong, and various Asian publications in the early 1990s, 12 additional U.S. television network affiliate stations in May 1994 (Bagdikian, 1989, pp. 806, 808; 1992, p. 241; Reuters, 1994, p. 4).

For a time, the most successful of the political-media marriages was that of Berlusconi. In 1994, he used his empire of the three largest private television networks in Italy, 34 magazines, two publishing houses, and several newspapers, in addition to his control of Italy's largest supermarket chain, real estate, insurance and advertising interests, and a professional soccer team, to successfully lead a rightist electoral victory, which made him the country's prime minister, in partnership with the neofascists. Berlusconi's achievement came at a time when his media corporation, Filinvest, was tainted by corruption charges (the national *tangentopoli,* "Kickback City," scandal) and $2.2 billion in debt and amid a growing government confrontation with the Mafia, a crime syndicate with which he is reputedly associated. The future of his financial holdings and his coalition with the neofascist National Alliance Party (led by Mussolini's granddaughter) are among the stakes in Italy's shifting political tide (Cowell, 1994). But his coalition broke up in December 1994, and Berlusconi was forced to resign.

New mergers, acquisitions, and sell-offs continually change the specific makeup of conglomerates, such as these global multimedia operations, but also continue to enlarge the centralized control or influence over messages, images, lifestyles, cultural standards, social values, and political norms. One study that discussed, in part, the political economy of the blockbuster film *Batman* showed how it cemented the biggest communications merger to date, Time, Inc. and Warner Communications, and created a financial bonanza for the new conglomerate. Time promoted the film in its magazine pages while Warner milked it through box office receipts (Warner Films), sound track sales (Warner Records), a separate crossover album and a single recording by one of the background musical performers, Prince (on contract with Warner Records), music publication (by Warner), and a video (by Warner) and sold it to MTV (in which Warner indirectly had stock interest). Noting that *Batman* was less a film than a product line, the study helped illustrate the relationship between economics and the production of culture (Meehan, Mosco, & Wasko, 1993).

Communications Are Us

With TNCs, including the global communication enterprises, so much in control of the means to speak and be heard, the First Amendment seems increasingly hollow with each new television sitcom, tabloid news show, and *People* magazine-style program format. Still, the democratic space open to public discourse about serious issues is found in varying degrees and political orientations—in desktop publishing, public access radio and cable TV, college media, alternative news publications, electronic bulletin boards, and even within the occasional cracks in the empire of established mass media. The revolutionary writer and political theoretician Antonio Gramsci (1891-1937) envisioned the possibilities for revolutionary change even within the dominant ideological apparatuses (the legal, political, governing, media institutions) of society, as long as the intellectuals and artists within had "feeling [for] the elemental passions of the people, understanding them and thus explaining and justifying them in a particular historical situation" (cited in Joll, 1978, p. 131).

In the United States, Gramscian alternatives to the hegemony of mainstream media include *Extra!, Lies of Our Times, The Nation, The Progressive, Z* magazine, *In These Times, Rolling Stone, Village Voice, Columbia Journalism Review, Mother Jones, On the Issues, Tikkun, The Utne Reader,* Paper Tiger Television and video productions, the program *P.O.V.* on PBS, many community cable stations around the country, Alternative Radio, Pacifica Radio, and a number of independent film, radio, and television producers and radio stations. An impressive assortment of books that have done much to educate people about the diverse cultural resources in the United States and that are critical of reactionary political tendencies are widely available in commercial and university bookstores. In the past few years, electronic bulletin boards and conference groups have been trading information that counteracts the official distortions of establishment politics carried in the mass media.

Realistically speaking, however, virtually all of these alternative sources of information have relatively small and overlapping audiences. Yet, for those skeptical of official "truths," there is no shortage of critical counterpoints, although not packaged in the slick gloss of the mainstream media. The power of the corporate media to defend the ideological hegemony of ruling interests and their superior technological means of communicating ideas and images represent the principal obstacles to public understanding of social issues. Corporate communication institutions have a history that needs to be understood if new public intellectuals (educated people dedicated to democratic change) are to make a difference in their attempts to construct socially responsible information alternatives.

And transnational communication conglomerates show no signs of becoming dinosaurs. One of them, the Atlanta-based news corporation Cable News Network International (CNNI), started in 1980, is a growing institution of worldwide persuasion. It transmits television news in Japanese, Spanish, Polish, and English (and soon French) all over the world, half of its international programming coming directly from its U.S. cable news station, with 28 news bureaus (and 500 radio stations) reaching some 130 million people. During the Persian Gulf War, the whole world was glued to CNNI's version of events, a measure of influence that no president could ignore. Another global phenomenon, the American music television network, MTV, with its high-tech music and electronic image culture, reaches 215 million homes in 75 countries (McPhail, 1993).

Managing the Global Media

The First Amendment of the U.S. Constitution gives citizens the right to speak but does not guarantee the right to be heard or to receive information. And given the audience reach of modern mass media, very few indeed have much of a voice or access to information that actually matters. Although more than 25,000 newspaper, magazine, television, book publishing, and motion picture outlets are in the United States, the majority are controlled by just 20 corporations. Almost all of them can be classified as "economic conservative," according to Ben Bagdikian, a former Pulitzer Prize winning journalist, *Washington Post* editor, and professor at the University of California, Berkeley (Bagdikian, 1992, pp. ix, 4-5).

Concentration in media ownership has been an ongoing pattern in the newspaper business. The *New York Times,* for example, from the perspective of ruling elites the "newspaper of record" in the United States, is a major corporation in its own right. By 1989, the *Times* owned 26 daily (including a few years later, the leading New England daily *Boston Globe*) and 9 nondaily newspapers, along with five television stations and two radio stations, many magazines (e.g., *Family Circle, Tennis, Golf Digest*), and one third ownership in the *International Herald Tribune.* The *Times* is also an owner of several Canadian and U.S. paper mills, which protects the corporation from rising newsprint prices (*Left Business Observer,* 1989). Starting in the 1970s, the *Times* began moving toward a more commercial look, with a six-column format and an increased advertising ratio. Was the British press tycoon Lord Thompson merely being frank when he declared his definition of news to be "nonsense used to fill in the space between ads" (cited in Bissio, 1990, p. 85)?

From its magisterial gothic logo, to its claims of publishing "all the news that's fit to print," to its full-page ads for luxury Fifth Avenue department stores and industry public relations, to its board of directors who interlock with IBM,

J. P. Morgan, Ford, American Express, New York Life, General Dynamics, Merck, and many other Fortune 500 companies, the hugely profitable ($160 million by 1987) *New York Times* is a voice of ruling-class interests (*Left Business Observer,* 1989). African Americans and Latinos receive little attention as regular citizens, much less as role models, in the *Times* or other mainstream media, and when the press does choose to pay heed, it rarely fails to focus on matters of race. Organized labor is typically noticed only when strikes occur and is rarely given adequate space to explain its justification for disrupting the normal flow of production and services.

Women likewise continue to be either largely invisible in the mass media or greatly underrepresented as respected role models. In a 1994 Freedom Forum (Gannett chain-sponsored association) study, men received 75% of front-page American newspaper references, and that was during a period when the Nancy Kerrigan–Tonya Harding skating soap opera was a national media obsession. Moreover, men wrote 67% of front-page articles and 72% of pieces on the op-ed pages. Of the reporters in the press, only 21% were women; on the three major television networks, CBS had 32%, ABC had 17%, and NBC had 14%. Compared with men (82%), women were given positive portrayals as authorities only 51% of the time, the rest mainly as sports figures or entertainers.

In foreign affairs, the concentration of media is revealed in its consolidation of bias—generally protective of imperialistic ambitions of nation state and business interests. In the recent past, on the one hand, the American media managed to miss the scoop on the Iran-*Contra* scandal, which came to light only after the story appeared in a Lebanese publication. On the other hand, the American media were quick to cover the 1983 story of KAL flight 007 as a case of cold-blooded communism when a Soviet aircraft was ordered to shoot down a South Korean commercial jetliner that it believed was a military spy plane overflying its territory (a *New York Times* editorial headlined it "Murder in the Air"). Yet, the same editorial page offered only sympathetic treatment for the U.S. naval officer who gave a similar order against an Iranian passenger plane shot down in commercial airspace 5 years later while other New York papers defused the horror of the incident by warning of Iranian retaliation (Lee & Solomon, 1990, pp. 278-283).

For those interested in more balanced or more critical perspectives on world and national events, alternative sources are certainly available. A group called Project Censored, based at Sonoma State University in California since 1976, reports important stories that most mass media ignore or refuse to cover. The top of the list of underreported stories in 1993, as determined by a panel of judges from among distinguished writers, publishers, editors, academics, public inter-est associations, and the mass media, was based on a report of the United Nations Children's Fund (UNICEF) that revealed that of the cases of murdered children in industrialized countries, 90% occur in the United States. One might think that in a country that so often touts "family values" as a national virtue, such a story would receive a glut of mainstream media attention.

A second story, covered in the alternative publication *Extra!* but missing from the major media concerned the facts of life of U.S. intervention in Somalia, pointing out that four leading U.S. oil companies at the time were holding exclusive concessions to drill the black gold in the Somali countryside. Some other missing stories from the 1993 Project Censored list included the resumption of U.S. Army biological warfare testing, the Chernobyl coverup, the CIA's involvement in the overthrow of the Aristide government in Haiti, and a major education report on the failure of the school voucher system. Some of these stories did get spot coverage in some large or medium-sized urban newspapers, but lacking sustained attention by the television networks, the large circulation weeklies, or the influential daily press, it becomes more demanding of people's time and energy to locate and substitute critical alternative media for those on which they regularly have come to rely.

Smaller and more flexible media, such as stationary and mobile telephone, telex, electronic mail (e-mail), data transmission, facsimile, citizens band radio, and post, offer more extensive access for creative and interactive communication, but not with the distributional power of print and broadcast media. Although the number of channels for communications has increased, definite limits remain to what can be said. The monitoring of telephone, actual and potential, imposes controls on what can be transmitted over the wire, and the same is certainly true for electronic bulletin boards and other forms of communication. Television and the press occasionally report on the arrest of an individual for sexual solicitation or computer software illegally sent over e-mail, but these stories also tell us that someone is watching. James Bamford's book *The Puzzle Palace* (1982), about the operations of the National Security Agency (NSA), revealed that international carriers and other private transnational communication corporations routinely recorded and relayed information on Americans' overseas telephone, data, and messages to the highest and most secretive branch of government security, some of which the NSA, in turn, passed on to the FBI. In 1982, a federal appeals court ruled favorably on the legality of these intercepts.

Control over newly emerging communication technologies follows the same patterns as the older media. The computer services provider Prodigy, for example, is a joint venture of IBM and Sears; one of its competitors, Compuserve, is a subsidiary of the financial accounting firm H & R Block; and the Delphi Internet network is owned by Rupert Murdoch's Australia-based News Corporation. This raises a number of concerns as diverse as the cost of access for ordinary users and the issue of censorship. Prodigy, which promotes itself as a "family-oriented" service, regularly monitors and removes messages it considers offensive (Feder, 1991, p. E5).

Transnationalized Division of
Production and Consumption

The last half of the 20th century has seen enormous changes in the organization of world capital, moving toward much greater integration of production systems and labor forces. Telecommunications and computers have been central to this transformation by greatly reducing the constraints of time and space. Over the centuries, regional specialization in the production of goods and services, as was once the case with English cloth making and Chinese silks, has given way to a far more differentiated and complex division of labor. Exchange markets have developed in layers upon layers in every region of the world. In the post-World War II era, TNCs have led the way in relocating manufacturing and service industries to both the major industrial countries and to the Third World, making commodities more indistinguishable as to their points of origin.

Japanese soy sauce is largely made from American soybeans, whereas Honda and Toyota automobiles driven by Americans were probably made in the United States. The IBM desktop computer was assembled mainly with components from east and southeast Asia. Motorola television sets are actually made by Matsushita Electric. Telephones that Americans use were probably made in Singapore or Hong Kong, clothing perhaps came from China, and winter fruits and vegetables probably from Mexico, and Central and South America. By 1992, Sony owned CBS Records and had a 20% share of the U.S. film market, bringing out *Bugsy, Terminator 2, Prince of Tides,* and *Hook* through its control of Columbia Pictures and Tri-Star Pictures (Barnet & Cavanagh, 1994, p. 135).

The development of transoceanic cables, microwave, satellite systems, and wired and fiber optic cable, all manufactured by a few transnational telecommunications corporations, have greatly expanded information channels while greatly reducing their cost. Communication and information technologies provide a pipeline for preserving economic advantage over nation states and medium and small-scale business, particularly in Third World countries. These invisible weapons make it far more difficult for nationalist interests in the Third World to mobilize in defense of their borders and their sovereignty and to be able to respond in practical ways to problems of poverty, unemployment, and basic social infrastructure. Overwhelmingly, transmission of information over these circuits is transacted by a few powerful governments and a few private business corporations, particularly international banks. AT&T data show that long-distance telephone calls from Third World countries overwhelmingly terminate in the leading industrial and former colonial countries.

The New York Clearing House Interbank Payment System (CHIPS), representing 11 Big Apple banks and many others around the world, is the largest computerized dollar trading network in the world. CHIPS is linked to its British counterpart, CHAPS, and a satellite-based banking system for global electronic funds transfer called Society for Worldwide Interbank Financial Transactions (SWIFT), which ties together some 3,200 institutions in 73 operating countries on a 24-hour basis. SWIFT grew from some 77,000 daily transactions transmitted in 1977 to almost 2 million in 1993. This has greatly facilitated the transborder trading of securities and other financial services, along with arbitraging in the tax and interest rate markets—by the mid-1980s to the tune of $4 trillion daily (Fox, 1993, p. 25; Mulgan, 1991, pp. 28, 232)! The Société Internationale de Télécommunication Aéronautique (SITA) is an international system of leased lines used for coordinating information by the airline industry. Visa and Mastercard use similar networks.

The global shopping mall, the great icon of transnational consumption culture, is a powerful socializing force that "takes no prisoners." Cultural differences certainly remain between societies, but the packaged images of a Michael Jackson or a Madonna and Western–influenced urbanization in Third World countries create communities of lifestyles less bounded by national borders. The spread of material values and the speed of communicating them have given private business ventures, both great and small, new lessons in profiteering. Every radical tendency on the left (e.g., the women's movement, civil rights, ecology, peace issues, disarmament) and on the far right (e.g., racism, fascism, anti-Semitism, fundamentalism) and even middle-class consumer protection is exploited for its commercial possibilities. These become transliterated into the vernacular of television advertising, billboards, T-shirts, broadcast news, rock music, sitcoms, and other marketable formats that dilute or trivialize the original content of their meaning and power. Outside Cuba, the image of "Che" Guevara is appreciated more as a wall decoration than as a symbol of revolutionary struggle for national liberation, anti-imperialism, and socialism, an icon easily substituted in North American lifestyle by a travel or Kurt Cobain poster.

OLD WINE IN NEW BOTTLES: REGULATION AND DEREGULATION OF TELECOMMUNICATIONS

On the surface, a tremendous change has occurred in Western cultural and communication practices in the past few decades. The gradual breaking up of the AT&T monopoly in long distance and terminal equipment markets, the expansion of cable TV networks, the uses of home videography, recording, and playback, and so on have opened options for the creation and consumption of information and entertainment but have not led to the end of drudgery, history, or ideology that some had hastily predicted (Bell, 1960; Fukuyama, 1989; McLuhan, 1964). In the policy area, starting in the 1960s, the FCC opened private-line microwave markets with its "Above 890" decision, which induced long-distance telephone competition between MCI and AT&T. The Hush-A-Phone and Carterfone decisions benefited smaller companies seeking to break AT&T's telephone equipment manufacturing monopoly.

The most momentous change in recent telecommunications policy was the 1984 divestiture decree against Bell Telephone, pushed by the Justice Department's antitrust division. This led to the separation of AT&T from its 22 regional companies. which later merged into 7 regional Bell operating companies (RBOCs). AT&T's domestic long-distance service market share, challenged by companies like MCI (now partners with British Telecommunications and partially owned by IBM) and Sprint (originally owned by GTE and United Telecom) fell from more than 90% to 66%, but its revenues, generated by lower real (inflation-adjusted) prices, grew substantially (Bolton, 1993, p. 127). At the same time, the decline of Bell Labs, which had dominated telephone equipment R&D for decades, contributed to increases in the country's trade deficit. Apart from AT&T, the major beneficiaries of these regulatory changes were TNCs, especially those in financial services, which were able to significantly reduce their international communication costs and thereby move more of their business out of the United States. One study found that inflation-adjusted costs of an international telephone circuit declined from $22,000 in 1965 to $30 in 1985 (A. Posthuma, cited in Wilson, 1995, p. 212).

AT&T's breakup also freed it to compete in new international sales, services, and investments, including those in the computer and other nontelephone industries. Since 1984, AT&T has taken a large share of Sun Microsystems (and Java software), formed a joint venture with its long-distance rival GTE, and acquired the Paradyne Corporation prior to its establishing a credit card business in 1990. Its purchase of Easylink, part of Western Union before the latter's demise, joined with AT&T Mail, made AT&T the largest e-mail operator in the country. One of AT&T's spinoffs from the divestiture, Ameritech, bought up CyberTel in 1990, a major cellular telephone operation, while AT&T saved the best pickings for itself by merging with the biggest cellular phone company, McCaw Cellular Communications, Inc., in an $11.5 billion deal. In 1991, AT&T picked up NCR for $7.5 billion, the fifth-largest computer firm in the United States.

Other RBOCs have also been merging with nontelephone communication enterprises, with an eye toward the huge cable TV markets in the United States, the United Kingdom, where they are already the largest investors, France, and Israel (Bolton, 1993, p. 136). In 1993, Bell Atlantic began plans for a $33 billion merger with the then largest U.S. cable company, TCI, which would have given one company control of telephone or cable lines to 40% of U.S. homes. Although the deal ultimately fell through, other mergers of this magnitude are almost certain to occur in the currently weak antitrust environment, especially after the 1996 Telecommunications Act (Reed, 1995, p. 52). Since the Act's passage, AT&T has aggressively announced plans to move back into the $90 billion local telephone market that it gave up in 1984, at the same time breaking up into three companies, moving into Internet services and satellite broadcasting and dumping its computer manufacturing acquisitions (Lander, 1996; Norris, 1995).

Overseas, AT&T formed a manufacturing joint venture with Philips (Netherlands); bought up 25% of Olivetti, the Italian computer and business machine manufacturer;, and bought 20% of Italtel, Italy's leading telecommunications equipment manufacturer. AT&T also took over or acquired part of various communication corporations in the United Kingdom, Spain, Venezuela, and the Ukraine while setting up plants in Ireland, Japan, South Korea, Singapore, Thailand, and Taiwan (Bolton, 1993, p. 136). Virtually all of its subsidiary Western Electric telephones are now made in Asia. Overseas investments have had profound impacts on jobs and labor organizing in the United States.

The breakup of AT&T greatly weakened the unionization of its manufacturing and service employees, led by the Communication Workers of America and its affiliated unions; layoffs before 1982 were almost non-existent. Between 1986 and 1989 alone, some 200,000 jobs were lost, with announced additional cuts of 15% to 20% of the workforce in long-distance services between 1994 and 1996. Other telephone companies from November 1992 to February 1994 forced payroll reductions of 85,000 jobs. In early 1996, AT&T announced plans for additional layoffs of 40,000 workers, 70% to take place that year. This is part of the massive downsizing of the U.S. workforce by the Fortune 500 industrial corporations, resulting in 4.4 million layoffs from 1979 to 1992 alone (Andrews, 1994a; Barnet & Cavanagh, 1994, p. 417). U.S. corporations tend to view profitability more in terms of increased productivity (more output, fewer workers), rather than through improved product development, better work relations, job retraining, and increased sales.

Before 1984, some 80% of telephone company employees were unionized. By the late 1980s, unionization was below 60%. Despite the breakup's huge impact on the telecommunications unions, AT&T informed the labor leadership of its acquiescence in the divestiture only 3 hours before the story went to the press (Bolton, 1993, pp. 125-131). In Europe, the largest private-sector employer is the telecommunications conglomerate, Siemens (Germany), and the largest public-sector employer is Deutsches Bundespost, the German post office (Mulgan, 1991, p. 35).

In the Communications Act of 1934 (Section 201), the FCC was created "to make available, so far as possible, to all the people of the United States a rapid,

TABLE 5.3 The Top 25 Telecommunications Firms (ranked by market value)

Ranking–February 1996	Entity
1 (1)	AT&T
2 (2)	GTE
3 (3)	BellSouth
4 (4)	SBC Communications
5 (5)	Ameritech
6 (6)	Bell Atlantic
7 (8)	NYNEX
8 (9)	MCI Communications
9 (7)	US West Communications Group
10 (10)	Airtouch Communications
11 (12)	Sprint
12 (11)	Pacific Telesis Group
13 (NR)	WorldCom
14 (14)	Alltel
15 (NR)	U.S. Robotics
16 (23)	Frontier
17 (20)	Tellabs
18 (NR)	MFS Communications
19 (15)	DSC Communications
20 (NR)	Panasat
21 (16)	U.S. Cellular
22 (25)	Stratacom
23 (26)	Nextel Communications
24 (22)	Paging Network
25 (19)	Telephone & Data Systems

SOURCE: "The Top 1000 Ranked by Industry," *Business Week,* March 25, 1996, pp. 106-145; March 27, 1995, pp. 104-149.

NOTES: Values in brackets = 1995 ranking. NR = Not ranked.

efficient, nation-wide, and world-wide wire and radio-communication service with adequate facilities at reasonable charges" (cited in Bolter et al., 1990, p. 83) and were supposed to serve, in the words of the commission, the "public interest, convenience and necessity." The principle of providing "universal service," under which AT&T had held a regulated "natural monopoly" position in local and long-distance telephone services and equipment in exchange for affordable residential and business access (some 96% of the population by the time of the corporation's divestiture) was superseded by a deregulatory approach that was expected to induce greater diversity and competition in communication

Pat Buchanan, Fred Barnes,
John McLaughlin, David Gergen,
Robert Novak, William F. Buckley,
George Will

Sam Donaldson, Mark Shields,
Michael Kinsley, Morton Kondracke,
Al Hunt, Jack Germond,
Hodding Carter

TV's Spectrum of Political Thought

technology product markets (see Table 5.3 for the top 25 telecommunications firms).

The Reagan and Bush administrations pushed for deregulation in other telecommunications sectors as well. In 1987, the FCC, under Commissioner Mark Fowler, proceeded, with administration backing, to lift the "fairness doctrine." This meant that television stations no longer had to air diverse opinions on controversial topics. Also eliminated was the "equal-time rule" that had required the stations to give candidates for public office an opportunity to rebut their opponent's policy statements. Under Reagan, the FCC raised the number of television and AM and FM radio stations a single company could own from 7 to 12 of each; numbers increased in 1992 during the Bush administration to 30 AM and 30 FM radio stations. Depending on the number of existing stations, corporations could also raise the number of AM and FM frequencies in a single market from one to three of each in the new ruling. This ruling provided a grand opportunity for large broadcasting companies, including CBS, Capital Cities/ABC, Group W-Westinghouse, and Infinity Broadcasting, to increase their radio holdings in large urban markets like New York, Chicago, or Los Angeles.

With major broadcasting companies consolidating control of their urban markets, the options for station rivalry, and whatever the diversity of content that springs from it, are diminished. The ability of smaller and alternative, including minority-owned, stations to hold on to audiences and the advertising dollar is also threatened. Broadcasting stations already have time-brokering agreements by which one station leases its airtime, essentially becoming a "robot transmitter" for another station. And even "public" broadcasting has looked little different in recent years, both in terms of the lineup of its commercial backers and the generally conservative programming that dominates its airtime. With Republican efforts to remove all federal funding from the Corporation for Public Broadcasting, noncommercial stations are under greater pressure to take on more advertising support, and some of the smaller market stations could disappear altogether.

It is expected that remaining legal restrictions limiting services offered by the RBOCs will be lifted, allowing them to compete more vigorously with cable TV, long-distance telephone, information networks, and other telecommunications services, presently banned under the "modified final judgment" in the AT&T divestiture case. Although the Bell Atlantic–TCI merger failed in 1993, Time Warner, the largest media monopoly, second-largest cable operator, and one of the leading cable program producers (including its HBO), combined with another one of the "baby Bell" companies, U.S. West, the same year in a plan to build an interactive network capable of carrying movies, games, data, video catalogs, and telephone calls. A year later, Time Warner began negotiations with GE in an attempt to take over its NBC television network, which would give the media conglomerate control over one of the big three television networks, its cable operations (CNBC and America's Talking stations), its satellite-delivered programming operations in Europe and South America, and other media ventures, possibly with GE retaining control of NBC's owned and operated local stations. Earlier, TCI and Time Warner, together representing 40% of the cable market, had insisted that as a condition for carriage, GE's CNBC all-news cable channel would have to alter its format so as not to compete with CNN, which each of the two cable giants partially own (Reed, 1995, p. 53).

Many new cross-ownership deals are expected throughout the 1990s that would reconstruct and make the market for communication services far more complex, if not truly competitive. The huge advantage of the wealthy RBOCs, with their regional monopoly positions and nearly universally wired telephone networks, and the emerging power of the largest cable corporations constitute some major threats to setting up a public-oriented national information system. RBOCs providing television in areas formerly dominated by broadcast and cable companies may be able to deliver programs, like telephone numbers, to the home on request, rather than through preselected menus of over-the-air (OTA) and cable systems, but it is far from clear whether the new format will better enlighten people or save them money. Even if people did prefer the more à la carte approach, would the telephone companies provide access channels for public-focused program producers even in the limited way that cable operators do?

In 1993, the Clinton administration, especially Vice President Albert Gore, began promoting the idea of an "information superhighway" that would extend the country's communication infrastructure and supposedly protect public access. Most of the new networks in electronic communications have been put together by the private sector, especially in cable, computer, and telephone, the key interest and advisory group on the "information highway" being the Information Industry Association (Apple, AT&T, Compaq, Control Data Systems, Cray Research, Data General, Digital Equipment, Hewlett-Packard, IBM, Silicon Graphics, Sun Microsystems, Tandem, and Unisys being the major players). And although the FCC is constrained to operate in the "public interest" and "for the maximum benefit of all the people of the United States" (*National Broadcasting Co. v. United States,* 1943), the practical definition of those terms have, by default, been appropriated by the corporate communication sector (Schreibman, 1993).

Democrats and Republicans in the House and Senate took early initiatives to break down restrictions on cross-ownership of television and telephone services as a way of encouraging competition through private investment. How the public interest will be protected in this private sector free-for-all is far from clear and is particularly threatening to e-mail, data, and bulletin board networks on Usenet, World Wide Web, or Internet, whose users have enjoyed easy and inexpensive access until now (see Table 5.4). Despite all the benevolent claims made about

TABLE 5.4 Some Critical Political User Groups on the Internet

World Newsgroups	War and War Toys
alt.current-events.bosnia	alt.war
soc.culture.bosna-herzgvna	alt.war.vietnam
soc.culture.croatia	alt.desert-storm
soc.culture.mexican	rec.aviation.military
soc.culture.korean	sci.military
talk.politics.china	alt.engr.explosives
soc.culture.japan	alt.politics.org.un
talk.politics.mideast	

Intelligence/Spying	Political Activism
alt.politics.org.covert	alt.activism
alt.politics.org.cia	alt.activism.d
alt.politics.org.nsa	misc.activism.progressive
sci.crypt	soc.rights.human
alt.politics.org.fbi	alt.government.abuse
alt.politics.org.batf	alt.whistleblowing
alt.law-enforcement	alt.discrimination
	alt.politics.equality

Journalism	
alt.journalism	alt.society.civil-disob
alt.journalism.criticism	alt.society.civil-liberties
alt.news-media	alt.society.anarchy
misc.writing	alt.society.revolution
alt.freedom.of.information.act	alt.society.civil-liberty
alt.society.foia	

Peacenet (Country Studies)

Africa	*Asia*	*Central America*	*Europe*
africa.forum	hr.asiapacific	carnet.general	baltic.news
africa.news	reg.china	nicadrl	env.europe
bitl.africa	reg.seasia	reg.elsalvador	list.nato
	tibet.information	reg.mex.news	yugo.antiwar

SOURCE: Minnick (1995, pp. 61-62).

deregulation and privatization of the "electronic frontier," many expect that the information superhighway will be operated with very expensive toll booths. The real public opportunities to use the information superhighway, in any event, are far from certain.

One strong advocate of public interest access to information networks is an organization called Computer Professionals for Social Responsibility (CPSR), which has been pushing "universal access" approaches to information technology. The communications industries generally do not share CPSR's sense of social responsibility and tend to prefer strictly fee-based entry, especially if congestion on the Internet becomes a problem. This will induce a kind of electronic triage between high-roller types, such as large businesses with video conference requirements, and ordinary users with less money to put on the table but with no less need to communicate long-distance. Already, those with more expensive and faster modems have an information transmission cost advantage over those stuck with machines of slower bits-per-second connection capabilities.

Competition for control of the superhighway also pits the RBOCs against the cable TV companies and value-added computer service providers (e.g., Prodigy, America Online, The Well, Compuserve). Several major computer service providers are already under the control of huge corporations (e.g., Dialog by Lockheed, The Source by Reader's Digest). For most people, information services like Nexis–Lexis, Dialog, or Dow Jones are financially way out of range. And at the present rate of technological convergence and corporate mergers, it appears almost inevitable that ability-to-pay principles will prevail over the universal access concept, if only to help communication corporations recover from the enormous debt burdens that takeovers incur.

In September 1994, a sign of the times appeared with the news that the largest television shopping company, Home Shopping Network, was going to buy Internet Shopping Network, possibly sealing the fate of the Web to commercial designs. The U.S. government, which initiated the Internet and originally prohibited it from being used for sales or advertising, gave way to market interests and began to transfer its own involvement to the private sector, alarming public-sector users, which included the fiber optic backbone managed by private corporations (e.g., IBM). Now privatized, it is expected that the prevalent on-line users will be commercial marketers, communication and business-oriented database companies, financial and stockbrokerage firms, and other for-profit operations. By late 1994, more than 20,000 companies were on-line, and others, such as CommerceNet, a consortium of 25 major electronics companies, financial houses, and banks, were waiting to start up Internet business transactions, marking another triumph for deregulation and privatization ("Internet Ventures," 1994).

Communications industries, as a whole, appear to favor deregulation of the information infrastructure, which, as recent experience in the airline industry would suggest, will probably end in additional mergers, acquisitions, reduced competition, rising prices, and more expensive access. The result would be a polarity of rich information enclaves and poor information ghettos. Commercialization of information access via the Internet, for example, means that many groups will be priced out of educational opportunities. Among America's poor in 1987, 25% did not even have telephone service (Mosco, 1993, p. 44). For African American and Latino children, who have only 40% of the rate of computer ownership as European American children and who are far more likely to use school computers for little more than typing and rote exercises, this will compound the inequities in childhood development.

A very different sort of access problem exists, as anyone who has used the Internet system knows, in that an enormous amount of clutter is circulated that passes as information. This is almost impossible to avoid if electronic bulletin boards are to be relatively open to users. The number of interest groups proliferate on the system at a rate that creates an extremely fragmented set of information channels. One history professor at the University of California, Berkeley, Carla Hesse, compares this scenario to the worst excesses of the French Revolution, which can only confuse the frequent user who is trying to make sense of the world. Another problem cited by Hesse is the lack of accountability of much of the information that is listed, often anonymously, a kind of *samizdat* of messages and digital images that "can be seamlessly altered and rearranged" (cited in Markoff, 1994a, pp. E1, E5).

POLITICS AS PUBLIC RELATIONS: LEAVE IT TO DEAVER (AND ASSOCIATES)

Although the mass media continue to report surveys suggesting that Americans are disgusted with politics as usual, they tend not to report that such fed-up feelings can be credited to the media themselves. As one former national Republican Party chair noted, "You sell your candidates and your programs the way a business sells it products" (cited in McGinniss, 1993, p. 162). Ever since Eisenhower hired Young & Rubicam and Nixon brought in J. Walter Thompson to polish up their images, television political advertising has become the principal conduit for presenting not challenging ideas, but crafted images. Over the years, political advertising has been marked by more style over substance, customized messages over consistent arguments, irrational over rational appeals, studied symbolic manipulation of audiences (via polling and voter research), playing to the media (news-focused sound bites), and other control techniques. Perhaps the French were not far off the mark when, in 1694, they were first in defining advertising (*publicité*) as "a crime committed with many witnesses" (Baudot, 1989, p. 6).

An information technology intrusively well known to modern-day U.S. politics is polling. Day by day, even hour by hour, changes in the public mood can be simulated in computer programs originally borrowed from 1950s Pentagon "wargaming" technology. By the 1960s, business was running such programs with demographic data to test variables for marketing detergent and soft drinks. One developer of market simulation models was Richard Wirthlin, who went on to start up his own company, Decision Making Information, in 1969. A year later, he joined Ronald Reagan's California gubernatorial campaign, for whom, along with other Republicans (including George Bush), he would work for the next 20 years as pollster and chief marketing analyst.

Wirthlin's "Political Information System" database and working team targeted "up-to-the-minute attitudinal survey work, fixed demographic information, historical voting patterns for every county in the country, on-going assessment of political party strength in each state, and subjective analysis" (Nelson, 1989, pp. 74-75). Under this system of political surveillance, "daily and weekly gathering of polling data became the basis for every decision and public statement or action" (Perry, 1984, p. 169). Political candidates are packaged and sold like boxes of breakfast cereal, and this serves as a staple of the country's electoral process. The cost of organizing elections on this basis also helps weed out less well-heeled and independent candidates, reducing the number who might otherwise attempt to seek office without big corporate financial backing. Electoral processes of this sort discourage potential leaders with messages and values not easily accommodated within the style and lexicon of the mass media.

Highly attentive to his media advisors and inner circle of public relations experts, President Reagan was anointed by a deferential media as the "Great Communicator." He appeared polished and folksy in scripted situations and adept at delivering one-liners embroidered into his speeches. On the spot, however, as captured in press conference transcripts, Reagan's responses typically were more incoherent and inaccurate, prompting a two-volume book series by Mark Green and Gail MacColl, entitled *There He Goes Again: Ronald Reagan's Reign of Error.* Closest to Reagan was another media star, his wife, Nancy, who spent much of her first ladyship bailing him out of unpleasant gaffes and awkward political moments. With the help of Wirthlin's polling skills, Nancy Reagan herself was turned on to the issue of teenage drug use, which fashioned a public image for her, particularly after she launched the well-publicized sound bite "Just say no."

Reagan's media adviser and deputy chief of staff, Michael Deaver (1981-85), was a master at stage managing the presidential image and particularly the "photo opportunity." As advance man for Reagan's travel agenda, Deaver prepared the backgrounds, the angles, the symbols to cast his boss in heroic poses, always with an eye to the television and press cameras and the nightly news. As the speechwriters carefully crafted the great communicator's spoken words to register the sound bite of the day, Deaver worked assiduously as part of the White House media management team to produce the images. Deaver's job was to get the pictures right and to keep the president as far from reporters as possible (Regan, 1988, p. 248). Reagan almost always worked from choreographed moves and prepared scripts, but on the occasions when he felt compelled to ad lib, the president was a cannon of incoherent phrases and misstatements of fact. After Reagan secured a second term, Deaver returned to private industry as a consultant for, among others, Philip Morris, cashing in on his former role as a member of the White House inner circle.

One of the meanest and sleaziest presidential media campaigns was the 1988 Bush-Dukakis contest. Despite an early wide lead in the polls, Dukakis was outmaneuvered in the deployment of media images and ability to get across a resonant political message. Bush's public relations consultants, largely shunning substantive discussion of issues and mentored by 8 years of Reagan's light touch, turned on the Massachusetts governor's technocratic personal style. The vice president's advertising team, led by Lee Atwater and Roger Ailes, held nothing back in the art of political chicanery. They created television images of their candidate in front of the polluted Boston Harbor (posing him as an environmentalist after the Reagan administration slashed funding to clean up the environment), using flag saluting as a "patriotic" issue (calling Dukakis a "card-carrying member of the ACLU"), formulating a "wimp" image of Dukakis (in a mocking ad about his commander-in-chief credentials), and playing on the public's anxieties about crime and race (insinuating Dukakis's flaccid liberalism) with the notorious ad featuring "Willie" Horton (an African American who committed a rape while on furlough from a Massachusetts prison).

Richard Nixon, who held the Capitol press corps at arms length during his political career, learned to make use of the media and political advertising to destroy his Democratic Party opponents in 1968 and 1972, as Lyndon Johnson had done in 1964 (most memorably with the infamous "Daisy spot" television ad featuring a countdown voiceover as an atomic blast cuts into the image of an innocent young girl pulling petals off a daisy, implying that his Republican opponent, Barry Goldwater, had a trigger-happy nuclear finger; Ginsberg, 1986, p. 161n). One of Nixon's 1968 ads, playing on the fear of violent crime and emphasizing his "law and order" campaign theme, used film footage of a middle-aged woman walking alone at night along a city street, the only audio portion being the sound of footsteps followed by a voiceover narration of violent crime statistics in the United States, as the video fades to "NIXON." Hubert Humphrey's backers, facing defeat in the 1968 election, countered with an attack on Nixon's choice for vice president, Spiro Agnew. The ad showed a television screen that read "Agnew for Vice President?" with a sound background of raucous laughter, followed by a slide reading "This would be funny if it weren't so serious" (Diamond & Bates, 1992, pp. 159-161, 169).

In 1992, Bill Clinton overcame a negative media image pursued by the Bush reelection group that started with such "character issues" as his alleged pot smoking 20 years earlier and marital infidelity and moved on to an attempt to stick the former Arkansas governor with a "slick Willie" persona. Clinton stuck to the economic issues, aided by a long recession, and had himself booked heavily on the national talk show circuit (47 appearances to Bush's 16; Bennett, 1996, p. 71). Ross Perot captured 19% of voter turnout that year on the strength

of his personal fortune and a Paul Harvey–style conservative, Midwestern populism, remarkable, given that he had no authentic party machinery behind him. In the end, Perot's renegade approach undermined the incumbent, added interest to the electoral process, and helped Clinton get elected.

One mastermind of campaign strategies during the Reagan and Bush years was Sig Rogich. A Nevada–based advertising executive, Rogich's previous clients included Frank Sinatra (whom he helped obtain a gambling license) and Donald Trump. He also served as a boxing commissioner, a director at Bally's Casino, and a media advisor to help clean up the image of the Stardust Hotel, then linked to organized crime. Rogich was brought to the 1988 Bush campaign to design attack ads against Dukakis.

Bush dominated media iconography that year by hammering away at "liberal" strawmen and using the "l" word (*liberal*) as synonymous with someone hopelessly out of touch, if not subversive—in hopes of planting on his opponent the political kiss of death (Bennett, 1996, chap. 2). By the 1980s, almost every major contested national race involved the use of media consultants and partisan pollsters. Even city hall and state legislature candidates were hiring the services of polling and broadcast advertising experts (Hagstrom, 1992, pp. 4-5).

Television has been central in the transformation of U.S. elections. Into the 1950s, the two major political parties were still choosing their presidential candidates in "smoke filled rooms," and party conventions continued to hold a lot of meaning. The shift to state primaries gave the public somewhat more and the party apparatus somewhat less of a voice (party activists nonetheless still dominate the choice of nominees) but also required presidential candidates to barnstorm the country in search of convincing victories, heavy financing, delegate votes, and the attention of the mass media. By 1972, Nixon had spent $61 million and McGovern $30 million in the presidential campaign. In 1984, Jesse Helms spent more than $13 million and Jay Rockefeller almost $8 million in their respective Senate races in North Carolina and West Virginia. From his vast cash reservoir, Perot spent $35 million on television ads alone in a losing 1992 presidential effort. In 1994, California Republican tycoon Michael Huffington poured out $28 million from his personal vault in a failed bid to "buy" a U.S. Senate seat (Bennett, 1996, p. 60).

More than anyone, the big media institutions cash in on the spoils of paid political advertising and the almost endless spectacles of the long campaign season that unites politicians and news organizations in symbiotic relationships. The Bush and Dukakis campaigns in the 1988 presidential election spent at least $60 million in television commercials. By 1986, the country's major elections already ran up costs of $1.8 billion, of which two thirds was spent on television advertising and in which private political action committees (PACs, of which the biggest was the National Conservative PAC), outspent the party organizations by three to one. Like most media commercials, the blurbs for candidates for elected office pander to rapid-fire and often subliminal images, stereotypes, caricatures, and 10- to 15-second sound bites, with very little time or information on which to make rational conclusions (Dye, Zeigler, & Lichter, 1992, p. 195; Nelson, 1989, p. 77; Volti, 1992, p. 181). The short "spot" political ads that mark U.S. elections can say very little of substance and, therefore, often "go for the jugular" with phrases that stick in the public consciousness.

By the 1990s, the major political parties and movements were grafting new information techniques onto their political and ideological agendas. Computer profiling of past candidates' speeches enabled political opponents to trace any apparent contradictions in their rhetoric as an instrument for publicly discrediting them. Tycoon politician Ross Perot went to network political "infomercial" formats in 1992 and 1996, while the successful right-wing radio talk show personality Rush Limbaugh used his syndicated television program for naked self-promotion and ideological assaults on President Clinton, liberals, and the Democratic Party. Two failed candidates in 1994 tried to revive their political careers by entering the talk show business, Oliver North (who lost a Senate race in Virginia) appearing on more than 120 television stations by mid-1995, and former Governor Mario Cuomo (New York) trying his luck on radio.

Even the former (often ridiculed) vice president and media critic Dan Quayle could not resist the image-building possibilities of broadcast advertising (as a way of informally initiating a 1996 run for the presidency), presenting himself in a televised advertisement for Frito–Lay potato chips during the 1994 football Super Bowl (parodying his celebrated inability to spell *potato* during an embarrassing 1992 media campaign event). Syndicated radio talk show personality Don Imus quipped that a "potato chip commercial is a step up for him" (Dowd, 1994, p. 6). Quayle eventually withdrew from the 1996 race.

In 1993, Paul Weylich, a conservative spokesperson, started up the 24-hour satellite- and cable-delivered National Empowerment Television program, following on the success of archconservative Representative Newt Gingrich's (R– Georgia) public access show on American Citizens' Television. (Gingrich has been one of NET's stars.) The same year, Lamar Alexander, a presidential candidate in 1996, began a monthly Republican Exchange Satellite Network broadcast. The following year, the Republicans began a weekly, hour-long television news magazine format program, *G.O.P.-TV,* hosted by the party chair, Haley Barbour, to showcase conservative leaders and issues, recruit new members, and solicit financial contributions (Berke, 1994, p. 20). President Bush's former media advisor, Roger Ailes, served concurrently as executive producer of Rush Limbaugh's nightly television program and president of GE-NBC's CNBC Cable Network until he moved on in 1996 to run Rupert Murdoch's Fox Broadcasting 24-hour cable news channel. TV *über alles?*

Radio talk show hosts, most of whom are conservative, have some of the biggest megaphones in American political discourse. Some have parlayed their name recognition for political opportunity by entering electoral contests (Howard Stern in the New York governor race in 1994; Mike Siegel as a candidate for the 1996 governor's race in Washington). Senator Jesse Helms (R–North Carolina) and Congressional Representative Robert Dornan (R– California) launched their political careers as conservative talk show personalities. Some 800 talk show hosts on American radio have a "bully pulpit" for steering the political agenda, particularly when they choose to start a bandwagon effect in influencing the public mood. With the pressures of hosting a daily talk show, it is much easier to be trendy than original or independent.

Mass media and the structural biases within U.S. politics limit its democratic reach. Can this change? Will electoral rules dramatically be restructured, such as by eliminating corporate campaign contributions or giving free airtime to minority parties (as is done in many European countries)? Can the "market" make chip-based technologies available to ordinary people and provide them with the means of producing news media and creating their own audiences? Can camcorders, faxes, computers, modems, and VCRs help people see the ideological manipulation used against them by the big media? Perhaps, but without critical understanding of the political economic basis of power and how the corporate aristocracy undermines democratic initiatives and personal sovereignty, small media enthusiasts are likely to simply create ersatz versions of the big media or, out of frustrated ambition, sell their talent to the highest bidder. Why does it take a Hollywood celebrity like Meryl Streep, the author William Greider asks, to make the case for eliminating the industrial carcinogen Alar from apples, when the so-called Environmental Protection Agency had known about its effects for many years? Greider seems easily convinced that even though media ownership is becoming more concentrated in fewer hands, wider access to media technology will create "TV guerrillas" who will air provocative works that established media dare not touch (Greider, 1992, pp. 307-330).

THE FIRST CASUALTIES OF WAR

In a liberal-democratic society like the United States, says Massachusetts Institute of Technology (MIT) professor Noam Chomsky, the combined interests of the state and organized economic concerns prefer (and require) the instrument of propaganda, rather than coercion, to bring the mass of people into line with establishment objectives (Chomsky, 1991, p. 5). The mass media, especially network television, serve as the government's and corporate business's main pipeline in creating public consensus. They do this by giving the widest possible coverage to a narrow list of approved and reliable "experts" who know all too well who butters their bread. It is not that establishment elites necessarily agree on the substance of most issues, but they do agree on the parameters of the debate and will close ranks in marginalizing radical analysis from the likes of Chomsky. Information detrimental to the image of state or capitalist power or sympathetic to socialist critiques is conscientiously elided.

In the first "televised war," the American mass media initially fully supported the government's invasion of Vietnam, and, indeed, two of its personalities, network anchors Chet Huntley and Walter Cronkite, had made propaganda films for the Pentagon on the threat of world communism (Kellner, 1990, p. 51). If the media contributed to undermining the war effort after the National Liberation Front's 1968 Tet Offensive by showing graphic images of U.S. soldiers in hopeless combat, it was not because they chose to educate the public about the country's imperialist objectives in Vietnam. It was more that the media could no longer ignore the growing resistance to the war at home and U.S. losses and demoralization in Vietnam without losing their own credibility. Their occasional leaking of certain hidden truths about the war (e.g., that Marines burned down peasant homes and murdered civilians) was not a critique of the war's immorality, but simply a statement about the government's "blundering efforts to do good" (Herman & Chomsky, 1988, p. 173). At all times, they ignored the most educated and articulate views of antiwar critics, the sufferings of the Vietnamese and other peoples in Indochina, and the history of and reasons for Vietnam's struggle for independence.

Unfortunately, government leaders learned few lessons about the U.S. invasion of Indochina that would change the ideological temper of its Cold War foreign policy. The postwar review became a way of finding scapegoats for the lost conflict, the "liberal" media being the Republicans' first suspect and the military never crediting the Vietnamese themselves for overcoming U.S. military-industrial superiority. And yet the Vietnamese did just that, albeit at a severe cost. U.S. bombing, measured at more than twice the tonnage dropped on all theaters of World War II combined, could not bend the will of the Vietnamese struggle for independence. When the United States developed and dropped "people sniffer" devices (the "XM3 Personnel Detector," an instrument that originated in food processing research), to try to detect enemy troop body ammonia, the Vietnamese placed buckets of urine in trees to saturate the sensing elements and direct the bombing that followed to unpopulated locations. Advanced technology could not overcome a popular will to be independent.

Perhaps the greatest "pushbutton fantasy" (Mosco, 1982) of the "communications revolution" was the media spectacle of the Persian Gulf War (1990-91). U.S. television networks and CNN collaborated in presenting images of a picture-perfect electronic battlefield to households all over the world. Pentagon–organized news pool reporters, sometimes draped in camouflage or Israeli–supplied protective masks to create a sense of being at the front, transmitted military-censored commentary and video mainly from the confines of a U.S. military installation in Saudi Arabia. The "patriotic" Gulf War coverage was the media's atonement for supposedly having bred the "Vietnam syndrome," what the conservative writer Norman Podhoretz described as "the sickly inhibitions against the use of military force" (cited in Herman & Chomsky, 1988, p. 236).

The mainstream media, especially television, covered the techno-war in the Gulf ("Operation Desert Storm") as if to put closure on the Vietnam syndrome. A military-media spectacle of "smart" weapons and other high-tech fireworks was made to look like a sterile, victimless operation. It was exceptional for journalists, Bill Moyers being one of them, to show pictures of the human devastation, and that was well after the war was over. The network anchors behaved, not as the dispassionate news reporters they claim to be, but more like cheerleaders for the local football team, conveniently ignoring the U.S. government's complicity in building up the Saddam Hussein regime in the first place. Pro-military "experts," including former chair of the joint chiefs of staff, Admiral William Crowe, were hired as paid consultants to the four networks, while critics of the war were discreetly avoided.

CBS commentator Charles Osgood referred to the bombing of Baghdad as "a marvel" and "picture perfect assaults"; ABC anchor Peter Jennings saw "brilliance" in the U.S. laser-guided bombs and the Scud as a "horrifying weapon"; NBC anchor Tom Brokaw referred to the possibility of prewar negotiations as "a nightmare . . . the worst possible scenario," and he asked on the first day of the bombing: "Can the United States allow Saddam Hussein to live?" Dan Rather, of CBS, initially the most cautious of the network anchors in repeating government disinformation, eventually joined in the pro-war chorus out of concern for the ratings competition, and he effused sentimentally about the effects of the war on the families of U.S. (but not Iraqi) soldiers (Kellner, 1992; Naureckas et al., 1991).

David Brinkley, on his weekly ABC–TV news discussion program, interviewed Defense Secretary Richard Cheney on what the news host called "Saddam Hussein's phony peace offer"; ABC's Cokie Roberts asked U.S. commander General Schwarzkopf, "Wouldn't it be smarter to institute a draft and get 18-year-olds?" CNN correspondent Peter Arnett was the lone U.S. television reporter in Baghdad throughout the bombing and gave a few rare glimpses of the destruction for which he was denounced as a quisling by many in media and political circles. The right-wing U.S. Senator Alan Simpson (R–Wyoming) and the White House press office attacked his report that the U.S. had bombed a milk factory and both branded him a conduit for Iraqi propaganda. Opponents of the war could not articulate their dissent, except in street demonstrations that the media virtually ignored. A few oppositional print and radio commentators lost their jobs as a result for expressing their views. Michael Deaver, Reagan's master image maker, said about the war coverage, "If you were going to hire a public relations firm to do the media relations for an international event, it couldn't be done any better than [how] this is being done" (Naureckas et al., 1991, p. 14; Taylor, 1992, p. 45).

A messianic message denouncing "appeasement" and of saving the world from the "new Hitler" (Hussein) was repeatedly voiced by President Bush as part of a public opinion strategy to demonize the enemy, a page borrowed from the rhetorical performance style of Reagan. It was reiterated again and again by the mass media, stirring up an a Manichaean good-versus-evil crusade, and Hussein was the perfect foil for media melodrama, American–style. Bush's earlier military defense of Hussein (and subsequent refusal to negotiate with him), the pressure he put on Congress not to sanction the regime's abuses against the Kurds, and his family's significant oil interest in the region all were missed by the media in the rush to war. The mainstream media also never explained the colonial origins of the Iraqi and Kuwaiti states, the historical frictions between them, and other relevant facts that would complicate simplistic comparisons with Hitler. The demonology was reminiscent of William Randolph Hearst's notorious order to a newspaper illustrator as the media mogul was trying to use "yellow journalism" to stir up public sentiment for intervention against Spain in Cuba: "You furnish the pictures. I'll furnish the war."

Much of the media's unrestrained enthusiasm for the invasion was captured in a kind of schoolboy excitement over the "Patriot" versus "Scud" symbolic conflict. Although neither of the missiles was effective in its military objectives, they both had some success as propaganda weapons. Iraqi forces, desperately trying to instigate a spontaneous Arab coalition against both Saudi Arabia and Israel, fired the Scuds wildly at both countries and, except for one hit inside Tel Aviv (possibly deflected by a Patriot) and another on a U.S. barracks in Dhahran, Saudi Arabia, killing 28 soldiers, it was an ineffective counterforce. Iraq began its pullout from Kuwait that same day. The U.S. military's exaggerated claims about the Patriots got full play in the American media even though no one in the news pool had actually witnessed the "facts" they reported.

Closed to critics of the war, the pool failed to challenge U.S. military claims about the rationale and conduct of the war, making the news establishment look more like a government lapdog than a watchdog. As a result, one of the many deceptions fed to an unwitting public was that the Patriot antimissile missiles had perfect intercept ratios against the Russian–made Scuds. In fact, the Pentagon would later have to confess (long after the war and its propaganda faded from consciousness) that the Patriots had no such success as earlier claimed. One MIT military weapons specialist reported that virtually every Patriot failed in its target mission and that the partial hits caused considerably more damage than the poorly targeted Iraqi weapons would have inflicted if left unimpeded (MacArthur, 1993, p. 162). The military-media collaboration was so complete that Assistant Secretary of Defense for Public Affairs Pete Williams, who managed the military's war debriefings for the media, moved on in his career to become a general assignment reporter for NBC News (owned by the big defense contractor, GE).

The media also carried false military reports that every fixed and mobile launch site for the Scuds was destroyed on the ground, when in truth only a third of the fixed and none of the mobile stations had been knocked out (Miller, 1992, p. A21). When incontrovertible evidence appeared that the U.S. military had bombed an infant milk formula factory, the American media defended the lie that it had been a biological weapons plant. When the U.S. bombed a civilian air raid shelter, the media refused to criticize the military's prevarications about it being a strategic military communications center. And when the 15-year-old daughter of the Kuwaiti ambassador to the United States (also a member of the ruling royal family), without identifying herself, told Congress and the American press in October 1990 that she had witnessed occupying Iraqi soldiers break into a hospital to steal incubators and dispose of the infants, a completely fabricated tale, the media printed the story as fact and without investigation. None of the reporters chose to question or make known the identity of the extremely partisan source.

Part of the public deception, described by Douglas Kellner (1992) and others, could be credited to the Republican Party public relations firm Hill & Knowlton, which was hired by the Kuwaiti royal family for $10 million to help mold American public opinion and sympathy for the emirate. The PR firm had pretested audiences to find out what agitated Americans the most and used the findings to package desirable propaganda for their client. Craig Fuller, George Bush's friend and former chief of staff when Bush was vice president, was president of Hill & Knowlton at the time of the Gulf War and took part in the deception, along with Ronald Reagan's former inaugural committee cochair, Robert Gray. A key element in the political swindle was the passing on of video news releases produced by Hill & Knowlton to broadcast news stations, a rather common practice in the media industry. Fuller next went on to become the top public relations executive for Philip Morris, "where he is promoting the fortunes of a real killer—the tobacco industry" (Bleifuss, 1994, pp. 73-74).

Hill & Knowlton tutored and steered the Kuwaiti ambassador's daughter through congressional hearings on human rights, never publicly revealing who she was. Nonetheless, Bush, Vice President Quayle, military commander General Schwarzkopf, and others in government repeated the incubator story to drum up support for invasion. CNN International, other news media, and Amnesty International were also fed and subsequently, without confirmation, carried the tragic tale and other fictitious allegations about Iraqi atrocities. Hill & Knowlton also doctored videos made in Kuwait and coached witnesses to revise their accounts about the Iraqi soldiers in an effort to convince the United Nations to join forces with the United States. The firm perhaps drew inspiration from the PR genius Edward Bernays, who asserted in his book *Propaganda* that "the conscious and intelligent manipulation of the organized habits and opinions of the masses is an important element in a democratic society" (Bleifuss, 1994, p. 77; Kellner, 1992, pp. 67-71).

It has been said that the first casualty of war is truth. Hardly any media in the world could have been more compliant in suspending their

158 ISSUES IN THE "INFORMATION SOCIETY"

professional ethics and spreading disinformation than the American news pool during the Gulf War. CBS and NBC refused film footage of civilian casualties in Iraq from two Emmy Award winners, Jon Alpert and Maryann DeLeo, who had long-term independent news supply contracts with the networks (Robinson, 1995, p. 107). Arguably, wars may be special periods when different dicta apply than during times of peace. But when one nation is at war as often as the United States, when the enemies are often weak and vulnerable nations, and when the outcomes are so one-sided (the U.S. forces repeatedly referred to the Gulf War as a "turkey shoot") that the victor inflicts horrific suffering and destruction on the vanquished, especially on civilian populations, acquiescence by the media can be read as collaboration in imperialism.

In 1991, Baghdad and the southern Iraqi city of Basra became the most bombed regions in as short a period as any other in history. As many as 150,000 Iraqis (perhaps more), mainly civilians, died directly from the assault or as a later result of the war because of damage to food, water, and health supplies. Under a less one-sided global political order, where the United States did not dominate international legal organizations as it does, American network anchors might very well have come under their jurisdiction for fomenting "propaganda for war," a crime under international law (e.g., the International Covenant on Civil and Political Rights, Article 20).

A critical media source, Fairness and Accuracy in Reporting (FAIR), noted that, during the Gulf War, only "about 1.5% of network sources and 1% of airtime were protesters [mostly at rallies], about the same number as sources asked about how the war had affected their travel plans," even though on the eve of the invasion about half the country preferred negotiations or sanctions to a military solution. Of some 878 on-the-air sources, only one, a spokesperson for Physicians Against Nuclear War, represented a peace organization, whereas seven commentators were Super Bowl football players. FAIR noted that the television networks were locked into support for the war, in part, because of their close links with the military-industrial complex.

The chair of Capital Cities/ABC sat on the board of Texaco Oil; CBS had board members from the Honeywell Corporation (a big defense contractor) and the Rand Corporation, which undertakes sensitive military research for the government; NBC's parent company, GE, produced $2 billion in weapons contracts of almost every kind, including the Patriot and Tomahawk missiles, for the Gulf War and had the royal family of Kuwait as one of its major stockholders. GE anticipated that contracts for the rebuilding of Kuwait were worth "hundreds of millions of dollars" (Naureckas, 1991). Given the billions that GE makes from the human and physical devastation of war, it is somewhat ironic that its advertising slogan is "We bring good things to life."

LEAVING I.T. TO EXPERTS

Illusions about the liberatory character of communications extends to the whole range of information technologies (IT), comprising a seamless web of subsystems linking microelectronics to computers, television, telecommunications, machine equipment, accounting instruments, e-mail, electronic publishing, word processing, building and graphic design, surveillance, record keeping, and many other applications. Collectively, they keep tabs on an enormous complex of data about societies and their citizens. For some who imagine great social potential in information and communication resources, there are reveries of digital technology returning to the hands and minds of the individual the personal creativity and artisanship lost in the industrial revolution and factory system of mass production. But what many miss is that technologies are, in essence, no more than the embodiment of social power and design and that the possibilities of individual fulfillment are constrained by the objectives of corporate planners, who think more and more in transnational terms and are less and less accountable on a personal basis.

In reality, TNCs have been growing while real wages (wages that factor in cost-of-living increases) have been shrinking. "Sooner or later, and quite probably sooner, the increasingly mechanized society must face another problem: the problem of income distribution" (Wassily Leontieff, cited in Noble, 1986, p. 26). Unless people have the means to purchase beyond their basic subsistence, it is pointless to talk about the liberating potential of technology. If the principal motivation behind technological change is profit, then as portrayed in Fritz Lang's 1927 grim film classic *Metropolis,* workers become disposable commodities in the search for reduced production costs and greater managerial control of the workplace through automation.

The image of technology as autonomous, that the logic of its development and use is beyond political and social value conflicts, is a core assumption of technological "expertise." The more that political and social intervention and accountability are removed (as in "deregulation"), the more the scientific and technical experts can function without interference. In its pure version, this outlook assumes that the appropriate direction, design, and uses of technology are self-evident, that technology itself is neutral, a mere tool, with respect to values and outcomes, and that the form of ownership of technology is irrelevant. Competing power interests enter the picture only *after* the development of technology.

Critical analysis views it differently. From the critical perspective, autonomous technology arguments are misinformed, exaggerated, or outright deceptive; human agency must never be overlooked regardless of how functionally sophisticated technical instruments may be. As one socially conscious professor of electrical and computer engineering argues, the "ideology of technology" represents "purposes quite remote from explaining reality to members of society. It fails to take political power and economic interests into account and thus masks their predominant role," at the same time that it "conceals the existence of specific, powerful corporations whose activities in pursuit of their interests are major factors in the problems of our contemporary society" (Balabanian, 1980, p. 10). Leaving technological inquiry and decisions to apolitical scientific or technical experts means that their own particular biases and self-interest or those of their sponsors will determine what sorts of research questions are asked, the methods employed, and the outcomes that result. In one set of hands, technology yields nuclear weapons, assembly lines, the South African pass system, skyscrapers, the genetically altered tomato, pollution, and Chernobyl; in another set of hands, we get mass transit, environmental protection, lower infant mortality, and vaccine for polio.

Most computer systems engineers, for example, would probably prefer to protect their expert status by designing applications software that is sufficiently arcane to require a permanent staff of on-site specialists to debug or adapt it to institutional specifications. In a similar way, the language of law and of bureaucracy is made obscure and inaccessible so as to preserve the indispensability of the respective professional groups engaged in legal and bureaucratic discourse. As Michael Shapiro notes, "Murray Edelman's insightful analysis of 'the political language of the helping professions' shows how the linguistic usages of a profession disable its clientele" (Shapiro, 1981, p. 162). Technical language, therefore, has not only explanatory but also political use value.

To use another example of how technique embodies a political economy, it is now quite clear that U.S. automobile manufacturers understood the superior safety features of airbags and the lives that would be saved and were capable of employing them in auto designs in the 1970s but did not do so out of economic considerations. It was business wisdom in the 1940s and 1950s that "what's good for General Motors is good for America." Yet, when General Motors took over the Pacific Electric Railway in the 1920s, a 1,200-mile network in Los Angeles of electric streetcars and interurban rail, it was not for the purpose of alleviating congestion or harmonizing human life with nature, but rather to kill fuel-efficient mass transit in favor of GM buses (powered by Standard Oil). This it accomplished, turning Los Angeles into a major market for its mass produced automobiles (Balabanian, 1980, p. 13). The consequences of a freeway culture for smog-inhaling Angelenos, one of the poorest designed cities in the United States, have gotten only worse over time.

It took a public intellectual (a person who offers her or his knowledge for the use of the public sphere) like Ralph Nader to point out that automobiles were *designed* to be unsafe and to show why government "consumer protection" was necessary. This led to the government instituting the first automobile safety standards (see Nader, 1965). When the then young lawyer was accused by the auto industry of opposing the consumer's "freedom of choice," Nader's response was that the auto industry was merely defending one's "inalienable right to go through the windshield" (Isaac, 1992, p. 127).

In the communications field, after television network news media complained that the "fairness doctrine," which required broadcasters to air balanced presentations of important and controversial issues, had forced them to cut back on public affairs programming and that deregulation would encourage more programming of this type, the FCC agreed and in 1987 repealed the requirement. Relying on "free market" principles and the industry's wisdom, public affairs programming (excluding the shock entertainment-oriented news magazine formats), in fact, has all but disappeared from commercial networks, allowing a "sloughing off of these less profitable areas" (Herman, 1995a, p. 93). Nader has fought the FCC decision ever since.

Even on public television, the de facto heir of public affairs programming, policy experts usually are represented as politically neutral in the sense that disagreements arise largely over various empirical observations about relatively innocuous issues, rather than over explicit ideological conflict. In the private sector, expertise generally is evoked on the basis of assumptions that science and technology are detached from partisan interests (including their own career advancement), which leads to attempts to depoliticize public issues (Nelkin, 1991, p. 276). Experts rationalize their politics (interest alignments) by resorting to "efficiency" claims and "cost-benefit" types of assessments—technical fixes that can never account for the wide range of potential social and ecological consequences. In fact, their empirical "neutrality" conceals real political choices. With rare exceptions (e.g., the public intellectuals), technical experts are employed by people and institutions with deep pockets for whom they lend an aura of scientific authority. From a critical perspective, they are "hired guns."

Political consultants, for example, are supposed to be on top of the latest methods of polling that provide campaign managers with techniques for selling their candidates. The expert's proficiency with information technology helps the politician's team customize the message for the demographic unit (audience) that he or she happens to be addressing at that moment. Computers facilitate the flexible production of propaganda that can match up audience profiles with appropriate "talking points" in the candidate's standard speech, much as advertising copy is adjusted for targeted consumer groups. "Politicians relate to potential voters," wrote Nicholas Garnham, "not as rational beings concerned for the public good, but in the mode of advertising, as creatures of passing and largely irrational appetite, to whose self-interest they must appeal" (Garnham, 1986, p. 48).

Even outside the realm of electoral politics, the expert empowers sponsors of research with information that wins contests for political or economic influence. With the aid of experts and sophisticated computer programs, the Weyerhauser Corporation was able "to develop a detailed plan of intensive forestry in the 1960s," cultivate tree plantations, expedite timber selling permits, and justify the cutting down of old growth forests—the habitat of the spotted owl and other endangered species. But the use of experts and computers also benefited affluent environmental organizations like the Wilderness Society and the Sierra Club, which did their own number crunching and used disagreements among experts to make ecological counterarguments and prevent further erosion of the virgin forests (Dietrich, 1992, p. 232). Experts came out on both sides of the nuclear power issue, usually depending on where their income derived. Ordinary citizens do not normally have the means of employing "experts" or advanced technology to fight their battles.

Beneath the facade of the dispassionate, nonpartisan truth seeker, technical experts conveniently often serve as authorities for stripping politics (and regulation) from policy. At the same time, expertise represents the refinement of the division of labor, a highly specialized and knowledgeable resource in the administrative or production process, and hence is "itself a product of the development of technology" (Street, 1992, p. 128). Although experts may be highly conversant in certain aspects of the built or natural environment, they typically do not grasp the larger, especially social, issues involved, yet their knowledge is often called on to defend policy decisions, as if technical know-how were sufficient to reveal cause and effect. The implication, invoked more and more frequently, is that political disputes are irrelevant to science and that process (and its marketplace corollary, supply and demand) is what matters—*how* to get at policy decisions, but *not who* decides.

ARTIFICIAL INTELLIGENCE: THE LITTLE ROBOT THAT COULDN'T

Critical theorists argue that the knowledge and intelligence underlying "expertise" are not clear-cut, objective terms about which people will agree, but rather ultimately depend on the dominant *values* and biases of a community or society. If, for example, an expert herbalist from rural China, someone credited with great medical intelligence and knowledge in her or his own community, were to show up and offer medical assistance in one of the high-tech medical facilities in Boston or New York City, those in authority would probably promptly escort the visitor out of the hospital, if not have the person arrested for attempting to practice without a license. In parts of China, in contrast, where technically advanced seismographic instruments are not available, a highly respected "expert" may be someone capable of carefully observing animal behavior and other natural evidence to predict an imminent earthquake. In the United States, natural or holistic approaches to healing have a difficult time gaining acceptance by insurance companies and the medical establishment as even an auxiliary form of health care.

Intelligence, rationality, and expertise are thus contextual, rather than "objective" (in the scientific sense) and are related to the social or political economic definition of the problem at hand. For that reason, different groups and communities continually fight over where resources should be invested—research on nutritional and holistic health and preventative medical practices or on more industrially based pharmaceutical technology, for example. Another example is the conflict between advocates of federal spending on public education versus those pushing for cuts in all forms of social spending. In the end, politics (who wields more influence) of one type or another will decide.

Technical and economic rationality has a logical nexus to the idea of mechanical reasoning and what has become the computer science subfield of artificial intelligence (AI). Robots that can "recognize" objects and move them about and computers that can "play" chess are certainly impressive displays of information processing, but are these demonstrations of intelligence? Or is AI, as some would argue, simply an oxymoron, a contradiction in terms? Can any*thing* that is artificial (synthetic) be *intelligent*—that is, be imbued with the ability to reason like humans and to modify its "thinking"? In 1996, an IBM supercomputer, "Blue Blue," programmed to compute more than 200 million moves per second and fed game strategies by a U.S. grandmaster, won a single in a six-game match from world chess champion Garry Kasparov. IBM lauded the outcome as a great victory for AI (McDougall, 1996, p. A16). A year later, aided by several grandmasters, a more advanced machine took the rematch.

AI is based on the computer programming of complex algorithmic statements (a series of if-then rules) and is supposed to make inferences from relevant and alternative information before rendering conclusions and suggesting courses of action. Enthusiasts for AI generally assume that humans are "thinking machines" whose reasoning patterns can be replicated in computers to solve problems. Human problem solvers are used as models in "expert systems" for programming AI devices. One enthusiast, Pamela McCorduck, even wrote a "semi-official history" about AI called *Machines Who Think.* So far, except for industrial robots, programmed for routine mechanical tasks, relatively little has been achieved to justify the high expectations in this branch of computer science, although applications have been found for such areas as assembly, medical diagnosis, chemistry, and chess games. In its beginnings, AI was directed to the task of reducing labor, extending managerial control, and expanding corporate profit (Perrolle, 1987, p. 16).

As discussed in Chapter 1, automation had long been a threat to working people, and the digital technologies introduced in the 1970s and 1980s only intensified their anxieties. AI embodies a threat not just to reduce the human skill components in both labor and management occupations but also to replace these job categories altogether with computer-guided assembly and administration. Robots have already replaced many thousands of workers in Japanese and U.S. factories, and computers have made redundant the supervisory duties that many middle-level managers used to perform. In the field of AI, social scientists and computer scientists (the artificial intelligentsia?) see a future where there are no limits to the complexity of computer decision making, with "rational" calculating capabilities far ahead of humans. Even if this were possible, one might ask, to what ends would "thinking machines" be put?

Quite apart from the social and economic substance of this way of thinking about computers is a still more problematic question: What is human reasoning? To some leading AI experts, such as those who work at MIT, human reasoning is a phenomenon that can be understood by breaking down decisions into discrete rules of rational introspection, as if thinking necessarily followed sequential paths. A classic test of AI is the computer that can "play" chess and win against very advanced human players. It is only a matter of time, some have predicted, before a computer chess program will be ranked number one, outwitting all the grand masters of the game.

Others, such as the philosopher Hubert Dreyfus (1992), see only grand delusions in the views of the AI community. Dreyfus grants that certain types of calculations wherein the human equivalent of rote memory is involved can be done and done faster and more reliably by computers than by people but that the intellectual processes of the brain, on which AI theory is modeled, is simply not knowable. AI mastery includes such problems as mathematical calculations or naming the capitals of countries or games that have limited rule possibilities (e.g., tic-tac-toe), in which interpretation or meaning is not at play. In interpreting language or games with creative strategies, a computer would be at a total loss because it is not capable of deciphering connotations whose meanings come from an immediate context. For example, in a sentence like "The bull is charging," how would a computer know whether the speaker intended to utter a statement of fact, of alarm, of amusement, of something metaphorical (not literal), of doubt, or of refutation, of another statement? The tone of the speaker, the speaker's relationship to the listener(s), the surrounding circumstances when the utterance was made, and an unlimited array of other contextual considerations would convey meanings and intentions that an inanimate machine could never consider.

Back to the chess game example: Dreyfus argues that computers can only calculate rule options (if this move, then that move) toward set objectives. But chess is more than a game of fixed rules. The chess master learns as he or she plays the game, picking up on new creative move possibilities that unfold through discovery and that provoke sudden bold maneuvers, which a computer would be incapable of "seeing" because its recognition patterns are preprogrammed. A person who beats a computer by using a particular strategy can beat the same computer again and again following the same pattern, because the computer, left on its own, simply cannot learn from its mistakes.

The implication of this is that human reasoning cannot be bested by even the most sophisticated calculating machines, even if computers are very useful in the mundane requirements of data sifting and where value selection or interpretation is not cybernetically defaulted. Humans appear to reason by way of gestalt, sudden breakthroughs in understanding their social and physical environment through spontaneous perceptions of the whole experience, for which formal logic is only a limited tool. Moreover, what we call "reason" does not imply discovery of objective "truth," because reasoning is always co-joined with personal motives, value orientations, socialized ways of seeing, and intentions that are never universal. As romantic an illusion as the idea of a human-invented life form may be, as in Frankenstein's monster, it is only in the end just an illusion. The notion of a "thinking machine," either as fantasy or ruse, in practice imposes a set of values and power interests of some people over others, all the while couched in marvelous value-free neutrality. The word *cybernetics,* which forms the foundation of modern information theory, derives its meaning from Greek and translates as "governor." It is not that information machines, computers, will ever govern, but that the illusion of such will be cultivated through impressive displays of cybernetic "intelligence." The philosophy that says a system of formal rules is what constitutes knowledge and intelligence divests those who reach understanding by other means—namely, everyone. It also presents us with the last frontier of the industrial world—the subordination of human intellect and sensibilities to the "rationality" of machine technologies and those in command of them, a more dystopian than romantic vision of the future.

The limits of AI can be illustrated in a research experiment at the University of California, Berkeley. Pandora is a newspaper-fetching robot that supposedly figures out that it should wear a raincoat during precipitation. Assuming such a robot could be programmed to retrieve a delivered newspaper, even with a capacity to respond to some visual and audio cues, certain existential problems occur. If a downpour and thunderstorm happen just before the paper is retrieved, should the robot go out in the rain and risk damage? If yes, should it take the chance of soiling the floor of its residence with its mud-soaked "feet"? In the case of a heavy soaking rain, does the condition of the newspaper make it worth retrieving at all? Can a robot exhibit "common sense" (see Rose, 1984, chap. 4)?

A human would respond to these contextual matters, many psychologists would argue, with gestalt-type rationality, not in linear (if-then) progression assumed by AI advocates, but in integrated and fluid patterns of interpretation combining intuition, past experience, sensory mechanisms, and logic and where the whole is more than the sum of its parts. Claims about advanced (e.g., "fifth generation") computers being vastly superior versions of the brain assumes a kind of human wiring that is not established as anatomical fact. If we do not yet understand how the human mind, human reasoning, and human intelligence actually work, how can we "create intelligence"? asks one computer scientist, Joseph Weizenbaum (1985).

Weizenbaum, himself one of the original architects of AI, has also been a leading skeptic about what he sees as the exaggerated claims about AI reasoning potential and other computer decision-making applications. At MIT's Artificial Intelligence Laboratory from 1964 to 1966, he developed a computer language analysis program, which he called ELIZA (from the "Pygmalion" character), as a way of showing how easy it is to demonstrate computer "intelligence"—as defined by the celebrated British mathematician and theoretician Alan Turing. The test of intelligence, according to Turing, was whether a machine could respond to typed questions with reasonable answers. Weizenbaum set out to construct a Turing machine.

He designed a "conversation" between a person and a computer, based on Rogerian psychotherapy, that would make it appear that the computer was actually thinking, when, in fact, all it was doing was constructing sentences based on keywords typed by the human partner. The Rogerian technique is to get the patient to take the initiative in conversing with the therapist. Thus, the ELIZA network carried conversations such as the following:

Caller: Are you really a psychiatrist?

ELIZA: Why are you interested in a psychiatrist?

Caller: I'm not really interested in a psychiatrist.

ELIZA: You say that you are not really interested in a psychiatrist.

Caller: That's right. I never had hang-ups about parents or stuff like that.

ELIZA: Tell me some more about your parents. (cited in Penzias, 1989, p. 144)

This "conversation" involved no actual reasoning or understanding at all on the part of the computer, but simply a series of patterned responses and stock phrases. Weizenbaum himself wrote a book to make precisely that point (Weizenbaum, 1976). Yet, many of those who communicated with ELIZA refused to believe that it was nothing but a machine, and some were seriously distressed when the experiment was discontinued in the early 1970s. ELIZA made Weizenbaum only more critical about the potential for social manipulation through the use of technology. He would later (1976) write:

The human is unique by virtue of the fact that he must necessarily confront problems that arise from his unique biological and emotional needs. . . . No other organism, and certainly no computer, can be made to confront genuine human problems in human terms. . . . Computers can make judicial decisions, computers can make psychiatric judgments. They can flip coins in much more sophisticated ways than can the most patient human being. The point is that they *ought* not be given such tasks. They may even be able to arrive at "correct" decisions in some cases—but always and necessarily on bases no human being should be willing to accept. (cited in Perrolle, 1987, pp. 231, 233)

One apprehension about surrendering to AI was depicted in the Stanley Kubrick film *2001: A Space Odyssey* (based on the book by Arthur C. Clarke). In this very imaginative 1968 film, an onboard spaceship computer, called "Hal," on which everyday operations are dependent, reveals in a climactic moment that "he" has a will of "his" own, "refuses" to be shut down, and "acts" to take over political control of the craft. Although such a story line exaggerates the capacities of computers, it does raise the question of how far computer "decision making" has already been assimilated as a core assumption about everyday governance in Western society.

Neil Postman, a critic of popular culture, fears that the United States has degenerated into a "technopoly," in which the dominant ideology "subordinates the claims of our nature, our biology, our emotions, our spirituality" to the sovereignty of the computer "by showing that it 'thinks' better than we can." Actually, despite huge investments from Fortune 500 companies, the limited outcomes of AI have disappointed most of its early enthusiasts. AI research is concentrated at four universities: Carnegie-Mellon, MIT, Stanford, and Yale. The arrogance of computer scientists is captured in the quip of Professor Marvin Minsky at one of the big four, MIT, who enthusiastically comments on AI information-processing units: "If we are lucky, they will keep us as pets" (Postman, 1993, p. 111).

This kind of thinking, though it has its adherents among some in the science community, is regarded as naive, if not dangerous, by others. An important element in understanding the ideology behind the "brain as computer" metaphor is the fact that large science and technology research institutions like MIT rely heavily on the support of military and corporate (often defense focused) funding. Through the 1980s, MIT, for example, received about two thirds of all its external funding for military-related research. The computer science community has little cross-fertilization of ideas with the social sciences, especially from critical scholars. In AI, 69% of all electrical engineering basic research, 90.5% of its applied research, 54.8% of computer science basic research, and 86.7% of its applied research came from the Department of Defense (DOD) in 1983, according to the Congressional Office of Technology Assessment (cited in Athanasiou, 1985, p. 31).

The major AI user is the U.S. military. Its communication system uses about 17 million lines of programming (Mulgan, 1991, p. 115). In 1983, the Defense Advanced Research Projects Agency (DARPA) undertook what started as a 5-year, $600 million "strategic computing initiative" (SCI) to give more explicit military direction to the AI program in the United States, with applications for "autonomous tanks" for the Army, a "pilot's associate" for the Air Force, and "intelligent battle management systems" for the Navy (Athanasiou, 1985, pp. 28-29). SCI became a core component of the Reagan administration's strategic defense initiative ("Star Wars"), a ballistic missile-based shield in space that is supposed to locate and destroy on-warning any attempt to launch missiles from anywhere on Earth.

Critics at the time argued that such a system—relying on 10 million lines of computer code without ever being fully tested and which could not knock out all 10,000 or more Soviet missiles—has too high a potential for an annihilating accidental nuclear war, and even if it could work, would prove to be part of an offensive, not defensive, system. Others have argued that the cost of building such a system, estimated as high as $100-200 billion per year (*New York Times,* April 11, 1986: cited in Ziegler, 1990, p. 280). would deprive Americans of such basic necessities as affordable housing, public education, job training, Head Start, school lunch programs, transportation improvements, environmental protection, and other social needs, in addition to the lost advantages of foregone investments in nonmilitary research and development (R&D).

COMMUNICATIONS FOR
MILITARY COMMAND AND CONTROL

One enduring tradition in U.S. foreign policy is the pursuit of the Monroe Doctrine, a political tenet, dating back to an assertive but insupportable proclamation of President James Monroe in 1823, that the Americas south of the United States are a special interest area and protectorate of their northern neighbor. Since then, the United States has continuously intervened in Latin American affairs, installed puppet governments, trained its military and police forces, and exercised political and economic hegemony throughout the region. After World War II, successive U.S. presidents extended the doctrine of direct and covert intervention to protect U.S. interests throughout the world, as in Iran, Italy, Greece, Guatemala, Brazil, Cuba, the Dominican Republic, Vietnam, Korea, Chile, Grenada, Panama, Iraq, Libya, Lebanon, the Philippines, Cambodia, Laos, and several other countries. About 90% of all U.S. arms sales (with its 70% of world market share in 1993) goes to unelected and dictatorial regimes, even though 96% of American people polled oppose weapons sales to such undemocratic governments (Allen & Closson, 1994, p. 1).

Included within the arsenal of U.S. weapons for political and economic destabilization, assassination, manipulation, and disinformation was the use of the airwaves for transmitting propaganda and instigation to overthrow governments the U.S. opposed. Radio Swan, owned by the United Fruit Company, was used by the CIA and Cuban exiles to prepare anti-Castro Cubans for the ill-fated CIA–directed Bay of Pigs invasion in 1961. In the early 1950s, the U.S. government distributed radio sets in the Philippines that received only one frequency—the one used to carry campaign news for the presidential candidate, Ramon Magsaysay, whom the Eisenhower administration was backing. Magsaysay's chief credit was that he had been a faithful anti-communist in the war against the Huks.

Radio Liberty and Radio Free Europe, organized by the CIA in the early 1950s, beamed broadcasts into eastern Europe and the Soviet Union to encourage uprisings against their governments. In the 1980s, Radio Marti and TV Marti were organized with Congressional approval to try to provoke the overthrow of the Castro government. The anti-Castro broadcasts were projects of Bay of Pigs veterans and organizers of earlier clandestine Cuban exile radio stations in collaboration with the Reagan and Bush White House. Radio Free Asia was approved by Congress in 1994 as a weapon for destabilizing the governments of China, Laos, Cambodia, Vietnam, Burma, North Korea, and Tibet. These government propaganda organs are joined in the war of persuasion by at least five evangelical stations that operate outside the U.S. mainland in Guam, Latin America, and the former Soviet Union.

Among the most secretive communications applications are those of the military. As much as half of the radio frequencies in Europe and the United States is reserved for military use, while communication satellites used by the military far outnumber commercial craft, about 75% of the total in orbit (Barnaby & Williams, 1983). Information about the heavy use of U.S. space shuttle flights by the military is censored. Disclosure about the real proportion of government spending consumed by the Pentagon is also withheld from the public. In fact, the whole conduct of the Persian Gulf War was shrouded in deception and misinformation about the necessity for military intervention, and the major U.S. media were strategic instruments in the propaganda aspects of the triumph. One study found that television was vital to the Bush administration and the military in conjuring images of a sanitary techno-war, which disguised one of the most violent, destructive, one-sided, and deceitful assaults on a nation state in modern history (Kellner, 1992).

The military-industrial complex and the employment of communications and information services extends far beyond the conduct of war to the full-time preparation for war. In the area of computer research at the top four academic institutions, more than 80% of federal funding comes from the DOD (and 71% of all federal academic computer science research), which deprives the federally funded National Science Foundation (NSF) of much-needed support. Major universities with strong computer science programs, such as MIT, Carnegie–Mellon, University of California, Berkeley, and Stanford, are heavily dependent on military research funds, which means that professors and their graduate students in such programs will be hard pressed to continue functioning without working on projects directly or indirectly related to weapons development. About a third of what became a $1 billion SCI went to universities. In 1985, the NSF could fund academics at those institutions an average of $31,000, at a time when the DOD was offering the same schools $279,000 per cooperating faculty member (Selvin, 1988, pp. 563-564).

Not all science researchers choose to cooperate. Albert Einstein often spoke out publicly and passionately against the militarization of science (for which he came under secret FBI surveillance). During the U.S. war in Indochina, many research academics working on military contracts opted to abandon the lucrative opportunities for advancement and status to find work in less destructive areas of science. One of those in the advanced section of the computer field dealing with AI, Joseph Weizenbaum, turned critic of the war in Indochina and of those in the scientific fields who continue to ignore the human consequences of military research. Weizenbaum found that U.S. military computers employed in Vietnam were programmed to lie to government officials and thus instigate policy decisions based on disinformation while shielding the chain of command from personal accountability (Weizenbaum, 1981a, pp. 559-561).

Norbert Wiener (1894-1964), often thought of as the founder of the science of cybernetics, had also warned of the repressive possibilities of computers in the hands of military and commercial interests. The "triple constriction" he cited on the potential of information flow for social progress included profitable communications driving out the less profitable, domination of the information industries by the rich, and the inevitable allure of communications to the power hungry (Wiener, 1948, pp. 161-162). Although he was a child prodigy (earning a Ph.D. in mathematics from Harvard at age 18) and a brilliant mathematician, Wiener never lost sight of the embedded relationship of technology to politics.

"COMPUTER DEMOCRACY"?
GRASSROOTS COMMUNICATORS,
POPULISTS, PIRATES, AND HACKERS

As in the periods just after the two world wars, when communication technology development was monopolized by the military, the post-Vietnam War era has seen a relative shift in research priorities to civilian and commercial applications, although the military-industrial complex has far from disappeared. From the general user's perspective, some of the best military communication hand-me-downs have been e-mail, electronic bulletin boards (bbs), and other computer user networks. Text and image communication of these types may lack the dimensionality and sensory impact of face-to-face discourse but has certain advantages over telephony, telegraphy, telex, and facsimile (fax) in terms of multipoint reach, information transfer costs, storage, and response time options. At the same time, electronic communications lack the qualities of emotion and intent conveyed in interpersonal communications and the documentation status of fax. One additional aspect of e-mail is that it brings together groups of individuals previously unknown to one another, linked by some common interest.

Some writers (Rheingold, 1991) see electronic communications forming "virtual communities" of shared interests, overcoming spatial, social, and cultural barriers that inhibited their communicating in the past. Indeed, it is entirely possible for e-mail users to communicate quite regularly without being aware of one another's class, gender, ethnic, or racial identity, or disability for that matter. Virtual reality communities, in which physical representations stand in for the real participants, give the sense of being able to become creators, instead of passive observers, of media content. For some, this excites the idea of computers having democratizing power. Although there is some value in such a belief, social relations on the whole are not changed simply by having technological instruments at people's disposal. Class relations of power are not so easily broken.

This is true whether technology is employed in peaceful or aggressive overtures. If technical capability were the decisive factor in human relations, U.S. attempts to "communicate" its will through saturation bombing in Indochina or British force against the American revolutionaries would have had different outcomes because, clearly, the victors in each situation were not those with the superior technology. When political will is unified, even inferior technological instruments can overwhelm the better-armed aggressor. Mao Zedong, the intellectual and military leader of the Communist Party–led revolution that turned China into a socialist state in 1949, understood this practical wisdom.

Recognizing that the Nationalist (Guomindang) forces of Chiang Kai-shek (Jiang Jieshi) had a certain advantage in U.S.–supplied equipment, Mao urged his followers, during the course of the long and protracted revolutionary struggle, that it is important to think dialectically about how to turn bad events (e.g., Japanese imperialism in China) into positive outcomes (e.g., a national liberation movement; Mao, 1968, pp. 125-129). Regardless of one's ideology, this is probably a wise and practical injunction. For those who view Western society as aimless, protective of selfish interests, and increasingly restrictive of inalienable rights, it is possible to work for alternative and democratic values by engaging the structures and instruments of control in ways that foster more humane social relationships.

In the 1980s, a new form of electronic messaging began proliferating that took advantage of a Vietnam War-era military-based technology and turned it into an instrument of popular and grassroots communication. The Advanced Research Projects Agency of the U.S. DOD had introduced in 1969 a packet-switched network (a form of sending fixed-length "bursts" of digitally disassembled data blocks), transmitted as interspersed bits spaced among multiple message components and reassembled at the other end. Commercial, scientific, activist, interest-oriented, and academic user groups eventually got access to this system, starting up such e-mail networks as Bitnet and electronic bulletin boards for the exchange of information and ideas on computers, games, human rights, the environment, eroticism, and what have you. This later led to the creation of the globally interconnecting network called Internet and its World Wide Web software applications.

As of 1993, the Internet linked some 400,000 host computers, 11,000 networks, and an estimated 15 to 20 million users worldwide (10-12 million in the United States) in some 102 countries, with the numbers rapidly increasing each year (Chapman & Rotenberg, 1993, pp. 5, 16). Although large corporations (IBM, Sears, H&R Block, Lockheed) dominate videotex services (interactive information system by computer), a major challenge has come from the activist community concerned with issues of peace, human rights, ecology, democratic movements, and social justice, which united several computer networks to form an Institute for Global Communications (IGC), based in San Francisco. IGC, in turn, is linked to sister networks in other parts of the world, such as GreenNet in England, NordNet in Sweden, and Web in Canada, as major nodes and a host of smaller interlinked nodes in many other countries, some 15,000 subscribers in 90 countries collectively managed by the Association for Progressive Communications (APC).

This form of internationalism is joined with other modes of communication—videographers and filmmakers, recording artists, cable TV producers, amateur radio users, and others (Frederick, 1994). Thus, the generation of people who stood in opposition to U.S. military intervention in Vietnam and other places or who were otherwise alienated from government and transnational capitalism had now converted a "bad thing" (war technology) into a "good thing" (communication for peace, justice, and environmental protection). Many believe that electronic communications can help reignite protest against politics as usual and shift the public agenda toward progressive social policy issues.

Antonio Gramsci, a renowned Italian socialist theoretician imprisoned by Mussolini, argued that ideology and the control of dominant ideas in society are not fixed, but rather are contested terrain. Gramsci believed that every person has intellectual skills but that only established institutions designate for society the ruling intellectuals. Truly independent public intellectuals can separate themselves from the dominant industrial-financial class and take active leadership for the elevation of mass consciousness and moral political reform (Joll, 1978, pp. 117-134). Gramsci might have agreed that the Internet represents a democratizing potential capable of bypassing the major channels (e.g., government, mass media, church, schools) of cultural hegemony—that is, the ways ruling ideas are ingrained while oppositional ones are banished from public discourse. The underground distribution of Tom Paine's militant writings helped inflame Americans against British colonial rule. But although also text-oriented, e-mail may be simply too visible to effect radical change, especially given the existing means of establishment surveillance against political suspects.

Some computer networks are making a difference, however. One on-line user system catering to progressive political, educational, public policy, and environmental concerns is Peacenet, a part of the APC. Since 1986, it has operated on a nonprofit basis through partner nodes in several other countries, including Britain, Brazil, Canada, Sweden, Australia, and Nicaragua, linked over telephone lines to social activists in these and many other countries. Through this computer network, individuals and groups (e.g., Greenpeace, Physicians for Social Responsibility, Amnesty International) are able to establish communities of interest and to communicate on or close to a real-time basis. This is particularly useful for organizing on-line or in-person conferences, mobilizing protests such as for human rights emergencies, sharing current information on issues like AIDS or violent conflicts in some region of the world, accessing specialized databases, or for putting together publications by transmitting manuscripts electronically. On-line systems like Peacenet also offer opportunities to become educated about other organizations or to "meet" members on the network.

While progressive groups are making good use of on-line bulletin boards, far right organizations are at least as active in setting up networks for causes of their own. In the wake of the collapse of the Soviet Union, neo-Nazi organizations have sprung up throughout Europe, and e-mail, cellular telephones, and fax machines have become their most effective instruments for sharing propaganda, fund-raising, and popularizing their worldviews. Electronic technologies enable them to bypass regulations that, for example, in Germany make it illegal to deny the Holocaust. These media can quite easily be used in anonymous fashion to avoid police detection. In 1993, right-wing groups were held accountable for nearly 2,000 crimes, including 8 murders (17 murders the previous year). The Thule network is a neo-Nazi computer mail system with some 2,000 users and shares with its members a hit list of anti-fascists in an effort to strengthen its goal of "purifying the race" (Neuffler, 1994).

Those with access to e-mail networks certainly have new opportunities for sharing information and ideas, for better or for worse, in a relatively unconstrained environment. The differences between e-mail and regular telephone communication involve cost (bulk usage on the former providing economy of scale pricing), multipoint, multiuser reception (telephone mainly used for one-to-one voice communication, e-mail for sharing written messages with one or many geographically dispersed respondents on a real-time, interactive mode or convenient, stored basis), and document transfer capability (e-mail allowing the sending of computer files). (Social organization networking is covered extensively in such journals as *Whole Earth Review, Technology Review,* and *Scientific American.*) Often, e-mail exchanges circulate ideas and ways of seeing that are not well covered in much of the detail of mainstream media. Many users report having regular communication with citizens living in countries where they are not otherwise able to publicly discuss certain sides of existing political situations. It is also common to find people on e-mail networks who share common interests but with whom one might never otherwise communicate. Such a celebrated romance through correspondence of poets Robert and Elizabeth Barrett Browning in today's busy world may sometimes occur over e-mail, parodied in a "Doonesbury" serial.

At the same time, it is important to keep in focus the political and economic realities. Among these realities are the ventures of opportunity-seeking commercial interests and the concerns of those who would restrict public communication in the cause of security, moral guardianship, or property protection. The author of popular, radical critiques of American mass communications, Herbert Schiller, argues that corporate capitalists have perennially encouraged the U.S. government to subsidize new infrastructures before commercial interests will move in—until the risk factors are overcome and the market is well established—practices defended in the name of free enterprise (Schiller, 1993a).

Most of the risky R&D costs associated with the development of the computer, transistor, semiconductor, microprocessor, microwave, fiber optics, satellite vehicles, launching capability, and e-mail, in fact, were underwritten by the federal government. Including the military, the federal government is the largest single purchaser of information equipment in the world, spending $25 billion a year and more than $200 billion during the past decade on computers and related services alone (Pear, 1994). In 1993, the Clinton administration, following the logic of pure market capitalism, supported the selling off of newly available

radio frequencies at public auction, the first time that broadcast franchises have been sold to the highest bidder. As Schiller puts it rather sardonically, "Why not put the Great Lakes, the Rocky Mountains and the national parks on the [auction] block?" When the sale was completed in 1995, the big winners were AT&T, the RBOCs, Sprint, and the cable giant TCI (Schiller, 1993a, p. 64; 1996, p. 84).

What is at stake is not simply a few frequencies, but the defense of the "public sphere," the responsibility of government to reserve social space in the common interest (see Chapter 2). The degree of access now afforded by the Internet has been undermined by Congressional actions in privatizing a system that had been subsidized by public NSF funding. The NSF, with a minuscule grant of $12 million per year, funded much of the research conducted on the Internet, along with a computer backbone system (NSFNET) in the United States for such users, and it set national policy for the system. The private sector (IBM, MCI, and Merit, Inc.) managed the technical aspects of the system for NSF (LaQuey, 1993, pp. 27-28). With the loss of public funding, the Internet by the early 21st century could become available only to those of financial means, pulling away from less-endowed educational institutions and millions of ordinary users.

The Internet is part of a vast satellite-delivered telecommunications complex that routes millions of telephone, banking, television, billing services, e-mail, weather information, oceanographic data, CIA spy information, and planetary mapping transmissions on a daily basis. Geosynchronous orbits of communication satellites, at 22,300 miles above the equator, require space vehicles to move at a speed sufficient to keep them positioned over the same point relative to the earth. Communication satellites receive transmissions, amplify them, and reflect them back to earth to parabolic antennas, making them vulnerable to natural and human interference. Many accidental, as well as intentional, disruptions of satellite communications occur, and for many years the United States and the Soviet Union practiced jamming of international radio as part of their Cold War struggle.

Individual rebels with knowledge of electronics may also occasionally sabotage the best-laid plans of the information elites. In 1986, for example, a disgruntled backyard satellite dish owner and hacker, John MacDougall, broke into an HBO movie transmission by bouncing a message off its Hughes Communication Galaxy I satellite on its channel 23 and forcing the regular program off the air. Its millions of viewers suddenly saw substituted on their television screens the following message: "GOODEVENING HBO FROM CAPTAIN MIDNIGHT $12.95/MONTH? NO WAY! (SHOWTIME/MOVIE CHANNEL BEWARE!)." Not only had MacDougall twitted one of the monarchs of the airwaves, Time Inc.'s Home Box Office, but he alerted the whole U.S. military establishment that its impregnable system of communication, many of its satellite positions available as public information, was vulnerable to interruption or jamming by ordinary foot soldiers (Goldberg, 1986, pp. 26, 29).

Regular pirate broadcasters also abound, several hundred operating in the northeastern part of North America. Although most cater to alternative music not picked up on industry-controlled pop charts, much of it is very political, ranging from anarchism to the extreme right. One report notes that one pirate transmitter, referred to as "He Man Radio," caters to macho listeners with uninhibited tales of male conquest (Binder, 1994). The pirates are no match in shortwave outreach to the U.S. government, however, which leads the world with, at it peak in the early 1990s, more than 2,500 hours of weekly transmissions or to Britain's BBC "World Service," which may have an even bigger listenership.

The most broad-based form of piracy takes place in the household, university, and workplace, where communication equipment, such as VCRs or computers, routinely duplicate rented, cable, or off-the-air programs or software for their own collections, involving some $60 to $70 billion in losses for patent owners. In China, illegally duplicated CD-ROM software from IBM, Microsoft, and Lotus amounts to some $830 million in lost revenues to those companies (Parker, 1995). Photocopying for non-commercial use by teachers and students is another increasingly policed but nonetheless commonplace copyright violation.

By the mid-1980s, already more than half of all database, spreadsheet, and accounting programs and nearly half of all word-processing programs were pirated (Barnet & Cavanagh, 1994, p. 145). For the recording industry, the creation of the compact disk offers near invulnerability of duplication, except back to tape. It is only a matter of time before that technical barrier will fall to mass piracy, inasmuch as relatively few people respect the multimillionaire status of pop stars or computer, film, and television industry executives. Although there are always cracks in any empire of power, pirates without a political program for change represent but a tiny blip of interference on the screen of corporate-dominated profits and politics.

TELECOMMUNICATIONS:
WHO RULES THE (AIR)WAVES?

Powerful international development agencies, such as the World Bank, a brain-child of the Franklin D. Roosevelt administration and still headquartered in Washington, D.C., have long tried to convince Third World countries that telecommunications investments bring economic development (e.g., see the work of three World Bank technocrats in Saunders, Warford, & Wellenius, 1983). The essential argument is that poor countries have low telephone densities (per 100 population) and wealthy countries have high densities, that telecommunications triggers economic activity, and, therefore, that Third World countries would be well served to invest billions in such an infrastructure, taking World Bank loans and advice and securing the financial support of other foreign commercial banks for that purpose. Other research on this issue suggests that Third World countries, in fact, are more likely to expand their economies and their communication facilities *without* World Bank intervention and that, where the Bank is involved in telecommunications, its real rationale is to create investment opportunities for transnational corporate manufacturers and end users while fostering debt-dependency on borrower nations and shifting internal investment from the public to the private sector (Bello, 1994; Sussman, 1987).

The basic flaw in the World Bank's linking of development to telecommunications investment is that First World (leading industrial) countries were *never* underdeveloped relative to the rest of the world, and, therefore, telecommunications had nothing to do with their original state of economic self-reliance. The corporations that dominate global telecommunications markets in the 1990s were the very ones that had control at the beginning of the century: AT&T (U.S.), Siemens (Germany), Philips (Netherlands), Ericsson (Sweden), IBM (U.S.), and Cable & Wireless (U.K.). Japan's NTT and its major electronics partners (NEC and others), which were started up or expanded into major telecommunications entities only in the postwar period, are the exceptions.

Telecommunications has been central in providing TNCs in the West and Japan with the means of subjugating the budding capitalist aspirations of the Third World and appropriating the latter's human and natural resources for First World development. That modern telecommunications is associated with wealthy and powerful nations has more to do with the instrumental value of information technology in establishing an international division of production and labor and the utility of communication satellites, oceanic cables, computers, and the like in permitting centralized control over far-flung decentralized geographic operations.

When the telegraph, telephone, and radio formed the foundation of telecommunications infrastructure in the United States and western Europe, user groups were virtually all domestic, stimulating further internal development, wealth accumulation, national integration, and other spread effects. The telegraph was used to speed up and expand urbanization and decentralized development to various parts of the United States; the telephone was appropriated by farmers and urban residents for their own communication and information web; and television and radio became a common popular culture and vernacular, linking America's east and west coasts.

This is not the experience of countries like the Philippines, Indonesia, and Egypt, which turned to external sources and forms of telecommunications technology that delivered few services to the vast majority of their citizens. The colonial era that preceded their independence impeded broad-based development, integrating and promoting only those sectors (e.g., cotton, sugar, petroleum) that complemented the metropolitan countries' own development. Hence, unlike the comparatively widespread and even development within the now leading industrial countries, the focus of telecommunications investment in the Third World has intensified the latter's underdevelopment.

In the Philippines, the list of major users of the country's telecommunications facilities reads like a who's who of the international *Fortune* 500. On television, one sees reruns of canned American television series, with such familiar sponsors as Nestle, Procter & Gamble, Coca-Cola, and Colgate Palmolive (for extensive documentation, see Sussman, 1982). Long-distance telephone and data circuits on the oceanic cable and satellite connections are largely reserved for these and other transnational firms that can justify the cost through economies of large-scale usage. Subsidiaries of U.S., Japanese, and western European corporations provide the path to upward mobility for educated and enterprising Filipino business professionals, which only reinforces their estrangement from the underfed, rural peasants and urban workers who struggle for existence on poverty pay scales of less than $4 a day (and declining in purchasing power). This framework for "development" is premised on a theory of "trickle down" economics: Let the rich get richer, and some of it will filter down to the masses. The Third World relationship to the information society is more fully discussed in Chapter 7.

CHAPTER 6

Privatizing the Ether

Regulation, Deregulation, and Information Apartheid

With the crusade for economic "liberalization" during the Reagan and Bush era, it became almost an article of faith that an unimpeded private sector could carry on the affairs of society more economically, more efficiently, and more equitably than any public ownership or state-regulated alternative. "Reaganomics" led to a wave of deregulation of industry that, in effect, meant the government was going to ignore enforcement of antitrust legislation, withdraw much of the regulation of such industries as transportation, communication, and banking, and cut social spending while boosting defense outlays. Public support for education was pared at a time when U.S. achievement test scores were declining. The administration sought military advantage over the Soviet Union by introducing new weapons systems and actively intervening in areas of the world where Soviet–aided states held power, as in Nicaragua, Grenada, Angola, North Korea, Afghanistan, and Cuba, or where radical nationalist regimes were in control (Libya, Iran, and Panama).

In communications, broadcasting megamergers and takeovers included the capture of one of the big three television networks by one of the biggest electronics firms in the United States. RCA, a major military contractor itself, was acquired (with NBC-TV) in 1985 by the conglomerate General Electric (GE) during Ronald Reagan's second term in office. Reagan had been host of *General Electric Theater* for 8 years and a GE public relations personality in the

1950s. In the 1980s, GE was one of the major defense contractors, had a major stake in Reagan's "Star Wars" space-based missile delivery system, and supported many conservative political causes (including sponsorship of the right-wing *The McLaughlin Group* news analysis on Public Broadcasting). William French Smith, a personal friend and lawyer of Reagan, joined the board of directors of GE after serving as Reagan's attorney general. When Smith was attorney general, the Justice Department waived antitrust restrictions and approved the GE-RCA merger (Lee & Solomon, 1990, pp. 76-77).

In the Reagan years, the deregulation of telecommunications (and the privatization of telecommunications in other industrial countries) raised questions about public accountability. Ordinary telephone users saw basic rates rise rapidly after deregulation, which helped defray the costs that regional Bell operating companies (RBOCs) were paying for fiber-optic trunk lines. The largest corporate users, some 1% of total hookups but accounting for one third of telephone company revenues (Mulgan, 1991, p. 153), represented the most powerful interests in search of better services and lower prices—and the thrust behind the Federal Communications Commission (FCC) decisions of the 1960s and 1970s that paved the way for the divestiture of AT&T. Deregulation has indeed been good for corporations in both regards: Long-distance rates have come down, and customized services have proliferated. Half of all long-distance traffic is taken up by just 5% of domestic and long-distance users (Office of Technology Assessment, 1990, p. 69).

Not entirely unexpectedly, the Clinton administration did not think differently about deregulation and privatization of telecommunications. Clinton's chair of the FCC, Reed Hundt, had previously served in a corporate law firm, Latham and Watkins, that mainly handled large telecommunications industry accounts. On *Talk of the Nation,* aired on National Public Radio (May 31, 1994), Hundt stressed marketplace competition over regulation as the solution to public interest concerns about the airwaves, cable, telephone, the "information highway," and so on. He urged the need to auction off newly available radio frequencies. One indication of the Clinton government's direction in telecommunications policy was apparent in its financial support, backed by a coalition of major electronics corporations, to Commercenet, a high-speed, business-oriented network on the Internet system for buying and selling goods, banking, engineering, and other data exchange, with security features built in to protect e-mail commercial transactions (Markoff, 1994b).

THE "ABILITY TO PAY" PRINCIPLE

The last major Congressional effort to monitor the growth of corporate control of telecommunications was in 1980 with the U.S. Senate report, *Structure of*

Corporate Concentration. This report documented the broad array of interlocking directorates between major banks, telecommunications, information processing and office equipment firms, and other leading industrial corporations. *Interlocking directorates* in this study referred to individuals serving as directors on two or more of the largest customer, supplier, and financial institutions that nominally are in competition with one another.

> For instance, directors of AT&T, IBM, Exxon, Xerox, Sperry and Eastman Kodak were all on the board of Citicorp. Directors of IBM, RCA, Sperry, Eastman Kodak and Exxon were represented on the board of Metropolitan Life. IBM, AT&T, Xerox and GTE were on the Bankers Trust board. AT&T, IBM, Eastman Kodak and Exxon were directors of Continental Illinois. AT&T, IBM, RCA and Exxon sat together on the board of Chemical New York. AT&T, IBM and GTE were all directors on Conoco. The most competitor interlocks were between AT&T and IBM. (U.S. Senate, 1980, p. 26)

Interlocking directorates represent an informal structure of collusion, allowing for market coordination and a restraint on competition. At the highest level of government, as well, especially at the cabinet level, are representatives of corporate interests whose jobs revolve between state policy making and chief executive officer functions of leading industrial enterprises. In the post-World War II period (1945-70), two thirds of the commissioners of the FCC went on to work in some capacity for the same communications corporations they once regulated—Reagan FCC chairs Mark Fowler to the Bell Atlantic Personal presidency, a cellular telephone company, and Dennis Patrick to the Time Warner Telecommunications presidency (Kellner, 1990, p. 130n; Mazzocco, 1994, p. 126). This could hardly have been the case had they taken independent and strong public interest positions as regulators. Under such an elite and co-opted system of regulation, the main struggles in industrial communications turn to capturing market position, rather than to responding to demands for making the media and information more publicly useful and accessible.

Since the early 1980s, communications and information have become increasingly commoditized. The once common architecture for various forms of distance communications (e.g., post, television, radio, messaging, telephone) has fragmented into differentiated, specialized, and value-added services for more affluent groups of users and traditional basic services for everyone else. Corporations are less likely to use ordinary mail for important communications, but rather the better value and less expensive bulk usage technologies of e-mail, computerized data banks, teleconferencing, fax, leased-line telephone, and when necessary, overnight postal delivery. Private, "cream-skimming" companies have taken over several niche markets in communications (e.g., overnight mail delivery, long-distance telephony, cable TV), taking advantage of the national backbone networks previously established by older corporations (the U.S. Postal Service; AT&T in telephony; NBC, CBS, and ABC in television). The niche markets have primarily benefited those with the ability to pay, whereas, for example, the U.S. Postal Service has been left, for the most part, with less profitable first class, bulk mail, parcel post, and airmail services.

Commercial interests have discovered a multitude of new communication possibilities for flooding information channels with exhortations to consume. But the functional sophistication of new means to communicate has encouraged the business world, especially those with access to cross-referenced data lists and consumer profiles of U.S. citizens, to target specific groups according to their "values, attitudes, and lifestyles." A credit card, a mailing address, even a driver's license objectifies the individual in a fusillade of advertising, public relations appeals, and solicitations of every imaginable type, building up mountains of paper waste in a techno-industrial society that was supposed to make paper obsolete. Culture is reduced to little more than an efficiently run marketing system in which products and citizens are merged into commoditized units of consumption and profit. Work and leisure become two sides of the same Taylorist project—the industrialization of everyday life. One writer, Vincent Mosco, has punned this reality as the "pay-per society," which he sees as a system that now charges for the most appealing forms of information and entertainment on a per-use basis. Special media events that used to be included in every television owner's viewing options, with advertising costs indirectly added to the price of products advertised, are now most likely to be available either on premium cable channels, pay-per-view programs, video rentals, or interactive computer home shopping. Much of the information that used to be available free in depository libraries as a benefit of the public tax collection must now be paid for in the form of expensive government documents or fee-based data bank access. Deregulation and the end of the universal service in favor of ability-to-pay principles opened up opportunities for telephone companies to make greater use of area-based calling charges for small-time users and more attractive bulk rate packages for corporate-scale customers (Mosco, 1988, pp. 4-10).

The Clinton administration, despite its public relations efforts on behalf of an "information superhighway," would not venture into regulatory turf so as not to offend "government downsizing" conservatives. Reliable indicators already suggest that computer services, such as America Online and CompuServe, are moving toward an ability-to-pay pricing structure and shunning periodicals that do not bring in large revenues, especially politically progressive publications that are not in vogue or do not have sufficiently large subscription bases. A director for on-line publications for CompuServe, for example, acknowledged this: "We're in business to make money. . . . The overall tone is going to be: You have to sell me on you" (quoted in Ness, 1994, p. 25).

Erik Ness also noted that Prodigy has kicked people off their service who contest proposed rate hikes and censored political discussion of which it does not approve. CompuServe rents its subscriber lists to businesses interested in its users' demographic profiles. In one of its trade publications, America Online boasted of its "upscale" subscriber list as a way of attracting new members "who pay up to $200 a month to enjoy hundreds of entertaining and informative services" (Ness, 1994, pp. 24-25). Presumably, few would come from the ranks of the working class or the poor.

All of this consumer behavior is monitored for future marketing and sales opportunities. Mosco (1988) writes, "How much money you have, what you like to buy, your views on capital punishment, your preference for president or for laundry detergent—the new technology is used to draw detailed marketing profiles of individual households for what is called (using appropriate military language) precise targeting of potential buyers" (p. 6). During the 1980s deregulation and merger mania, corporations, some with limited or no previous experience in mass media, began to soak up communication companies in multi-billion-dollar financial deals. These new media giants did understand the strategic importance of information enterprises in the current stage of capitalism. Many are transnational and have opened satellite, computer, telephone, data, and broadcast businesses abroad to create global economies of scale in information, advertising, and entertainment, merging existing markets in the West with newfound audience potentials among the affluent in the industrial and Third World countries.

Even if only one tenth of the Indian and Chinese people were part of the affluent class ready to buy into the material culture of the West, that would add more than 200 million new consumers for transnationally produced goods and services, including computers, television programs, video games, and other communication and information pastimes. The enchanting prospects of such new riches induce the industrial producers of these information and entertainment commodities to standardize mass market products so as to overcome the cultural resistance of far more heterogeneous user groups. This effort had led to more action-oriented programming (e.g., sports, war coverage) and simpler story lines in television and film; more universal symbols in video games; more generic, insignificant, and less critical newswriting; the use of mass marketing conventions in advertising; and avoidance of progressive social values (that occasionally creep into American media) so as not to offend conservative government gatekeepers in authoritarian state systems.

In early 1995, the *International Herald Tribune* (owned jointly by the *Washington Post* and *New York Times* companies) publicly apologized to the Singaporean government for publishing a commentary by an American academic teaching in that country that had suggested some "intolerant regimes" in Asia make use of a "compliant judiciary" to weaken domestic political opposition. A

Singaporean court ruled that both the newspaper and the writer were guilty of contempt against the state even though Singapore was never mentioned in the article. This brought about $25,000 in penalties and costs. The ruling also opened the way for a libel suit and a stiff financial penalty against the defendants by the former prime minister, Lee Kuan Yew, who has already won several major court awards against the press (Shenon, 1995).

Despite their occasional critical reports on right-wing police state activity, however, foreign media operating in such countries more usually provide information and entertaining diversions that do nothing to question the status quo. Although nation states may legitimately impose restrictions on the behavior of foreign corporations, including foreign media, acquiescence in the violation of universal standards of human and civil rights by the Western media, including the principle of independent judiciaries, amounts to tolerance of censorship. The *International Herald Tribune* is based in Paris but prints an Asian edition out of Singapore. Its public apology to a totalitarian regime (a demeanor it never assumed in communist party–run countries) and deference to state control effectively puts the newspaper in complicity with political repression. It does so because Singapore offers a technically suitable and cost-effective printing, news-gathering, and circulation center for regional publication, an arrangement the Post–Times joint venture will not turn down.

PERIPHERALS AND PERIPHERY CULTURES: ISSUES OF CLASS, GENDER, RACE, IMPERIALISM, AND TECHNOLOGICAL DEPENDENCY

One central objective in the industrialization of technology has been the continual transformation of work and work relations, discussed with respect to Taylorism and Fordism in previous chapters. The electronics industry not only brought changes in the means of communication but also led to automation of production technology, such that with computer-aided design, the manufacture of a "different type of part requires only new [computer] instructions instead of new machines. This flexibility not only extends automation to a wide range of new occupations and workplaces, but makes possible a thorough reorganization of any workplace in which it is applied" (Shaiken, 1980, p. 3). As automation increases, the worker is made into "a monitor rather than an active participant" (Shaiken, 1980, p. 12).

Digital and laser-based technologies have had enormous impact not only on the downgrading of skilled work but also on unskilled work. At checkout counters, workers mindlessly and monotonously move items across bar code readers or punch picture-coded electronic cash register buttons—performance skills that could be executed by monkeys. Where skill, status, and reward are so reduced, can personal alienation and its social consequences be unrelated to such technological designs? If these are the kinds of jobs that await America's new workforce, what incentives are there for youth to acquire advanced education or for those without the means for advanced education to assume positive attitudes toward legal work opportunities?

Other societal concerns about electronic technology are evident. Visit any video game arcade, and it is patently obvious that it is entertainment designed mainly for teenage and young adult males (the home video game Doom, a huge seller among males of all ages, is based on gratuitous violence against evil "demons"). One academic observer, Marsha Kinder, finds that such discrimination against girls occurs not only in immediate play area access but also over the long term because video games provide an entry into the world of computers (Starks, 1993). Arcade video games largely deal with blunt power relationships and simulated violence, themes of aggression that are adapted to the common socializing and interest of boys, even where females are featured as warrior types (as in one version of Mortal Kombat). The reluctance to develop thematic content more appealing to girls may partly represent the lack of imagination of the largely male video game designers but may also relate to the assumption that higher thresholds of violence bags more quarters than less aggressive content.

The formula of using vicarious brutality in video game design, now enhanced by more lifelike graphics, appears to be moving to more intense levels. In one game, Night Trap, a "woman in flimsy negligee" is dragged off from her bedroom by three men, who restrain her while a fourth man "plunges an electric drill into her neck." When questioned in the U.S. Senate about its moral values, an executive of Sega Enterprises, which marketed the game, defended the video depiction of sexual violence as intended only for adult audiences, suggesting that industry responsibility for encouraging perverse moral standards ends when the consumer is age 18 or over (Puga, 1993, p. 1). Public concern with video violence parallels similar kinds of criticism of films, television, recorded and radio-transmitted "gangsta rap," and other media that incite attacks on women and police.

The mass media do little more to highlight women as role models, although the improved status of women in print and broadcasting today, though not consistent, has been at least partly reflective of the general change in social attitudes about gender. One 1993 study found that women are referenced in the major news magazines (*Time, Newsweek,* and *U.S. News & World Report*) only 14% of the time and captured 34% of bylines and photo appearances—low, but still an improvement over data from a few years earlier. Results were similar in a study covering a wider range of 20 newspapers. The latter study found a parallel in American television coverage of the 1992 Olympics, for which women received far less attention than men even though they brought home more medals (Jon Bekken, communication via e-mail, February 21, 1993).

As two well-placed businesswomen found (Dawn-Marie Driscoll and Carol R. Goldberg), women's roles in society remain greatly underrepresented in the mass media. One study they cite of the ABC news program *Nightline* and another of the Public Broadcasting System's (PBS) *MacNeil/Lehrer News Hour* indicated that, over long periods of investigation, the two shows, respectively, featured women as experts only 11% and 13% of the time even though women-held government and corporate management positions have risen dramatically during the past 15 years (41% of all business managers by 1991, according to one survey). Women are also overlooked as business experts in the leading daily newspapers—and treated, in fact, only slightly better in stories written by women. The corporate outlook of media retains a patriarchal ethos (Driscoll & Goldberg, 1994).

THE PSEUDO WORLD OF DISNEY

One pillar of American mass culture, the Walt Disney Company, has often been one of the purveyors of false and distorted images of women, as well as of peoples of Third World origins and, for that matter, of much of history. The Cinderella of the classical written tradition is strong and rational; in Disney's animated version, she is passive and helpless. Snow White in the European tale is full of affirmative and assertive qualities; in Disney, she is acquiescent and simpleminded (Meehan, 1988, p. 103). In two of Disney's more recent animations, *Beauty and the Beast* and *The Little Mermaid,* the lily-white female denizens of valley and ocean are treated as more courageous and independent-minded but ultimately no less submissive to the archaic notion that marriage to a handsome white male aristocrat is women's most cherished fantasy.

Disney comics have always been dominated by a patriarchy of ducks, mice, dogs, and other animals (stand-ins, of course, for their human male counterparts). The organization's historical rightist orientation extended to its hiring practices as well. Up to the 1970s, at least, Disney organization management had very few Catholics and Jews and no women or African Americans (Kunzle, 1975, p. 21n). The progenitor of the empire, Walt Disney (1901-66) himself was known for his right-wing, anti-Jewish, and anti-labor politics, his cooperation with the McCarthy era House Un-American Activities Committee, and probable role, like that of Ronald Reagan, as a Hollywood informer for the FBI. According to one biography of Disney, "His studio was notably lacking in Jewish employees [quite remarkable given the level of available Jewish talent in Hollywood], and at least once he presented a fairly vicious caricature of the Jew on screen," which made him useful to extremists in the pre-World War II years who were portraying Jews in Hollywood as part of an "international Jewish conspiracy" (Schickel, 1968, p. 95).

Disney's animated film *Aladdin* is full of negative verbal and visual stereotypes and physical caricatures of Arabs, whereas the heroic "Arab" characters in the story line are made to look remarkably European. Disney's Donald Duck comics have long caricatured Third World peoples as ignorant noble savages and the "all-American" Donald as their fearless white liberator. Such images are little more sophisticated than the fantasy-filled but deeply condescending letters of Christopher Columbus to the Spanish throne, describing fecklessly naive indigenous peoples he encountered in his imagined Orient—in fact, the Caribbean. The difference some 500 years later is that although relatively few people have seen Columbus' letters, the Disney comics have circulated and been read by millions of people. By 1962, Disney comics were already selling 50 million copies per month in 50 countries and 15 languages (Kunzle, 1975, p. 14n).

The public relations office of the famous theme park Disneyland, opened in 1955, advertised that "you can encounter 'wild' animals and native [African] 'savages' who often display their hostility to your invasion of their jungle privacy. . . . From stockades in Adventureland, you can actually shoot at Indians" (Kunzle, 1975, p. 21). Disney himself was proud of his "Indian Villages" at Disneyland, describing them as "a forest of tepees, built exactly as the redskins made them, plus a tribe of Indians. . . . They're friendly, though, and will perform their tribal dances." A visitor to the theme park "could peer down the island" from above "Fort Wilderness" and witness "the burning cabin of some poor pioneer who was attacked by the redskins," Disney wrote in the 1950s. And if you needed respite from all that fear and excitement, "Frontierland" offered relief at its main restaurant, "Aunt Jemima's Pancake House," where an African American "Aunt Jemima" would stand all day signing autographs (Weiner, 1994, p. 133).

Millions of school-age children (and adults) learned about the origins of the U.S. war with Mexico through Disney's television fantasy filters. Disney accorded heroic image and status to people like Davy Crockett, who, more serious historians insist, never wore a coonskin cap and tried to surrender rather than fight at the battle for the Alamo. Disney's historical rewrite made noble characters out of drunkards, drug addicts, and land grabbers who organized the secession from Mexico. Disney also made an icon of Jim Bowie, an Indian hunter, who, according to at least one finding, may have led the command at the Alamo for the gold, stolen from Apaches, that he stashed beneath the fort—a historical claim credible enough to convince the city council of San Antonio in 1994 to give digging permits (Myerson, 1994). Texas would later be separated from a weak Mexico (itself independent from Spain only in 1821) by its non-Hispanic settlers before being added as a U.S. slave state in 1845 and inciting a national border war with Mexico from 1846 to 1848.

In Disney's monumental re-creation of the past, which included plans to add, by 1998, a $650 million, 3,000-acre theme park of American history, "Disney's America," there is no Realityland. This venture was not regarded as a harmless diversion by serious American historians concerned about how the myth master Disney would treat sensitive subjects like the Civil War, the Vietnam War, the issue of slavery, and the experiences of Native Americans. Pulitzer Prize–winning historian David McCullough, joined by more than 30 other historians and authors, called Disney's America a "plastic, contrived history" and a "sacrilege" that the park would be built in the area of 13 historical sites and 16 Civil War battlegrounds, including Bull Run (the first major battle of the War) and Manassas. Environmentalists also opposed the idea out of concern for the added pollution, already high in northern Virginia, that the project and surrounding development would bring. Disney demanded $160 million in state assistance to make it happen. This time, the Disney people did not get their way ("Disney Drops Its Efforts," 1994; Hohler, 1994; Janofsky, 1994a).

Despite Disney's $400,000 lobbying effort, pressure from historians, environmentalists, preservationists, and from real estate interests concerned about protecting residential land values succeeded in forcing the company to withdraw or relocate. Some 1,200 protesters had marched to Washington D.C., calling the Disney proposal a "Trojan Mouse." The result means, however, that another site in Virginia, or possibly another state, will be targeted. Had Disney not demanded as much public assistance from Virginia state and county officials who otherwise appeared anxious to comply, the company would have demonstrated that, indeed, *everything* is for sale, including a nation's historical memory (Peirce, 1994).

Within the inner chambers of Disney, the most tangible reality is its growing wealth. The company has 16 offices in 45 countries and is bent on further expansion. By 1993, its combined divisions (e.g., 4 theme parks, films, 400 Disney stores, consumer products, licensed products) produced $8.5 billion in revenues. In 1994, Disney was concurrently running a cable station (Disney Channel), a television station (KCAL in Los Angeles), a television programming division (*Home Improvement, All-American Girl, Boy Meets World, Ellen, Thunder Alley, Mike & Maty* [daytime talk show]; *Hardball, Blossom,* and *Empty Nest*), and three movie studios. In *Lion King,* it had one of the biggest grossing films of all time while it was concurrently running the Broadway play *Beauty and the Beast.* Disney's labor practices in Haiti and other countries in producing its various clothing lines earned it the opprobrium of being one of the ten worst corporations in 1996 (Mokhiber, 1996, pp. 11-12).

Disney contacted CBS Inc. in 1994 about the possibility of buying out that media empire. CBS is a major television network that has about 200 affiliate television stations, 7 stations of its own in major urban markets, a television programming division, television licensing agreements with more than 300 partners in 95 countries, and a videocassette marketing operation (Fabrikant, 1994).

Disney did not get CBS but, instead, captured a bigger plumb from Capital Cities, the world's largest broadcasting network, ABC, for the acquisition price of $19.2 billion, making Disney-Capital Cities/ABC the world's most costly media conglomerate. Only the merger of the tobacco and food conglomerate RJR Nabisco by the Wall Street investment firm Kohlberg Kravis, Roberts & Co. in 1989 for $30.6 billion was bigger. With ABC, Disney gained control of its 225 affiliates, 10 television and 21 radio stations, plus the cable channels ESPN, Lifetime, and Arts & Entertainment. Disney also took over ABC's seven daily newspapers, shopping guides, specialty books, and several magazines (Ramstad, 1995). Deregulation has allowed greater concentration of the media than ever before, akin to the U.S. "robber baron" era of industrial combinations in the late 19th century.

The Disneyland theme park located just outside Tokyo draws some 300,000 people yearly. After a nearly $1 billion loss in its first year of operation at its Euro-Disney outside Paris, Disney began to reveal some of its "moral pragmatism" by discarding its puritanical no-alcohol policy at the theme park. French unions sued Disney for violations of civil liberties when the company tried to impose on its prospective employees weight and height restrictions and prohibitions on beards, mustaches, long hair, and certain types of jewelry and, for women, short skirts, mascara, and colorful undergarments (Ritzer, 1993, p. 111). The company expects to cash in on its global reach by expanding its Disney store chain (268 in 1994) throughout Europe, Japan, and East Asia (Sims, 1994). It should not be surprising if sporting mouse ears becomes trendy at the Great Wall.

In the 1990s, Disney expanded into the world of children's formal education. The Edison Project is a brainstorm of Christopher Whittle, who founded the satellite-delivered Channel One advertiser-sponsored current events program, 10 minutes of "news" sandwiched between commercials for candy, soft drinks, and video games, for some 40% of the country's high schools. Whittle had earlier distinguished himself as a student at the University of Tennessee by creating "Time Savers," a kind of Cliff's Notes for courses taught at the university and for students who did not care to read the classics or even attend class. In late 1994, Whittle negotiated a deal to sell Channel One to a publishing house, K–III (*The Weekly Reader, New York,* high school videos and films, school book clubs, and school newspapers), for $250 million. K–III planned to add commercials, as other corporations were rushing in to take over remaining niches in the school curriculum where product placement opportunities could be pursued (Spillane, 1994, p. 600). The plan of the Edison Project was to create by 1996 a national chain of profit-driven private secondary schools with supposedly "advanced" curricula.

Disney, led by its chair, Michael Eisner, negotiated with Whittle to take controlling interest in the Edison Project and the lucrative Channel One—and away from its present owners, Whittle, the media conglomerate Time Warner, the electronics giant Philips, and Britain's Associated Newspapers Holdings. Control of the project would give Disney access to a potential audience of 8 million, with enormous opportunities for sale of education software and the vast line of other Disney character (e.g., Mickey Mouse) paraphernalia. Other interested stakeholders included Apple Computer, which would supply the schools' personal computer needs and thereby corner a larger market share, and McDonald's, which is interested in running the schools' cafeterias—and make its diet of fat-rich french fries and hamburgers a staple of the kindergarten-through-grade-12 set (Landler & Grover, 1993, pp. 22-23).

With expanding commercial sponsorship of education, what happens to the treatment of serious political, social, and environmental issues in the curriculum? Will non-unionized teachers in a privatized environment be free to teach young people about toxic wastes for which industrial corporations are responsible if an electronics corporation that uses gases, solvents, acids, and other dangerous chemicals in its production process is a sponsor or donor to school programs? Will students receive frank information about the nutritional quality of fast food, when McDonald's is a financial backer? Will the 12-minute Channel One news programs (including 2 minutes of commercials) that Whittle marketed to schools broaden children's understanding of racism and other forms of institutionalized discrimination in American society? And although privileged young people may get to attend these private schools, what level of official concern will remain for public schooling and for the 50% of both high school and college students who have to work to get by, thereby neglecting their education?

In more recent years, the Disney company has partially rehabilitated its discriminatory history and image by extending health insurance benefits to partners of gay and lesbian employees. In an industry in which gays and lesbians contribute so much talent, Disney had little other choice. Nonetheless, this move cast the company in a visible conflict with religious extremists within the Southern Baptist church who demanded a boycott of Disney theme parks and all of its products. It remained to be seen whether Disney would back down and rescind insurance coverage to its employees in deference to the religious right or, in the light of its global material interests, take the more pragmatic approach of embracing all groups on the basis of their industrial and purchasing power.

On the teaching of race and ethnicity, television in the United States has not had a distinguished record. With few exceptions, most black characters on American television are cast as pimps, prostitutes, thieves, murderers, drug users and pushers, nurturing mamas and nannies, or as simpleminded buffoons, "bugging their eyes, pouting their lips and looking stupid amid the sounds of white laughter. Any serious issue within the plot usually ends up so white-washed the initial purpose must have been to take the racial edge off in order to make it palatable to a white majority" (Spigner, 1994, p. 8). Simpleminded stereotyping captures the imagination of white sponsors more often than does intelligent scripting. Hence, the refusal of commercial networks to pick up the excellent television series (later carried by PBS) *I'll Fly Away,* about life in the south as seen from the perspective of a black housekeeper in a white household, who joins the cause of the 1950s civil rights movement.

Although most programs on commercial television do little to educate or incite critical thinking about any issue, the lack of attention to peoples of African, Asian, Latino, Native American, and other non-European ancestries has more to do with lack of representation within the media establishment. The *New York Times* reported that most of the 1,700 daily newspapers in the United States do not have a single African American, Latino, or Asian American journalist and that of the 5% of these ethnicities in the professional media, most are employed by minority-owned companies (cited in Parenti, 1993, p. 12). Entry for black television newscasters in the 1960s and 1970s initially focused their talents on sports, crime, and civil rights issues, but in the past decade the influential local news anchor roles have been opened in many cities to African Americans, as well as to Asian Americans, Latinos, and women. Media ownership seems to be the key variable on how blacks are portrayed. A journalist's study of 3,200 black neighborhood news stories in the Boston media area revealed that African Americans were portrayed in negative stereotype 85% of the time, whereas black-owned media largely focused on educational and entrepreneurial aspirations of the same communities and how the residents tried to get beyond government neglect and poverty. The two neighborhoods studied constituted 7% of the city's crime, yet 59% of the coverage of these communities was about black criminality (Lee & Solomon, 1990, p. 243). The fact that most news agencies have created "glass ceilings" for African American media professionals suggests that the white power structure is holding back black progress and effective ways of dealing with the complex problems of a multi-ethnic, multi-cultural society.

Although the first half-century of commercial broadcasting offered little more than insulting or patronizing racial stereotypes (e.g., *Amos 'n' Andy, Beulah,* Charlie Chan movies, The Lone Ranger and sidekick Tonto), in more recent years a few commercial television successes have portrayed African Americans in roles with which most Americans can identify. The conservative 1950s and early 1960s were a near "whiteout" for black television and film talents (with a few breakthrough exceptions like the Sidney Poitier roles). This was followed by a period of more visible but nonetheless degrading, stereotyped characterizations (*Sanford and Son, The Jeffersons,* and the "blaxploitation" movies such as *Shaft* and *Superfly*), as well as some more challenging and positive film roles (e.g., *Sounder, Lady Sings the Blues*).

The civil rights movement of the 1960s had reverberations in the mass media. Starting in 1965, Bill Cosby introduced in the popular *I Spy* series a significant shift in television's treatment of African Americans, allowing a talented black actor to take on a costarring and more fully developed character assignment, free of insulting or patronizing racial images. In the 1980s, *The Cosby Show* was a further attempt that probably succeeded in the "normalizing" of black home life, although it did so by casting the Huxtable family within an affluent, professional-class household and lifestyle. Nonetheless, the program writers managed to introduce story lines of unselfconscious friendships between the Huxtable family and characters of other ethnicities that nudged the culture toward racial harmony.

African Americans, according to some studies, watch more television (77.3 hours per week) and tend to rely on it more for information than do European Americans (50.1 hours per week), even though it minimizes their cultural status, thereby intensifying black people's sense of estrangement in white society (Barnet & Cavanagh, 1994, p. 156; Johnson, 1993, p. 272). Poverty and limited institutional facilities (quality schools, museums, library collections) for commemorating the contributions of African Americans to the common American experience narrow the cultural alternatives to television for most black youth. A few children's programs on public television (e.g., *Sesame Street, Ghostwriter*) have promoted images of casual inter-racial friendship, but this is otherwise exceptional on U.S. television. Marriages or intimate relationships between different ethnic groups, though increasingly common in U.S. demographics, are isolated on television, as if producers and advertisers consider it still too controversial—or perhaps believe that it requires scripting more intelligent than sponsors are willing to stomach.

It is widely assumed that computers will become increasingly important in gaining access to information. For the majority of African Americans who do not own computers, this is likely to mean another barrier that reduces their opportunities in the workplace and thus in earning capability. Although at some inner-city schools black students get hands-on experience in learning to use and program computers, most school systems in poor neighborhoods do not provide either the equipment or the staff training needed, compared with more affluent districts. Essential technological change, as occurred in the post-Civil War era, radically altered the relationship of people and their tools. In the age of computers, the permeation of technology into the private areas of life goes deeper than any time previously, and the consequences of computer illiteracy are likely to be profoundly destabilizing for individuals, for groups of citizens, and for society as a whole.

GLOBOCOPS: THE NEW PARTY LINE

The early use of communication technology for police and military surveillance, discussed in Chapter 2, was but a part of the huge arsenal of the nation state for the social control of citizens. In the United States, the origins of police control go back to 17th-century "night watch" forces in Boston and, later for the protection of private property, in other urban areas. In the south, plantation interests formed slave patrols; in the northern and western states, militias evolved for the defense of farmlands against Native Americans. By the late 19th century, police were actively being employed, often brutally, against workers, unions, and the working class in general, especially in large factory towns and cities. Their numbers grew parallel with the expansion of cities and industrial capitalism (Hirota, 1983, pp. 25-26).

America's "permanent war economy" (Melman, 1985), driven by the fear of a return to economic depression following World War II, brought spin-offs from military research and technology to the area of police security, with increasing paramilitary forms of social control. The police riot in Chicago under the hard-line mayor and Democratic Party boss Richard J. Daley during the 1968 Democratic National Convention in that city, following his refusal to issue a parade permit or allow demonstrators to sleep in public parks, gave vent to a violent backlash against the peace, new politics, and countercultural elements of the "Movement." This virtual "second front" in the Vietnam War led to the Kent State and Jackson State massacres 2 years later. In 1968, there were also massive demonstrations against the war and against political regimes in Paris,

Prague, and throughout Europe, while the Cultural Revolution in China and other political confrontations spread to the rest of Asia and the Third World.

Within this polarized political atmosphere, police forces in the United States and their natural allies, including those industries that profited from the sale of various military and police matériel, favored the diffusion of Vietnam War technologies, helicopters, infrared detection equipment, armored transports, S.W.A.T. teams, and various other "flexible response"-type strategies. Congress appropriated $1.75 billion in 1973 for the Law Enforcement Assistance Administration under the Justice Department that gave it "a repressive capacity unparalleled in history" (Hirota, 1983, pp. 26-27). Where real threats to property, community, or state security do not exist, the sheer magnitude of the costs of new equipment, such as modern and fully equipped police vehicles and computers for regional and national police records, actuates a corollary to Parkinson's Law—namely, that the level of surveillance rises to fill the computer space and expense allotted to it. In Britain, for example, use of the Police National Computer, started in 1974, introduced a heavy cost factor that forced police to meet a higher quota of spot checks on cars and their drivers without resulting in better crime-fighting performance, but rather more inconvenience to motorists (Street, 1992, p. 101).

A National Law Enforcement Telecommunications System allows police forces to hook up to federal computer terminals in various agencies, such as the Department of Justice, Drug Enforcement Agency, Treasury Enforcement Computer System, Postal Inspection Service, and State Department. The FBI's National Crime Information Center alone holds some 17 million records. Germany's Central Police Computer, called INPOL, has one of the tightest systems of state-run computer matching and surveillance of citizens, assisted with data from other state institutions such as Lufthansa Airlines, with plans to introduce a national ID card system, reminiscent of the security system set up by Hitler (Lyon, 1994, chap 6). Although the idea of such a network for tracking people who have broken the law may give comfort to some out of fear of criminal anarchy, it also gives rise to fear on the part of others who conceive of its authoritarian power to control legal dissident behavior. In October 1994, the passage of the Communications Assistance for Law Enforcement Act requires that telecommunications carriers and manufacturers "design their networks to optimize law enforcement's wiretapping ability" (Cassidy, 1995, p. 54).

There is a fine line between the legitimate role of police in the defense of people's security and its potential function in the service of repressive ends. Those who have experienced the latter in their pursuit of political and economic justice and democracy and those who know the history of police power in support of statist and private corporate interests have good reason to fear unbridled uses of security technologies, many of which were created in the context of fighting imperial wars. Computers are but one item in the armory of police technologies recently developed for public order, which include CS gas, water cannons, plastic bullets, riot shields, and others (Street, 1992, p. 52). The technification of police power distances people into broad categories of suspicion, similar to the ways that automated air wars since Vietnam have removed the soldier from seeing the "enemy" and thereby made violence more impersonal and less palpable.

Another real danger of police dependence on computers for information, apart from the potential for intentional abuse, is the reliability and authority of the instrument itself. One writer, Ian Reinecke, describes the case of a West Indian youth being arrested by London police on the basis of a computer record. After arrest, charge, conviction, and imprisonment, during which time the youth continually insisted on his innocence, the court would not listen. Reinecke warns that computers, especially police computers, have a higher standing in the legal system than most individuals (especially those already distanced from the arbiters of justice by race, class, ethnic, and gender prejudice) and are rarely disputed. After 3 months of incarceration, the police admitted that the West Indian youth had been wrongly imprisoned on the basis of mistaken identity, having had the same name as another youth they had been seeking (Reinecke, 1984, p. 222). Technology often provides the powerful with additional wedges of social space between themselves and the ranks of the powerless and the underclasses.

WHO'S ON-LINE, WHAT'S ON-FILE?

Compared with the popular knowledge and use of computers of some 15 or 20 years ago, the changes in everyday ways of doing things affected by electronic communication and information systems have been many. Automated tellers and checkouts, video game amusements, time-managed options for viewing videocassette films or taping television programs, improved sound quality of CDs, portable and automobile audiocassettes, facsimile transmitters, "virtual reality" types of entertainment, camcorders, and videotex information access and services are but a few artifacts that were not available to most people in the 1970s. And yet, it would be hard to make a persuasive argument that this new electromagnetic landscape, as it were, has made life easier or more enjoyable for most people in "developed" countries, even if there may be a few more "conveniences."

Many studies as well as anecdotal evidence suggest that people in the United States feel more pressed for time and more overworked than ever, which should not be surprising, considering that they work 320 hours per year (2 months) more than their counterparts in Germany or France (Schor, 1992, p. 2). Were computers developed for the purpose of making life easier and more interesting for most people? Or were they intended to time-compress more work out of people for the sake of greater productivity and profit without necessarily passing on these dividends to the workforce or the public? Indeed, Americans have recaptured the lead as the most productive workers in the industrial world and, at the same time, as among the poorest paid, with real wages (adjusted for inflation) continually declining. Are we becoming more liberated by technology or, as one analyst sees it, turning into a society of rulers and helots (Cohen, 1987)?

One report suggests that popular uses of the computer, aside from video games, has actually been very limited and somewhat of a failure in terms of the expectations of its promoters and the trade press. Despite great cost reductions over the years, two thirds of U.S. households still did not own a personal computer in 1993, and of those used in residences, most were by "students and people overworking at home." Of those households not using them for work brought home, only 17% had computers. This is a very low rate of diffusion for what has been touted as the hottest family item since television and the VCR (Saltus, 1993a).

The head of the Boston Computer Society, Bob Grenoble, admits to preferring a pen, a notepad, and a small calculator to a computer for most of his daily transcriptions, employing his home computers primarily for word processing (a kind of glitzy typewriter), games, and occasional accounting. Even the expected productivity gains in business applications because of its supposed efficiencies have been disputed in at least one major MIT business school study, which showed that the billions invested have little to show for it. Whatever impact computers have had is clearly a class phenomenon, as nearly two thirds of households with more than $100,000 income have home computers, the percentage steadily declining with lower income (e.g., one fifth of households with $20-30,000 income; below that level, only 11% of households; Saltus, 1993a). The more valuable applications of desktop computers are in business, government, and military offices.

If most people have limited access to or need for computers, this is not the case of empowered individuals and groups within the political economy. Technology is used to reinforce and reify the existing political and bureaucratic power assignments in society because, as Oscar Gandy argues, empowered institutions preserve their elite status by discovering the dysfunctional behaviors they have been put in charge to control. The classification schemes that institutions adopt, he says, ultimately define the agenda of what needs fixing. In the southern United States, for example, for 150 years it was virtually unheard of to find legal cases of white males raping black women, not because violence was not perpetrated on black women but because of the institutionalized prejudice that informed the practices of the legal system (Gandy, 1993, pp. 16-17). With computerized infrastructure at their disposal, institutions have more sophisticated means to transform information into a form or component of capital and into a more efficiently organized production-consumption cycle.

Financial Banks, Data Banks, Policy Groups, and Think Tanks

Data banks are systematic and generally large collections of stored information on various subjects or specialized database files, usually with restricted access. Several thousand databases were already on-line in the United States by the late 1980s, and by 1990, data flow cost the top 100 telecommunications users alone between $1 billion and $20 billion annually (Hamelink, 1994a, p. 32). One of the biggest on-line financial information services is Reuters, the British international news agency, founded in 1850, that parlayed its grip as a global news network into a financial bonanza with the opening of its Monitor Service business and money market information facility in 1973. The establishment of floating currency exchanges in 1971 created a windfall for the company as it began to offer both on-line currency and stock quotations. By 1985, Reuters' on-line news and financial services, valued at more than $500 million, accounted for 80% of its revenues. Its network of overseas correspondents encouraged the company to employ its global news network in the service of its more lucrative financial market concerns (Estabrooks, 1988, pp. 51-52).

Other major media enterprises that positioned themselves in the new business information environment included McGraw–Hill (one of the biggest book publishers and the owner of *Business Week* and 29 other specialty magazines), which went into electronic financial and data bank services; Reader's Digest, with the largest monthly circulation in the world, took over The Source, an electronic information and bulletin board service that contains a home shopping catalog; Dow Jones (*Wall Street Journal, Asian Wall Street Journal, Far Eastern Economic Review*) took on electronic news, airline flight information, and on-line stock quotations; and Knight-Ridder (newspaper publisher, cable station operator) started up an electronic business information division (Estabrooks, 1988, pp. 49-50).

Among the earliest data banks were credit bureaus, which began to operate in the 1890s to issue credit reports on individuals interested in obtaining loans, insurance, or other financial service. *Credit bureaus* are vast repositories of information on citizens, the vast majority of whom have no idea that such records even exist. By the end of 1989, credit bureaus held about 500 million credit files and issued some 450 credit reports in 1988 (Rosenberg, 1992, p. 203). The largest credit services include Dun & Bradstreet, which in 1984 purchased the market and media research firm A. C. Nielsen, and TRW, a major military contractor that builds spy satellites, which, among its other information services, processes U.S. Postal Service change-of-address forms.

Although citizens have certain rights with respect to their credit records (e.g., the Fair Credit Reporting Act of 1970), such as correcting false and biased information and, to some extent, limiting the distribution of personal information held by private and government sources, the proliferation of citizen data on present and past behavior, taste, whereabouts, friends and associates, forms of recreation, reading and spending habits, political preferences, and almost every other aspect of one's private life has never been greater. In September 1989, *Business Week* ran a story on privacy based on the successful effort of one of its editors, using his home computer, to receive personal information from a high-powered credit bureau about the financial behavior of then Vice President Dan Quayle and Illinois Congressional representative Richard Durbin. The results of the data search were relatively unobtrusive, compared with what can be learned about ordinary citizens. It points to the de facto noncompliance of the government and private sector in safeguarding the Fourth Amendment to the Constitution, which is supposed to guarantee citizens protection from illegal searches and seizures (Rosenberg, 1992, p. 205).

The methods employed by Nixon in the Watergate scandal would have to be regarded as adolescent by today's technical standards in political sabotage. With computers, electronic penetration can often do the job more efficiently. In 1990, state Republicans in the New Jersey Assembly used the skills of one of their staff members to break into the computer files of their Democratic rivals in a case that was subsequently covered up by both sides. The Republicans feared exposure of their involvement, and Democrats feared exposure of what the files contained (Kerr, 1990). This incident revealed how easily electronic spying can be used for political blackmail, dirty tricks, and further distortion of the electoral process.

Many ordinary citizens also fear government electronic surveillance of their personal lives. Yet, the degree of eavesdropping carried out by commercial enterprises is far more pervasive, much of it willingly subscribed to by consumption-conscious citizens forever in search of new products and gadgets to buy. With Caller–ID service in place, for example, commercial establishments can turn an innocent telephone inquiry into a "reverse directory," locating the person's telephone number and address and selling the data to a mailing or telemarketing list, which will lead to unexpected solicitations.

As *Business Week* put it, "Companies are collecting mountains of information about you, crunching it to predict how likely you are to buy a product, and using that knowledge to craft a marketing message precisely calibrated to get you to do so" (Berry et al., 1994, p. 56). Credit card transactions, mail and telephone surveys (often not represented as such), cash register receipts, warranty cards, participation in "lucky drawings," magazine subscriptions, mortgage payments, hotel registrations, even a driver's license, a marriage license, a car registration, or a hospital delivery starts the trail by which sales-hungry enterprises pinpoint potential consumers and flood them with lifestyle appeals constructed with the aid of computer profiling. You can run, but you can't hide from the marketing bloodhounds.

Long-distance financial markets were amplified by some closely concurrent developments. The most important was the expansion of transnational capital, especially after World War II and the international regimes that were set up in the wake of that horrendous conflict to promote global trade and investment (e.g., World Bank, IMF, GATT). Global capitalist actors, the transnational corporations (TNCs), created new demands on the major international trading currency, the U.S. dollar, which eventually led to a flood of greenbacks in the central banks of various countries, especially in western Europe and Japan. Alongside the global movements of capital, the restrictions on interstate banking and the separation of commercial and investment banking (e.g., the Glass-Steagall Act of 1933) withered in the 1980s. Telecommunications instruments, such as digital telephones, communication satellites, and computers, were developed as a result of the increasing demand for military and commercial information processing in calculating and sustaining national and imperial advantage.

The inability of the U.S. government to guarantee repayment in gold for remitted dollars through international banks forced the Nixon administration to take the U.S. currency off the gold standard and to allow its overvalued fixed exchange rate against other major currencies (e.g., British pound, German mark, French franc) to "float" in open market transactions. At the same time, government began to deregulate financial markets to encourage economic growth and attract foreign trade and investment. By this time, a web of communication satellites and oceanic cables were in place to facilitate such international currency and other financial paper exchanges on an instantaneous and 24-hour basis. With the proliferation of desktop computers, these processes were accelerated even further. Banks as a whole represent the biggest international user group for data transmission, and the latest in high-speed computers and telecommunications can be found on Wall Street (Forester, 1987, p. 225).

In a market-dominated society like the United States, public policy formation, public opinion, and ideological management rely heavily on information and interpretation provided by the mass media. Mass media are central institutions in normalizing the status quo and help certify those "experts" whom they call on to pilot public awareness on crucial domestic and foreign policy issues. The experts, in turn, come from the ranks of establishment circles, policy groups, and "think tanks" (policy discussion, research, and publication groups).

These include the establishment Committee for Economic Development, the centrist Brookings Institution, or wealthy, nonprofit foundations like the Ford Foundation, Carnegie Foundation, and Rockefeller Foundation, and increasingly from conservative or ultraconservative research organizations like the American Enterprise Institute (AEI), the Heritage Foundation, the Georgetown Center for Strategic and International Studies, the Hoover Institution, the Stanford Research Institute, the RAND Corporation, the Hudson Institute, and other groups that fund and publish the research of experts. Strategic to the formation of public opinion about business and politics, corporate- and foundation-financed think tanks like AEI provided many of the staff and ideas for the Reagan administration, along with the "militantly antigovernment-regulation, pro-big business organizations as the Institute for Contemporary Studies, the International Institute for Economic Research and the Institute for Educational Affairs" (Greenberg, 1989, p. 155).

One scholar on U.S. institutions, William Domhoff, who has written extensively on power in the United States, notes that "the media reinforce the legitimacy of the social system through the routine ways in which they accept and package events." Comments of business and government leaders are usually treated by the media at face value in such a way that "America's diplomatic aims are always honorable, corporate involvement overseas is necessary and legitimate, and revolutionary change in most countries is undesirable and must be discouraged whatever the plight of the majority of their citizens." Domhoff's view, however, is that the media's messages are often mixed but, more important, that their audience sweep matters little in terms of actual policy making but more in public opinion manipulation. The real power, he argues, takes place in the intersecting and complementary roles of corporate directors, foundations, think tanks, and government officials, who steer the public agenda in protecting the rules of private wealth accumulation. Expertise is drawn mainly from either the major think tanks or from academic circles funded by centrist or conservative foundations (Domhoff, 1983, chap. 4).

The analysis by Fairness and Accuracy in Reporting (FAIR) of the conservative political biases of *Nightline* and the *MacNeil/Lehrer News Hour* (discussed in Chapter 2) found that by "limiting the boundaries of acceptable views, the news media legitimize positions and certify 'expert' spokespersons while dismissing other positions and labeling their advocates as 'partisans.' Representatives from public interest groups and progressive academics are occasionally interviewed on domestic issues but almost never on foreign policy subjects. The researchers found that 60% of *MacNeil/Lehrer* guests were government officials and that the news program "serves as a veritable press agency for the views of U.S. officialdom—one that excludes the views of critics." Even though *MacNeil/Lehrer* is not formally "sponsored," it is conspicuously underwritten by TNCs; AT&T funded nearly half the news program's budget when it went to a 1-hour format in 1982 (Hoynes & Croteau, 1990, pp. 2, 3, 12, 13).

MacNeil/Lehrer (now called *Newshour with Jim Lehrer*) reflects no more and no less the policy elite view in the United States, even with occasional liberal trimmings. Who speaks on *MacNeil/Lehrer* echoes a narrow range of voices around the power configuration at the top, what Chomsky calls the "spectrum of thinkable thought," with the right-wing Heritage Foundation gaining a major presence during the Reagan and Bush years at the expense of the Brookings Institution. In foreign policy, most "outside experts" interviewed are likely to come from among elitist policy groups and think tanks such as the Council on Foreign Relations, which publishes the influential *Foreign Affairs* and runs small, insider discussion groups (of *Fortune* 500, government, military, and academic leaders), or the Foreign Policy Association, which does broader outreach through pamphleteering and sponsorship of international affairs programs on college campuses. In the economic arena, the most powerful groups are the Conference Board (founded in 1916) and the Committee for Economic Development, composed mainly of corporate leaders, who support research disseminated to business, government, and media, and is influential in placing members on the President's Council of Economic Advisors (Domhoff, 1983, pp. 85-95).

Interlocked with the communications corporations, the major industrial, business, and financial interests in the United States do not have to work hard at getting the attention of the mass media. In the business pages of the *Wall Street Journal* and other business publications, it is well known that much of the copy is simply composed and handed over by public relations departments. No conspiracies or backroom cabals are necessary for big corporations to exercise hegemony over public opinion; control is implemented at several levels of shared interests, even while intra-elite competition and dissent coexist. A widely cited 1973 study of the *New York Times* and *Washington Post* by Leon Sigal showed that 78% of their stories originated from government sources and that most of the others came from business contacts (cited in Herman, 1995b, p. 81).

A corporate elite consensus formed in reaction to Lyndon Johnson's Great Society programs. This consensus urged that a scaling back or at least a slowing down was needed against such public demands as civil rights of women, gays and lesbians, and ethnic minorities; welfare payments; public education; inner-city redevelopment; raising the minimum wage; reducing real unemployment; labor rights to organize; corporate social accountability; consumer protection; environmental and peace initiatives; judicial reforms; taxes on the rich; and other social projects. Even with a scattering of tepidly liberal dissenters in its ranks (none from the left), the mass media have been the main conduits in spreading conservative and elitist points of view.

Global Airlines, Assembly Lines, and Supermarkets

Telecommunications and the computer have been a boon to most large-scale industries but at the same time have raised the stakes of survival. In the airlines industry, which has expanded to handle increasing travel demand (400 to 500 million international travelers alone per year), computers have become central to the management of flight scheduling, passenger reservation systems, and streamlined services. And despite decreasing fuel prices, some of the biggest companies have crashed or been absorbed by the survivors. American Airlines' Sabre system for computerized reservations handles not only flight bookings but also hotel and auto reservations, tour packages, theater tickets, and other travel services, all of which bring fees to the airline-owned network. Foreign airlines complain that Sabre software discriminates in favor of American Airlines to their disadvantage (Rowan, 1993).

Another form of discrimination is latent in the Sabre software, and that is the kind practiced against travel agents. With every incoming call, agents are required to book a percentage of reservations, 26% at American Airlines, and to complete the written record of the transaction so that another call can come in within 13 seconds. Agents are electronically monitored the entire work period, which does not allow them to casually take breaks. Monitors secretly listen in to check the work pace and standard sales pitch of the agents. This eavesdropping creates pressure to manipulate the caller into a booking. At Air Canada, for a fee, callers can connect with less monitored special agents, thereby charging customers for the privilege of receiving relatively straightforward information (Garson, 1988, pp. 40-70). The Communications Workers of America estimates that some 15 million U.S. workers are surreptitiously monitored daily (Branscomb, 1994, p. 95).

At another tier of the communication labor force, computers and telecommunications have also facilitated the separation of production sites from marketing, financing, distribution, and retailing. For workers who manufacture the building blocks of the computer—the semiconductors, integrated circuits, and circuit boards—few will ever become citizens of the "information society." The physical conditions of the work, especially the contact with toxic chemicals used in the various stages of production of miniature components, frequently leads to anemia; heart irregularities; damage to internal organs, the spinal column, and the eyes; skin cancer; and pneumonia. Components production is also suspected of causing an immunity deficiency that resembles AIDS, as well as miscarriage, respiratory diseases, impotence, and various organ cancers. In the 1980s, the California Department of Health Services, suspecting contaminated groundwater from electronics production, found that Silicon Valley had 2.4 times the usual rate of miscarriages and 2.5 times the usual rate of birth defects (Perrolle, 1987, p. 37). And the compensation hardly makes it worth the risk: Entry assembly workers in 1987 made between $9,000 and $13,000 per year (whereas their counterparts in the Philippines, working under more severe conditions, earned about one tenth that amount; Hayes, 1989).

In the 1980s, IBM ran an ad featuring a look-alike Charlie Chaplin, the actor and socialist (1889-1977) whose classic film *Modern Times* derided the drudgery and alienation of industrial capitalism as seen through the eyes and experiences of one of its assembly line workers (played by Chaplin). In the IBM rewrite, the Chaplin impressionist is a happy, well-dressed pitchman for technology—the personal computer, which the ad calls "a tool for modern times" (Howard, 1986, p. 3). Today, fewer workers stand on assembly lines in grimy, poorly lit factories, doing the tedious chores portrayed in *Modern Times,* but the lines and columns of workers now punching data into computer keyboards, with lightning eye-hand coordination and their keystroke count recorded in internal memory to measure productivity, look much like the industrial workforce of old. For telephone operators, pressured to complete transactions within designated time limits, the "audio response systems" that give synthesized voice directory assistance merely speed up the pace of work, squeezing more call completions per minute out of each worker and saving the telephone company millions of dollars (Howard, 1986, pp. 20-21).

The infrastructure of communication and information technology has accelerated an international production system and division of labor, leading to the modern phenomenon in Western capitalist countries of the global supermarket. This means that Americans, the Japanese, and western Europeans can have fresh fruits and vegetables available throughout the winter as a result of agribusiness investments in the southern hemisphere and easy communications with the remotest farming sites in the Third World. These cheap labor havens for transnational investments bring the U.S. consumer the Mutant Ninja Turtle toys and Barbie dolls from cold, dreary workplaces in China, grapes picked by Chilean peasants, dresses sewn by rural Filipinas, coffee beans picked by Columbian *campesinos,* peanuts picked by stoop-backed Senegalese, and VCRs made in South Korean sweatshops. This international division of labor also means that the dirtiest and most polluting industrial tasks are often dispensed to exploited Third World workers and peasants, like U.S. oil-refining operations in the Caribbean, pesticide drainage in rural Thai villages from export agriculture, toxic ash dumping in Haiti and the Indian Ocean, and Japan's steel-sintering plants in the southern Philippines—not unlike the locating of industrial waste sites and incinerators in African American residential areas throughout the United States.

Locked into a desperate search for foreign investment in hopes of attracting new technology, jobs, foreign exchange, and exports, and often as a result of collusion between government and TNCs, many Third World leaders are willing to import polluting industries. For their part, transnational corporate CEOs and their government and financial allies express little reservation about exporting environmentally degrading forms of production. In an internal memorandum that was subsequently leaked, a vice president of the World Bank, Lawrence Summers, expressed his chagrin at the bottlenecks impeding the export of pollution. He argued:

Health impairing pollution should be done in the country with the lowest cost, which will be the country with the lowest wages. . . . I think the economic logic behind dumping a load of toxic waste in the lowest wage country is impeccable and we should face up to that. . . . I've always thought that underpopulated countries in Africa are vastly under-polluted, their air quality is probably vastly inefficiently high compared to Los Angeles or Mexico City. . . . Concern over an agent that causes one in a million chance of prostate cancer is obviously going to be much higher in a country where people survive to get prostate cancer than in a country where under-5 mortality is 200 per thousand. (cited in Dore, 1992, p. 85)

In the global supermarket's retail outlets, new information technology has transformed the methods of selling and marketing a whole array of commodities. The now ubiquitous use of universal product codes (UPCs, bar codes) and scanners allows merchants, advertisers, and manufacturers to solicit from customers, generally without their knowledge or consent, a wide array of data on their consumption habits: what they use (e.g., contraceptive products, junk foods, magazine purchases), brand names, size of products, spending patterns (their income range), and other personal information. In short, the purchase record forms a kind of customer "signature." If payment is transacted with a check, credit card, or bank card, the store now has an individual identity, address, credit record, and other data for targeting the customer and his or her lifestyle for future purchases or for selling the data to another marketing company (Toffler, 1990, pp. 102-103), individual, or institution. Combined with other records kept by government or commercial establishments, the inner sanctum of the individual the Fourth Amendment to the Constitution was intended to protect is routinely violated.

Jocks and Stocks: Selling Sport and Selling Short

Among the high roller telecommunications users are professional sports teams. Before the advent of television in the United States, professional sports earned most of its income from ticket sales and concession stands. Since the 1940s, revenues have soared from contracts with broadcast and cable stations. As the economist Andrew Zimbalist notes, CBS paid Major League Baseball $1.06 billion for the exclusive right to carry All-Star, playoff, World Series, and 16 weekend games from 1990 through 1993, whereas ESPN had an annual contract for 175 games at $100 million per year. Over the years, the share of televised games on "free" television has gone down while that of cable has gone up (and increasingly placed in higher "tiers" of cable channels, costing viewers more money to watch the "national pastime"). Cable is clearly the preferred transmission medium for team owners, as this means greater receipts, even if it comes at the expense of millions of people who cannot afford it. Moreover, the broadcast networks have covered themselves through ownership of cable sports coverage: ABC owns 80% of ESPN, NBC has a big share of Sportschannel America's regional networks, and CBS controls Midwest Sportschannel (Zimbalist, 1992, pp. 157-158).

There are other intersections between media and professional sports. Tele-Communication Inc. (TCI), which has a controlling interest in the American Movie Channel, QVC, Discovery, United Artists Entertainment, Home Sports Network, Prime Network (and its regional sports networks), Showtime, and much of Black Entertainment Network, also has a 22% share of Turner Broadcasting Systems (TBS), which, in turn, owned the Atlanta Braves. Time Warner, before proposing a merger with Turner in 1995, was already the second largest cable operator and world's largest media conglomerate ($13 billion in 1992 sales), with 21% ownership of TBS. In 1946, radio and television contributed about 3% of total revenues to Major League Baseball, but by 1990 the share was 50%. Other media also have their cut in profession sports: The *Tribune* owns Chicago superstation WGN, which televises Cubs and White Sox games, WGN radio, which also carries Cubs games, as well as WPIX-TV (which broadcasts Yankee games) and KTLA-TV (which broadcasts Anaheim Angels games). The media-sports complex extends to the National Football League, the National Hockey League, and the National Basketball Association (Zimbalist, 1992, pp. 48, 155).

Media are also intimately associated with financial markets. As increasingly big corporate entities themselves, the media have expanded the amount of attention given to Wall Street. Stock market and business reports are a regular feature of network television news and get even more specialized treatment on cable and public broadcasting programs dedicated to industry and financial news and analysis. Business and advertising have increased their print ratio in most daily newspapers, and some, like the non-unionized *USA Today,* with its section called "Money," are quite unabashed about their pro-business orientation. Beyond the wide array of print media, such as *Forbes, Fortune, Barron's, Business Week, Money, Wall St. Journal,* and numerous other business publications, many over-the-air and cable programs, such as *Wall St. Week, Nightly Business Report, Financial News Network, Adam Smith's Money World, Marketplace,* and other regular radio and television programs are dedicated to financial investing. Under corporate media control, the rich get richer and the poor get talk shows.

Certain forms of communication have become status symbols of the business culture in the United States—and increasingly in other countries as well. The cellular telephone is one of them, and the young up-and-coming executive strutting or driving with a portable phone attached to his or her ear is the latest mark of the arriviste, not unlike the way cigarette smoking once served as a movie cliché for virility.

Services like Prodigy (a joint venture of IBM and Sears) derives its main revenues from its on-line mail order catalog and other marketing services. Prodigy started out by offering free e-mail with a flat monthly fee but soon reneged on the deal. Later, it prohibited required users from contacting advertisers on their own or to indulge in language or topics (censorship) that offended the company's sensibilities. Prodigy's main competitors also include transnational enterprises: McGraw-Hill's BIX, H&R Block's CompuServe, and GE's Genie. One writer has raised the specter of these systems using their technical capabilities to surreptitiously download software that can monitor users' personal computer hard drives (Rosenberg, 1992, p. 326).

As manufacturing industries continue to shift from the production of U.S.–made goods into overseas operations, finance, and services, U.S. communications and discourse, too, have been swept by the exuberant devotion to money culture. The once casual way of stockbroking has been greatly transformed by the use of computer transactions. The National Association of Securities Dealers Automated Quotations (NASDAQ), a system of electronically transmitted buy-and-sell orders for "over the counter" securities, has been the fastest growing stock and bond market service in the United States. NASDAQ inspired a British counterpart, Stock Exchange Automated Quotation (SEAQ), to which it was linked in 1986. There is no question that computers have speeded up the pace of money and other financial paper transactions. On October 19, 1987, computer program trading of huge volumes of securities, involving automated investment decision-making programs, was, according to the Securities and Exchange Commission, the "largest single factor" for the "financial meltdown" that day. Program trading led to a 22.6% drop in stock prices in the New York Stock Exchange (the Dow Jones Industrial Average), the biggest stock market collapse in U.S. history (Associated Press, 1988).

On a global scale, the consequences could be more severe. In fact, the stock market crash of October 19, 1987, may have been precipitated by an earlier crash in the Hong Kong exchange, but in any event, the very same day saw stock market collapses in Chicago, Tokyo, London, Frankfurt, Paris, Sydney, and other major financial centers. At the time, more than $4 trillion per day were being electronically transferred by Western banks to take advantage of tax and interest rate differentials in different countries (Mulgan, 1991, p. 222). Instantaneous communications has added to the volatility of financial markets in recent years, making the global economy more vulnerable to sudden reverses, although current conditions seem to be considerably more stable than during the crash of 1929, which ushered in a worldwide economic depression. The concentration of financial and technological instruments in a handful of Wall Street and other major exchange markets leaves the vast majority of U.S. and other industrial state citizens extremely vulnerable and captive of the kingfish who play "casino capitalism" with workers' pensions and who have the capacity to wreck disaster on society.

Life, Liberty, and the Invasion of Privacy

As functional and useful as electronic communication devices can be in many instances, in others the same gadgets can serve as instruments of personal and political oppression. It is not, of course, the electronic listening, viewing, or recording devices themselves that are to blame, for who can blame an inanimate thing for making the very animate and intentional choices that sometimes harm people? Nonetheless, the presence of a "listening and watching" environment, a "surveillance society" (Gandy, 1993; Lyon, 1994), intensifies the climate of repressive potential; one might even envision the possibility of an electronics-based totalitarianism. Where people have strong reservations about the information society in which they live, this is the chief focus of their concern.

In 1992, it was reported that an engineer at the Olivetti Research Laboratory in Cambridge, England, now at the Xerox Research Center in Palo Alto, California, had developed a computer-based electronic badge that could track the whereabouts of anyone who wears it. It was designed for employers who want to monitor the movements of their employees within buildings or larger areas, not unlike the badges worn by the congenial ship crew on *Star Trek*. Some observers see such innovations in the workplace in less benign terms, as an intrusion on the limited free space of employees. An editor of the newsletter *Privacy Times* commented, "There's a lot of surveillance in the workplace these days. They could say you were in the men's room or the cafeteria too long or that you were sitting in so-and-so's office too long. It has the potential of changing the modern office into an electronic sweatshop" (Sloane, 1992).

The Department of Motor Vehicles in Oregon, as in many other states, sells personal information about drivers obtained from motor vehicle registrations to whoever wishes to purchase it. One computer consultant in the state decided to buy the information (for $222) and post it on the Internet. Although such information may be used by advertisers and other commercial interests to build consumer profiles (the make of automobile and address of its owner can be a very valuable asset) or by private detectives and other investigators, it could also be used by organized crime syndicates to track and make easy pickings of absent homeowners. With such easy access to driver information, it is simple to foresee an enraged driver, jealous lover, or impatient creditor taking down a license number, turning to a Web site and pursuing the imagined transgressor with murder in mind. Easy access to computer files reduces the temporal space among thought, information, and action at the expense of contemplation and reflection.

In the northeastern United States, a region usually thought of as vociferous in the defense of personal liberties, the fast-food chain Dunkin' Donuts, among hundreds of other businesses, allowed some 250 of its franchises to install surveillance cameras and microphones in the early 1990s. Ostensibly, the purpose would be for management to catch employees in the act of pilferage. The devices could not only videotape anyone within range of the checkout counters, however, but also pick up conversations up to 30 feet away. Media exposure of this situation, started by a small newspaper in Concord, New Hampshire, led to strong public reaction against what may have been a legal violation in several states of using audio surveillance without prior consent of customers and employees.

The effect of the taping was that private conversations between customers, perhaps privileged legal, business, medical, or occupational discussions, or between employees, perhaps about the conditions of work, could be monitored, leading to threats, punishment, exploitation, or blackmail. Character assassination opportunities, using surreptitious recordings, are limitless. Job pressure is intensified. Public pressure forced a quick retreat by Dunkin' Donuts executives, who promised to yank the surveillance equipment, although many other businesses continue to use and look for similar audio and video spy systems (Coakley & Murphy, 1994; Ryan, 1994).

E-mail is also used in the penetration of private space beyond what has already occurred with telephone, postal mail, and other media. In 1994, only one state, Michigan, had a law pertaining to harassment by stalking over the e-mail system, and because of it, a man was forced to stand trial for illegal pursuit of a woman by making persistent unwanted advances in a digital mode of expression. Because no physical contact was involved after an initial meeting between the plaintiff and the frustrated pursuer (an encounter that was mutually arranged by way of a video dating service), the Michigan court, and possibly other states, would have to wrestle with the boundaries of terror versus freedom of expression. In setting up a mailbox for paper or electronic messages, what rights does the subscriber implicitly surrender?

Communication technologies have been introduced in ways that also alter the private space in what was once considered the sanctity of home life. Few people would dispense with the convenience of the telephone, especially given the increased pressures on time allocation in the modern work environment. The near universal residential telephone network in the United States has become far too attractive a target of businesses wishing to use automatic dialing for inexpensive recorded advertising and sales pitching. The nuisance factor forces those with the means (17% of telephones in New York City, 40% in Los Angeles) to unlist their numbers (enduring the incongruity of a surcharge for *not* having their numbers published), the rest of the population putting up with commercial interruption of their leisure time (Tenner, 1991, p. 33).

Citizens nonetheless continue to buy into push button and other added telephone features, along with a host of additional communication apparatuses seen as convenient or entertaining: VCRs for television and video schedule management, the answering machine and voice mail for recorded messages and delayed response options, facsimile, e-mail, and other home computer applications. Then there are the host of video and computer games often proffered as "baby-sitting" diversions for children of overworked or distracted parents or guardians, perhaps a few other communication gadgets that serve recreation, amusement, security, or investment desires and, of course, the ubiquitous television (now cable-augmented), radio, and the various versions of audio recordings (what used to be the "phonograph"), and you have the typical household within the Western industrial universe. All of these often pleasurable or useful apparatuses are not without social and psychological costs, as they reduce in innumerable ways one's sense of private space and personal autonomy.

In 1989, for example, it was reported that the television ratings monitor group Nielsen Media Research had developed, together with the David Sarnoff Research Center in Princeton, NJ, a more reliable method for observing viewing habits. It was called a "passive people meter" and would use image recognition to record television preferences, including physical motion (e.g., leaving the room) and eye movement (e.g., napping). Other readings of such a device might not necessarily be so passive and, indeed, reminds us that the technological foundations, if not the resolve, toward an Orwellian society are already available. The people meter could be used not only by commercial interests for passive television watching but also by police or other political operatives to spot faces in a crowd of angry political activists (Carter, 1989). Given the surveillance predilections of the U.S. government, ostensibly in search of criminals and "terrorists," this is not too far-fetched an idea.

The aborted administration of Richard Nixon (1969-74), a man who ambitiously built a political career by smearing the reputations ("soft on communism") of political opponents and government officials, should be well remembered, among other ways, for having approved the Watergate incident, the break-in at the psychiatrist's office of Daniel Ellsberg following the release of the *Pentagon Papers,* and other intelligence "dirty tricks" operations. In the 1960s and 1970s, the U.S. government used police, military, FBI, and CIA agents in a massive infiltration of anti-war, anti-nuclear, African American, Latino, Native American, and other domestic political movements, as well as against counterculture bookstores, newspapers, street theater, rock groups, communes, schools, and even child care centers. (For accounts of this, see e.g., American Friends Service Committee, 1979; Churchill & Vander Wall, 1988; Glick, 1989; U.S. Senate, 1976; Wise, 1975; and the publication *Covert Action Quarterly.*) At least until the 1980s, the CIA kept U.S. and foreign journalists on its payroll, planting its own stories, disinformation, and other tools of psychological warfare, and counterpropaganda (to the Soviet propaganda machine) in newspapers and television news programs around the world, including the pages of the *Washington Post* (Hamelink, 1994b, pp. 24-29).

Although chided in Senate investigations in the 1970s, the FBI was back to its old dirty tricks in the 1980s under Reagan's imprimatur. Among its targets were legal liberal and left organizations. Freedom of Information Act documents revealed that, for 5 years, the Bureau harassed and broke into offices of political organizations and sometimes members' homes, such as the Committee in Solidarity with the People of El Salvador (CISPES), which opposed U.S.

military support to the murderous regime in El Salvador, and the progressive legal group that has been under FBI surveillance for 50 years, the National Lawyers Guild. Other FBI–targeted groups included the Southern Christian Leadership Conference (SCLC), the black clerical and civil rights group previously headed by Martin Luther King; the Roman Catholic Maryknoll Sisters in Chicago; and the United Automobile Workers (Shenon, 1988). During the Persian Gulf War, some 200 Arab Americans were interrogated by the FBI for no other reason than their ancestry. The FBI has admitted to supporting Philippines former dictator Ferdinand Marcos in his attempts to suppress Filipino American opposition to his regime, including the murder of two activists, in the United States. Names of Salvadoran political refugees deported from the United States back to El Salvador were provided by the Bureau to the regime's security forces (Parenti, 1995, pp. 149-157). Behavior of this sort reveals not only the politically intolerant and repressive conditioning of information gathering and police agencies but also the panoptic tendencies prevalent in large, complex, industrial societies.

The covert activities of the CIA abroad represent the export of police state behavior, involving support for counterrevolutionary movements and right-wing regimes; interference in trade unions, universities, mass media, political parties, and elections; secret wars; disinformation campaigns; police and military training; illegal drugs and arms shipments; and political assassinations. It is well documented that the CIA was involved in numerous coups d'état around the world and more recently in attempts to discredit Jean–Bertrand Aristide to prevent him from resuming the presidency of Haiti. Over the years, the CIA has had at its disposal the cooperation of hundreds of journalists, publishers, editors, and television executives. In the late 1970s, it was revealed in a Senate Intelligence Committee report (U.S. Senate, 1976) that the agency completely owned "more than 200 wire services, newspapers, magazines, and book publishing complexes" and had on its payroll "as spies and researchers" some 5,000 academics, along with thousands more from various organized crime syndicates, international drug dealers, Nazi collaborators, racists, anti-Semites, fascists, and various right-wing state leaders around the world (Parenti, 1995, pp. 157-162).

In the early 1990s, the advent of telephone caller identification technology raised the issue of protecting individuals' privacy from telephonic intrusion. The potential for abuse was significant. On the one hand, landlords or bankers could determine from what neighborhood inquiries about rentals or loans were coming. People who would otherwise use hotlines for seeking help for various physical or psychological problems, or battered women could be intimidated by the exposure of their telephone numbers. Social workers or health professionals might be afraid of calling clients from home. People with unlisted numbers would lose the value of that service.

On the other hand, many people appreciate this service for the protection it affords from unwanted or harassing telephone calls. The convenience of the telephone carries with it the penalty of random and often anonymous violation of one's private space. Some states have passed legislation requiring the telephone company to provide means for callers to prevent their identification from being transmitted, which would seem to negate the reason for employing caller ID technology in the first place. As in all aspects of public policy, the decision whether to permit caller ID will sacrifice the rights of certain groups in favor of others.

Much of the "information age" literature has exuded over the growth of e-mail. The fervor over e-mail as the *deus ex computor* of digital democracy does not hold up to the scrutiny of critical examination. Although e-mail offers new kinds of opportunities for engaging long-distance communities of interest and multipoint messaging, it is not without cost and serious risk. From a library users' electronic bulletin board, a notice was posted in 1993 that showed how easy it is to unobtrusively observe e-mail user behavior and even private messages themselves. Commands on the VAX/VMS computer system, such as "show users/full" will reveal all signed-on users at a given moment. Another simple "finger" command shows who is logged in at any terminal site or when a particular user was last logged. A "superuser" is one who has access to any file on the host system, and the practice of supervisors snooping on employee mail is widespread. *Macworld* reported in 1993 that 21.6% of companies surveyed admitted to such "electronic eavesdropping" practices (Branscomb, 1994, p. 93). The law is not clear as to who has legal rights to examine files of citizens or employees, but even if it were, enforcement would still be another matter.

Another sort of privacy trade off issue has arisen with the development of cellular portable telephones, a technology that has raised concerns about the ease with which eavesdroppers can listen in on conversations with store-bought radio scanning equipment. The National Security Agency (NSA), which is in charge of electronic intelligence gathering by the Department of Defense (DOD), and the FBI have both opposed making available scrambling and descrambling equipment that would help protect the sanctity of cellular telephone conversations, because they fear that such equipment might be used by elements on whom they routinely use surveillance. The NSA has also intervened in academic conferences organized for the exchange of knowledge in the field of encryption, preventing papers in some cases from being presented or published, whereas the DOD set rules in 1986 on prior review for conferences, meetings, and publication of all DOD–funded research (Shattuck & Spence, 1988).

American high-tech manufacturers have complained about NSA interference and attempts to impose encryption controls on their exports that do not threaten the invulnerability of government security. Manufacturers argue that the proliferation of encryption technology has made it impossible to contain the flow of this knowledge but that the standards used in defense equipment or weapons systems are of a completely different caliber from those for commercial applications. Encryption used in commercial fields, in contrast, is never fully protected from cracking. Some corporations discovered this with the help of a folk hero from the hacker community, known as Phiber Optik, who with his accomplices routinely broke into telephone company and credit bureau computers to get free telephone service, "liberate" celebrity credit accounts, and set up prank commands—which led to capture in 1992 and jail time (Gabriel, 1995).

In 1994, the Clinton administration proposed legislation that would require telephone and cable companies to put computer software on their systems that made them accessible to government wiretapping. The applications of the software, together with a government-designed encryption chip that would be installed in every new telephone and computer made in the United States, would facilitate government eavesdropping on private telecommunications and erode constitutional protections against illegal search and seizure. Not only voice communication but also the whole array of new shopping, entertainment, informational, and personal messaging is subject to potential spying. An additional concern is that the new equipment will force telephone and cable companies to pass on the added financial costs to customers, thereby asking the public to pay for the surveillance used against them (Andrews, 1994b).

Whereas questions of privacy and surveillance occupy much of the space in the critical literature on the information society, countries that have not reached the level of economic development of the West do not have the same concern about the pervasiveness of electronic intrusion. It is not that surveillance is a strange custom in the Third World, for there are certainly enough political and other kinds of prisoners to show that it is not. And, for that matter, the general attitude toward private space tends to be defined very differently where communal values are still very much intact. The main critique of the instrumental uses of information and communication technologies in Third World countries focuses on how they are used against them as a nation and as a culture by the foreign powers that wield them. This is discussed in the following chapter.

PART IV

THE GLOBAL
DIMENSIONS OF THE
"INFORMATION SOCIETY"

The "information society" is not simply an American or Western phenomenon. In fact, it is intimately tied to the countries of the Third World (Africa, Asia, Latin America, the Caribbean, and the Pacific), whose resources were appropriated by the Western powers and Japan during the hundreds of years of colonial rule. During the industrial revolution the Third World provided, and continues to provide at the end of the 20th century, most of the West's agricultural and mineral needs, in the process reorienting their land uses and economies for export to the capitalist powers. Much of the manufacturing assembly operations of transnational corporations (TNCs) have been relocated to the Third World to take advantage of cheap labor, big regional markets, and other favorable opportunities for investment. Without modern telecommunications, the ability of TNCs to operate plants overseas would be nearly impossible.

The Third World countries, in the meantime, have been flooded with the software products (e.g., television programs, films, advertisements, videos) of the communication technologies they helped build. This has raised concerns about the preservation of their cultures and their future prospects as independent producers of goods and services for the benefit of their own people. At the same time, as more and more American jobs and technology are shipped abroad, with the blessings of the U.S. government, the American workforce is reduced in size and opportunities to better their own lives, making them poorer in many respects and reducing the differences between themselves and the impoverished existence of most Third World workers.

CHAPTER 7

The Third World Meets the "Third Wave"

The *Third World* (also called the "South," the "developing countries," or the "less developed countries") is the nations comprising the African, Asian/Pacific, and Latin American/Caribbean regions, with the exception of those countries that were never colonies or dependencies of the industrial powers (e.g., Japan) or were created as white settler states (e.g., Israel, South Africa). On the whole, the Third World has remained economically far behind the leading capitalist economies, and the income gap with the First World (leading industrial countries) has grown in the post-independence period. Although expressions of solidarity in opposition to policies of the "imperialist" West have waned since the early 1980s, most Third World countries are no better off now than they were under colonialism—in many cases, even worse. The South has close to 80% of the world's population yet only 20% of the world's income and less than 10% of industrial patents (only 1% held by Third World nationals). At the same time, about 80% of the world's economic output and new private investment takes place among the leading 24 industrial states, the Organization for Economic Cooperation and Development (OECD).

Many Third World countries suffer periodic droughts, famine, epidemics, and extremely high infant mortality rates and, at the same time, annual population growth of from 2.5% to over 4%, even while 500,000 women die during childbirth. Some 100 million people (equaling the total populations of unified Germany and Scandinavia combined) are being added to the poverty list each year, and 1 billion people are under the category of "severely malnourished." Hundreds of millions more are deficient in the nutrients needed to support basic physical and mental development, and some 900 million adults are illiterate. About 14 million Third World children under the age of 5 die each year from hunger, disease, and poverty.

Much of these grim data has little to do with resources, but, instead, with historical inequalities in the ownership and distribution of land. Elites usually established their wealth through control or confiscation of irrigated areas dedicated to export crops as the poor were marginalized to the more arid and desolate regions or reduced to landlessness, tenancy, or wage labor status, all three conditions being widespread in countries such as Pakistan, Bangladesh, and Nepal. Export-oriented agribusiness made a small minority wealthy enough to enjoy Western lifestyles through their integration with world markets. This economic leverage enabled elites to perpetuate their political power while the poor majority were kept out of the governing and economic processes and retained as soldier pools in regional, national, religious, or tribal conflicts.

Not to overgeneralize, there are certainly great differences among Third World countries: 39 countries (with well over half of the world's population) having an annual per capita income of less than $425 (73 countries with less than $1,600) on one end and a few oil-rich Middle East countries and Singapore with mean incomes approaching or even surpassing those of the major industrial leaders. The "newly industrializing countries" (NICs) of Asia (Singapore, South Korea, Taiwan, and Hong Kong) have had impressive economic and export trade growth rates, which has brought an unprecedented degree of prosperity to millions of their citizens. China, still a very poor country, also has had remarkable growth in recent years. Argentina and Chile, with important natural resources and older industrial bases, are relatively stable economically, although their peoples have undergone long periods of dictatorship and repression in the postwar period, and in both countries the military stays close to the sidelines, the last attempted coup in Argentina occurring in 1990. Cuba, though poor, was able to provide universal health care and education before the collapse of its east European and Soviet socialist allies in the late 1980s. The impact of an enforced economic isolation and other acts of aggression by the United States brought the island country to near collapse by 1994.

Some 95% of the population growth over the next 35 years is expected to occur in the South, and most of these countries are experiencing economic growth rates well below their population increases. In many countries, well over half of the youth are malnourished. And although some countries, such as China and India, have made impressive gains in becoming self-sufficient in food grains, many Third World countries experience chronic food shortages and inattention to the plight, ignoring the need for agricultural credits and basic information infrastructure. (Some of the many good, basic-level studies on the history and socioeconomic conditions in the Third World are Baran, 1957; Barnet & Cavanagh, 1994; Barnet & Müller, 1974; Brown et al., 1992; Chaliand, 1989; Lappe & Collins, 1978; Moore, 1966; Nossiter, 1987; The Report of the South Commission, 1990; and Stavrianos, 1981.)

Apart from the unequal terms of commodity trade between most countries of the South and the North (major industrial countries) and out-of-control debt dependency that holds many of the former economically hostage to the latter, there is the lingering legacy of "cultural, media, and informational imperialism." This phrase refers to the flooding of the Third World with the media content of the West, including international news coverage, films, television programs, music, advertising, public relations, as well as educational materials, books, magazines, theater, and other lifestyle symbol transfers that encourage wasteful consumption of foreign goods and services, divided cultural identity, and emigration. A British scholar defined media imperialism as "the process whereby the ownership structure, distribution or content of the media in any country are singly or together subject to substantial external pressures from the media interests of any other country or countries, without proportionate reciprocation of influence by the country so affected" (J. Oliver Boyd-Barrett, cited in Mattelart, 1994, p. 177).

In Brazil, for example, almost all of the films shown on television come from Hollywood or other wealthy countries. Most other Third World countries have a similar immersion of imported cultural products, dominated by "a handful of global conglomerates with something to sell" and a content "drenched in expertly choreographed brutality" that conditions people "to believe that violence could improve their security" (George Gerbner, cited in Traber & Nordenstreng, 1992, p. 14). The problem of cultural invasion is not simply one of contact with foreigners, for all societies have experienced some degree of assimilation with external cultural values, but rather the extremely unequal terms of exchange on which these contacts are organized and the weak financial and technological foundations for developing modern endogenous arts and media.

The United States, together with France, Italy, Germany, and the United Kingdom, control up to 90% of all film exports. For lack of income or even electrification, television remains a limited pastime in the South (less than one tenth the number of sets per 1,000 population, compared with the North), but its middle classes are heavy consumers of American television programs. The United States has three times as many FM radio transmitters as all of the Third World combined. A handful of book publishers, mainly in the United States, United Kingdom, and Germany, control the worldwide distribution. In paper consumption (for books, magazines, and newspapers), the United States alone represents 40% of world total. Expenditures on advertising in the United States are greater than those in the rest of world combined, and American agencies prevail in foreign country advertising, including Europe (data cited in Frederick, 1993, chap. 3). And even though it redounds fully to the benefit of corporations, under U.S. federal law, TNCs can deduct the full cost of advertising from taxable income, which costs the government $3.3 billion annually in lost revenue (Canham-Clyne, 1995, p. 10).

Foreign culture exerts hegemonic power where the receptor country is asked to play on an unlevel playing field, and historical colonialism and imperialism have made survival a steep uphill battle for the Third World. Media and information transfers from the West impose Western language (most often English), typically spoken only by Third World elites. Media and language also convey cultural outlooks and values. Relatively poorly financed domestic cultural productions (theater, television, and film) often have a hard time competing for audience attention, particularly would-be affluent patrons, when the elites are drawn to technically slick foreign media. Science information transfers often convey data of interest mainly to their Western sources, thereby biasing the research agenda in recipient countries. Information and media transfers thereby exacerbate the dichotomy not only between North and South but also between Third World elites and the poor while, at the same time, reinforcing the status and legitimacy of Western-oriented leaders who wield the "look" of their foreign partners and patrons.

North–South contacts based on inequitable or "neocolonial" political and material relationships ultimately undermine local symbols, pastimes, cultural practices, and attempts to foster a national culture and economy. Under the present "free trade" agreements for the flow of media and information, many Third World leaders feel disadvantaged in their attempts to forge a stronger national identity. Without a national identity, they argue, people in poor countries cannot be mobilized toward a consensus on which to plan the future, leaving them vulnerable to the more powerful countries and more powerful foreign interests (e.g., military, transnational enterprises, religious groups, intelligence organizations). One form of cultural domination is documented in a 1980 study by Rafael Roncagliolo and Noreen Janus, who found that in 22 Latin American newspapers, TNCs employing transnational advertising firms took up 31% of advertising space (and a far higher percentage if very small and classified ads are not included; cited in Bissio, 1990, p. 85).

COLONIAL LEGACIES

Opposing points of view have argued that exposure to the Western media has a very positive influence in helping "modernize" Third World societies. A widely

cited communication scholar, Ithiel de Sola Pool, said, "Newspapers and radio enable people to conceive what it is to be a ruler, or a foreigner, or a millionaire, or a movie star" (Pool, 1966, p. 110). Daniel Lerner, his often-quoted colleague at MIT as one of the leading 1960s–era "modernization" specialists, declared, "What is required to motivate the isolated and illiterate peasants and tribesmen who compose the bulk of the world's population is to provide them with clues as to what the better things of life might be" (Lerner, 1963, pp. 341-342). Implicit in their analyses was that the peoples of the South lacked the individualistic motivations (what David McClelland had called the "need for achievement") necessary for development and that the North, especially the United States, would infect them with the germ of material progress through identification with American media, social-psychological, and cultural values (McClelland, 1961; see also Sussman & Lent, 1991).

Lerner, Pool, McClelland, and other defenders of the "free flow" of information often ignore the tangibly discriminatory effects of foreign information systems on Third World countries. First, it is important to address the historical-colonial circumstances under which modern communication technologies were introduced in the South. Radio stations in sub-Saharan Africa, for example, were set up "to provide information and entertainment to the colonial administrators and white settlers," privileging Europeans with political and cultural contacts with the metropolitan countries while transferring imperial culture to the educated administrative class of Africans. The black African nations were among the last to get their political, if not economic, independence from the European powers and today have the lowest rates of telecommunications and mass media facilities in the world, with, for example, an average of only two telephones per 1,000 people (even fewer if government and central business district connections are not counted). Much of the telephone traffic between African countries is channeled through Europe (Boafo, 1991, pp. 105-107).

Throughout Latin America, U.S. commercial radio made early inroads that, together with domestic private media interests, overwhelmed initiatives of noncommercial, public, and state-owned stations. Against the better-organized advertising and mass marketing groups, educational and cultural radio programming interests remained underserved. The Columbia Broadcasting System (CBS) and the National Broadcasting Company (NBC), both U.S. broadcasting corporations, ran the Cadena (Chain) de las Americas and the Cadena Panamerica, respectively, with affiliates all over the continent (Fox, 1988, p. 13). This provided avenues for American goods to penetrate the region and for American advertising agencies (e.g., J. Walter Thompson and McCann–Erickson) to create markets for its northern corporate clients. The long-term strategy worked, and Brazil, for example, which started with noncommercial radio, has a broadcasting system today that is saturated with advertising for both local products as well as for Coca–Cola, General Motors, Levi's jeans, Volkswagen, and many other foreign goods (Oliveira, 1993).

As part of the "Good Neighbor" policy of the Franklin D. Roosevelt era, the State Department backed pro-U.S. right-wing dictatorships throughout Latin America and the Caribbean, with the intention of securing good investment opportunities for U.S. businesses, reducing the "need" for military intervention and blocking the influence of Nazi Germany in the region. During World War II, U.S. companies advertising in Latin American media were subsidized by federal tax breaks while some 1,200 newspapers and 200 radio stations in Latin America were being fed daily news despatches from the State Department's Office of Coordination of Inter–American Affairs, headed by the oil and banking scion Nelson Rockefeller (Fox, 1988, pp. 13-14). The Rockefeller family interests also developed a supermarket chain and an in-house advertising agency in Latin America, in addition to a share of NBC through the Rockefeller–Morgan banking interests, making them direct beneficiaries of private commercial broadcasting in the region.

After World War II, television culture was quick to penetrate America's southern neighbors. Mexico, in 1950, was first to establish television, and as in the United States, commercialism superseded public or educational broadcasting alternatives throughout Latin America. All of the major U.S. networks, along with Time–Life media enterprises, had investments in Latin American television stations. This facilitated the sale of programs to Latino audiences. South of the border, *I Love Lucy, The Flintstones,* and *Batman* became well-known elements of the cultural lexicon at the same time that right-wing governments used coercive pressures to purge the media of progressive political content. Frank Sinatra, Dean Martin, and other American entertainers combined appearances on Cuban television with holidays at glitzy Mafia-owned resort hotels in Havana. Collaborating with the Central Intelligence Agency's (CIA) "health alteration committee" (hit squad), the Mafia would later help coordinate assassination plots against the Cuban revolutionary leadership that came to power in 1959 (U.S. Senate, 1976).

Cuba, Venezuela, and Costa Rica represented three major Latin American bridgeheads for U.S. television. By 1960, according to the study of Richard Bunce, ABC International was part owner of 3 Venezuelan television stations and managed another 8 of the country's 14 total. That year, ABC International also participated in the Central American Free Trade Area by setting up a Central American Television Network to distribute foreign programs and advertising to the region. This became especially timely when the revolution in Cuba ended U.S. commercial domination of that country's national television system, which until then had carried 62 U.S.-made serials. By the mid-1960s, ABC's stations were found throughout Central and South America, as well as in east Asia, the Middle East, Canada, and Australia, with distribution of almost 900 programs in more than 90 countries, reaching most of the television homes outside the United States (Bunce, 1976, pp. 79-83). ABC International and its program department, Worldvision, represented one early coordinated project of commercial television and advertising to set the world broadcast standard and become the beachhead of global capitalism in the southern hemisphere.

THE COLD WAR'S CHILLING
EFFECT ON MASS MEDIA

In Latin America, most of the governments during this formative period exercised tight political censorship of radio and press information, which, aided by generous U.S. military and intelligence assistance, tried to block nationalist and socialist groups from building popular movements. American media penetration, U.S. foreign policy, and dictatorial state apparatuses had the common objectives of protecting privileged foreign enclaves and keeping the world safe for capitalism. U.S. military intervention has been a security overhead cost to protect overseas U.S. investment and trade interests.

The extent of U.S. market domination of Latin America and other Third World regions during this period was unprecedented. Anti-U.S. feelings among Latinos came violently to the surface as anti-"Yanqui" demonstrations erupted during the visits of Vice President Nixon in the 1950s and Vice President Rockefeller in the 1960s. Locked into Cold War delusions of a Soviet–engineered "world communist conspiracy," official America was blind to Latinos' anti-imperialist anger behind these demonstrations.

By the 1980s, popular revolts against military rule worldwide and a wider understanding of the U.S. role in support of Latin American and other regional dictatorships (e.g., Marcos in the Philippines, the Shah of Iran, Suharto in Indonesia, the Greek junta, the apartheid regime in South Africa, military rule in South Korea, Pakistan, and Zaire), the perception of a diminished threat from the Soviet Union, and concern about runaway military spending brought more moderate outlooks in Congressional policy (though not in the Reagan White House). In Latin America, economic decline and huge debts to Western banks undermined military rule and led to civilian electoral politics, although the generals stayed poised to leave the barracks when needed—to "restore stability." With "isolationist" and "America First" Republicans more dominant in Washington's politics after 1994, the return of the Latin American generals may not be far off.

THE CHILEAN TRAGEDY

One Latin American country that resisted U.S. corporate domination of its economy was Chile under its popular unity president Salvador Allende

(1970-73). To help build up Chilean industrialization, Allende's democratic socialist program included nationalization of the big U.S. copper corporations, Kennecott and Anaconda, a move broadly supported by nationalists and most of the working class. International Telephone and Telegraph (ITT), a U.S. telecommunications corporation that controlled the main telephone utility in Chile, Chiltelco, and fearing nationalization, worked with the CIA, the U.S. Defense Intelligence Agency, and other U.S. businesses in the country, first to attempt to block Allende's election, and later to overthrow his government. ITT illegally funded opposition newspapers while the CIA bankrolled the press, a rival right-wing presidential candidate, conservative unionists, strikes, demonstrations, politicians, and anti-Allende conspirators in the Chilean armed forces, many of them trained at Fort Benning or in the Panama Canal Zone (Kolko, 1988, pp. 217-222; McCormick, 1989, p. 185).

An important plant for CIA propaganda was the conservative newspaper *El Mercurio*, one of whose former editors admitted that the paper's publishers "did everything we could to provoke a coup" (Constable & Valenzuela, 1991, p. 134). Meanwhile, ITT and the CIA collaborated on a draft program to destabilize the government. Before these facts were brought to light in U.S. Senate hearings in the mid-1970s, the American mass media had consistently failed to question the more aggressive aspects of U.S. foreign policy, whether in its assisted overthrow of elected governments (e.g., Iran [1953], Guatemala [1954], Brazil [1964], Guyana [1964], Indonesia [1965]) or its support for repressive dictatorships around the world. In Allende's Chile, the Nixon administration promised to "make the economy scream," while his national security adviser (and subsequent secretary of state), Henry Kissinger, defended U.S. destabilization operations against the reformist government with the arrogant assertion, "I don't see why we have to let a country go Marxist just because its people are irresponsible" (cited in Paterson et al., 1991, p. 589).

If the left was blocked from winning power through constitutional means and armed struggle was taken up as a result, so much the better—ruling circles in the United States could then "prove" that marxist governments are undemocratic. Failing to block Allende's election in 1970, Nixon ordered the CIA to organize a military coup against him and possibly his assassination, according to historian Gabriel Kolko, a strategy that initially failed. Protecting some $800 million in private American investments in the country, the U.S. government next encouraged the Chilean military (which it continued to fund) to step in (Kolko, 1988, pp. 217-222). The coup and Allende's death were followed by assassinations of other leftist leaders, book burning, heavy censorship of all media, political bans of various sorts, and the setting up of new media for political propaganda.

The United States had blocked the flow into Chile of all economic aid, government loans, or spare parts and induced the World Bank, the

Inter-American Bank, and private banks to follow its lead. Only the military continued to receive assistance. On September 11, 1973, Allende was overthrown in a military coup, and thousands of his supporters, along with Allende himself, were killed and many more tortured or forced into exile. With the junta, press controls, and the suspension of civil liberties all in place, Army Chief of Staff General Augusto Pinochet organized a wave of political jailing and violence against dissidents.

Meanwhile, U.S., World Bank, and International Monetary Fund economic assistance was quickly resumed, which weakened Chile's emerging independent manufacturing base and intensified its dependence on raw materials and processed goods exports (Bello, 1994, p. 44). It was a shocking but not isolated exhibition of U.S. Cold War extremism. The mainstream press in the United States helped legitimate the coup by characterizing the Allende government as politically corrupt and unstable, ignoring the fact that it was one of the few elected governments in Latin America at the time and that the military seizure led to massive civilian bloodshed.

Pinochet made himself Chilean president and remained in that position until 1990. Promising to "exterminate Marxism," he disbanded the parliament, banned political activity, and destroyed much of the political opposition. *El Mercurio* refused to fault the repressive dictatorship, and when one editor called attention to the murder of a prominent labor leader and criticized a government economic policy, he was fired. Television was also neutralized under Pinochet and reformatted to depoliticize and pacify the public with soap operas, military parades, and soccer coverage (Constable & Valenzuela, 1991, pp. 154-155). Following the 1973 coup, the American mass media generally and compliantly referred to the brutal dictatorship as the "Chilean government" but to Pinochet's democratically elected predecessor as the "Allende regime" (Parenti, 1993, p. 145). A new government under Patricio Aylwin in 1990 began a restoration of liberal democratic processes, and state-run television was sold off to private interests.

American television, movies, news wire services, music, and advertising continued to be pervasive features of Latin American mass media into the 1990s, although mainly in the form of cultural distribution and less in direct ownership. Even Cuba, under the socialist government of Fidel Castro and despite poor relations with the U.S. government, continued to air canned American programs on its state-run television stations and to exhibit American films in Havana's movie theaters. At the same time, Cuba has made great strides in developing independent and widely acclaimed films at its national film studio, ICAIC, and nationally oriented broadcasting programs. Nicaragua during the Sandinista period (1979-90) suffered under the Reagan government-organized and armed guerrilla terrorism of the *Contras* and broadened its control of the media as the offensive deepened. Unlike in Cuba, though, a private press, sometimes censored, functioned throughout the period. In the rest of the region, American media were still predominant, even with growing competition from Brazilian and Mexican television exports, the latter a major source of programming for the Spanish International Network in the United States.

Privately owned television in Brazil—in particular, the media conglomerate TV Globo—was able to function throughout the martial law period of 1964 to 1985 by working closely, politically and financially, with the generals, initially with the strategic partnership of Time, Inc. TV Globo played an important part in building legitimacy for the regime, exaggerating armed resistance to justify continuing military rule and heralding the country's economic "miracle." In return, TV Globo got the cooperation of the state telecommunications monopoly, Embratel, which carried its signal throughout the country. As in Chile, substantial foreign assistance and investment poured into the country despite its appalling human rights record but eventually stalled when Brazil began to stagnate under a nearly $100 billion debt by the mid-1980s ($117 billion by 1991, the Third World's biggest debtor nation). When massive political rallies began to materialize in Rio de Janeiro and Sao Paulo streets, *O Globo,* the newspaper arm of the Globo conglomerate, either ignored them or dealt only with their "entertaining" aspects until it became obvious that military rule was coming apart (Guimaraes & Amaral, 1988).

One often-celebrated "success story" of Brazilian media is the television soap opera, the *telenovela* (Rogers & Antola, 1985). This format has not only created a large audience in Brazil, crowding out foreign programs on prime time, but also has been exported to other countries in Latin America and to Portugal. As some would have it, this example refutes the thesis of U.S. "cultural imperialism." But, as others note, what the Brazilian *telenovela* represents, with its conspicuous merchandising of products displayed in each episode and raw marketing methods (including pretesting story endings on sample viewers to enlarge the audience), is a form of market manipulation of culture adopted from its mentor, the United States. The soap opera's main purpose is not to represent reality, educate, or inculcate humanist values, but to promote pacification and consumerist escapism. Rather than counteract the impact of homogenizing American material culture on the South, with few exceptions Brazilian television reinforces and helps internalize and internationalize these values by creolizing them with a thin veneer of local color (Oliveira, 1993; Sklair, 1991).

Although some argue that the "free flow" of information is an excuse for Western commercial and political control of Southern countries, others see it as a principle protected in the United Nation's 1948 Universal Declaration of

Human Rights, the 1966 International Covenant on Civil and Political Rights, and other international agreements. The declaration's Article 19 (restated in the covenant) declares, in part, "Everyone shall have the right to freedom of expression," including the "freedom to seek, receive and impart information and ideas of all kinds, regardless of frontiers," although the document also defends the rights of states in "the protection of national security or of public order, or of public health or morals." The precise boundaries of these provisions are not altogether clear. To what extent do communication corporations have rights to transgress borders without official permission? To what extent do states have the right to serve as guardian of a nation's cultural integrity?

In defense of the "free flow" concept, Pool argued that the decomposition of Third World culture that results from contact with the Western media can only indicate the failure of indigenous culture to respond to the emergent cultural demands of its people (Pool, 1974, p. 49). The assumption here is that cultural adaptation is a form of natural selection and can never be a response to coercion. The most forceful defense of "free flow" doctrine, however, is that the placing of restrictions on foreign media would leave people in the South entirely at the mercy of state-controlled media. In the extreme, there was a tendency in the United States to view statist interference with "free flow" as part of a Soviet conspiracy to destroy the "free press." Are such concerns humanitarian or mere smoke screens for the protection of transnational marketplace interests?

COMMUNICATIONS AS THE
"GREAT IMPERIAL BINDING FORCE"

All these positions contain elements of truth. One problem, though, is that advocates of "free flow" often overlook certain historical and political realities about the Third World and ignore some unpleasant facts about the evolution of Western societies. The West has had its share of state-sponsored violence, social disturbances, political repression, and official suppression of public information throughout the 20th century and earlier. The grip of fascism and military juntas in Europe, at their peak in the 1920s to 1940s, their continuity thereafter (Greece, Spain, Portugal, threatened in France), and the personal dictatorships in the name of socialism in eastern Europe and the Soviet Union show that political repression is in no way peculiar to the South. Nazi death camps, the Soviet Gulag, the nuclear violence unleashed in Hiroshima and Nagasaki in 1945, and the millions killed in Indochina from 1946 to 1975 during the French and U.S. interventions certainly rank, with the two world wars, among the worst atrocities ever committed against civilian populations.

Moreover, even if the Third World is on the whole more prone to political instability than most Western countries (overlooking Italy, which, from the end of World War II until 1983, had 42 different governments, more than one per year), it has had more to do with the distorted economic and social legacies of colonialism than with any innate political or cultural characteristics. During the "triangular trade" among Europe, Africa, and the Americas, peaking in the 18th century, England and France brought their own and imported Asian merchandise to the west coast of Africa, where they exchanged manufactured goods for human merchandise (slaves) to be transported as cargo to the "New World." At the plantations in North America, the West Indies. Latin America, and elsewhere, Africans were traded for raw materials such as cotton, sugar, molasses, indigo, and other tropical products grown by earlier slaves, which were shipped back to Europe and there processed by the machines of the early industrial revolution.

This international division of labor was replicated by other colonial powers in the Americas and both with and without the institution of slave labor in other parts of the world, linking the development of Europe and the Americas to the brutal exploitation of Africans and other Third World peoples (Frank, 1979, pp. 15-17). This distorted not only trade patterns in the African and Asian regions, which had prospered well into the 15th century, but also affected communication routes. Such areas as north and east Africa lost out to Western Europe as centers of social and economic exchange. Even now, telephone traffic between east and west Africa is routed through European capitals (Hamelink, 1995, p. 302), a lasting legacy of colonial communication policy, which a British governor of Victoria, Australia, called "the great imperial binding force."

In China, just after the British and French victory in the second Opium War (1856-58), in which drug trade was imposed on the Chinese people, the Manchus came under increasing pressure from the Europeans to allow them to set up telegraph lines. As Native American tribes in the United States understood the implications of the telegraph on the survival of their cultures, so too did many Chinese recognize that European telegraphy was an instrument for foreign political, economic, and cultural domination, although comprador merchants trading with Westerners tended to approve. When the Chinese government resisted, the Europeans set up lines anyway, as they did in the area from Foochow to Lo-hsing-t'a and Shanghai to Woosung (Headrick, 1991, pp. 56-57). When the Europeans got their lines, they insisted on transmitting only in foreign languages and in Morse code, a system not accommodating of the Chinese language. Eventually, the Chinese government, exhausted by the combined internal rebellions against the Manchu emperors and the unified front of foreign commercial interests, established its own lines, along with a code for transmitting Chinese characters.

Thus, the inherent instability of many Third World countries is related to a historical "development of underdevelopment" by which the colonized areas of the world were economically integrated into Europe, later joined by the United

States and Japan, providing the metropolitan powers with inexpensive raw materials and cheap (or slave) labor in exchange for expensive finished goods. By 1935, about 70% of the world's people were still living under some form of colonialism. Most of the Third World countries achieved their political independence only after World War II, and by that time, their economies, political practices, militaries, and scientific and cultural institutions were strongly linked to the former colonial powers, with the exception of those, such as China, Vietnam, North Korea, Cuba, and the early Indonesian republic, that delinked themselves from the First World. For most of the others, according to one official report of the nonaligned countries (those not part of U.S. or Soviet military alliance structures), their futures are "increasingly dictated by the perceptions and policies of governments in the North, of the multilateral institutions [e.g., World Bank, IMF, GATT] which a few of those governments control, and of the network of private institutions that are increasingly prominent" (The Report of the South Commission, 1990, p. 3).

As the dominant member of these multilateral institutions, the United States has heavily relied on political and ideological as well as economic pressures to keep non-capitalist, more independent countries like Cuba from pursuing their own, often radical development agendas. Demands for "free flow" of information not only serve certain idealist motivations but also the major market and foreign policy interests of the big powers. The first country to make use of shortwave reflection in the ionosphere, the Netherlands, sent government radio messages for the administration of its colonies in the East Indies. Radio was later used by the Axis and Allied powers as an instrument of international propaganda leading up to and during World War II.

Two early major works on the use of persuasive communications are those of Walter Lippmann (*Public Opinion,* 1922) and Harold Lasswell (*Propaganda Technique in the World War,* 1926), both arguing for the elite management of mass communication as a necessity of the modern industrial era. Following World War II, international communications development in the United States was heavily influenced by individuals, including Daniel Lerner and Wilbur Schramm, who had worked in psychological warfare and propaganda operations for the U.S. government. Lerner strongly believed in the "manifest destiny" of the United States to use international communication in the exercise of what he called the "persuasive transmission of enlightenment" (Lerner, 1969, p. 182). Radio Free Europe and Radio Liberty, beamed to the Soviet Union and eastern Europe, were early postwar projects of the CIA to "enlighten" that part of the world. The Cuba-U.S. "radio war" (as one-sided a "war" as the Persian Gulf War of 1991), starting in the 1960s, is a case study of how the "free flow" concept also can distort fundamental principles in international law and politics.

The Castro government (1959 to present) has been bombarded from the beginning by several waves of radio blitzkrieg, first by covert CIA transmissions in preparation for the Bay of Pigs invasion in 1961, and later by several different U.S. government-sanctioned or tolerated pirate and official transmissions intended to destabilize the revolutionary regime. Despite the evidence that the Castro government made use of its human and material resources and assistance from the socialist states to greatly improve the lives of the majority of Cuba's citizens, the United States continued to isolate and threaten the Cuban state. In 1985, the Reagan government created Radio Marti as a section of the Voice of America to try to bring order to the many inept attempts to weaken the Castro government through radio propaganda. By mid-1994, station transmission was increased to 71 hours per day over 17 various shortwave frequencies, each at 100,000 watts, to take advantage of Cuba's emigration crisis. Thus far, the broadcasts, like the added transmission, TV Marti, have not been successful in breaking Communist Party power in Cuba despite the country's dismal economic situation.

TV Marti is funded by the U.S. Congress and relies for its transmission on a tethered Navy balloon 2 miles high and close to the Cuban coast. In violation of international law (as declared by the International Telecommunications Union's International Frequency Registration Board, of which the United States is a member), TV Marti has used frequencies reserved by international agreement for Cuba's domestic use and in daily 4 ½-hour transmissions. One ironic twist is that Cuba imports far more television than does the United States, some 30% of its programming brought in from 26 countries, and close to half of its imports (42%) are from the United States. Several American human rights organizations, secular and religious, have tried to give substance to the "free flow" concept by illegally traveling (all tourist travel is banned) to Cuba and bringing publications to and from the island.

Most countries of the world have not missed the inconsistency of the U.S. government's approach in lecturing other countries about human rights. This has been made evident by the ovation given to Castro on his visits to Latin American conferences and at the 1994 inauguration of Nelson Mandela in South Africa and by the United Nations vote (101-2) in November 1994 to condemn the U.S. embargo against the island (Sussman, 1993). In 1996, European Union pressure forced the Clinton administration to back off from retaliatory sanctions against their trade with Cuba.

THIRD WORLD DEMANDS FOR A "NEW WORLD INFORMATION AND COMMUNICATION ORDER"

The non-aligned nations movement, led by Yugoslavia, India, and Algeria, with the support of the socialist states, declared in a 1976 summit in Colombo, Sri Lanka, that "a new international order in the fields of information and mass communications is as vital as a new international economic order" (cited in

TABLE 7.1 Global Media and Information Access, 1989 (per 1,000 population)

Region	Daily Newspaper Circulation, 1988	Radio Receivers	Television Receivers	Telephone Sets, 1981
Africa*	13	177	36	8[+]
Asia	81*	274 (E. Asia)	60	20[++]
		182 (S. Asia)		
Arab States	33	263**	90**	—
Europe	332***	701	372	241***
Latin America	86[+++]	342	140	55
North America#	253	2,016	796	840
Oceania	213	984	315	450
Former USSR	—	686	375	89
Developed Countries	336	—	—	445
Developing Countries	44	—	—	28
World	132	375	148	191

SOURCES: Adapted from Altschull (1995, pp. 297, 299, 301); Saunders, Warford, and Wellenius (1983, pp. 4-6).
NOTES: *Excluding Arab States; **Middle East, including Israel; ***Including former USSR.
+Excluding South Africa; ++Excluding Japan and Israel; +++Includes Caribbean.
#Not Including Mexico.

Frederick, 1993, p. 165). This meant the existing global structure of communication and information preserved the dominance of the former colonial countries in the dissemination of information relevant to future economic development—anything from the portrayal of world events in mass media to the ability to collect data on scarce natural resources or the political activities of dissident individuals and organizations. Ultimately, this, they argued, contributed to uneven economic development, Western exploitation of Third World human and natural resources, cultural and linguistic homogenization and hegemony, and the erosion of national sovereignty. Therefore, the nonaligned nations movement demanded a "New World Information and Communication Order" that would allow them to globally receive and distribute news and views on a more level playing field.

The individual countries of the South are at an enormous disadvantage in challenging the terms of trade and investment with Western countries when access to and dissemination of important information is extremely limited for them and readily available to capitalist states (see Table 7.1). An overwhelming proportion of memory-stored data is concentrated in the United States, 90% by one estimate (Gonzalez-Manet, 1992, p. 82). Control of information and information technology gives TNCs and their global subsidiaries the means to dominate market research and best understand and exploit market opportunities. It also guarantees TNCs continued control over patents of new technologies, which perpetuates technological dependency in the Third World.

The same is true in the area of national security. The U.S. government has commanding resources in disseminating its own version of events, including deliberate false propaganda ("psychological warfare"), in dealing with an enemy Third World country. U.S. private media, in cooperation with the government, operate worldwide information channels (e.g., CNN, UPI, AP, newspaper and television foreign correspondents, satellite transponders, telephone and fax, transborder data flow, airlines, tourism, film distribution, the U.S. Information Agency, Voice of America). The U.S. Information Agency, which sponsors transmissions into Cuba that favor the right-wing exiled leadership based in Miami, opposed requests by exiled President Aristide, whom the Clinton administration recognized as the legitimate leader of Haiti, to use a U.S. government channel to send radio broadcasts back to his militarized homeland even while it sponsored Cuban exiles without portfolio to transmit to Cuba (Quinn–Judge, 1994).

Collectively, as of 1986-87, AP, UPI, Reuters (U.K.), Agence France Presse (France), and Tass (USSR) dominated the worldwide flow of news, with 34 times as many words transmitted as the next five largest international news agencies and some 96% of the world's news transmission (Frederick, 1993, p. 128). Reuters Television (formerly Visnews), ABC–TV's affiliated World Television Network, and the American Cable News Network dominate the global flow of visual news.

Together, the Anglo-American grip on international news has the power to enormously influence global perceptions of world events. European and American media support for the Western military alliance (NATO), as in the Persian Gulf and Bosnia, was only amplified with the collapse of the Soviet Union. In recent years, several other news services, such as Knight-Ridder and the *Los Angeles Times* and *New York Times* services, have added a degree of competition to the big five press agencies, whereas UPI, now controlled by Saudi money and the Middle East Broadcasting company, has lost most of its print accounts and moved more into television news and data services. Few media outlets, especially in smaller markets, can function without reliance on these national and international news sources, which gives those individuals who own or control them enormous influence over the casting and "spin" of important issues of the day.

Consider the following imbalances in the international control and flow of information technology:

■ The United States, with less than 5% of the world's population, compared with 60% for all of Asia, has as many telephone lines; Manhattan has more than all of sub-Saharan Africa (comprising 49 countries); Italy as many as Latin America.

- Western countries transmit more than 80% of all telephone traffic, and nine of them have three quarters of all telephones; some 85% of the world's people have no telephones; the Third World together has less than 15% of the world's total lines.

- The Third World transmits among themselves less than 10% of the world's telephone, telex, or telegraph traffic.

- The United States and the former Soviet Union (now Commonwealth of Independent States, CIS) with a little more than 10% of the world's population, use more than half of the world's geostationary orbits (the Third World uses less than 10%).

- U.S. annual advertising expenditures is greater than all but 17 countries' gross national products.

- The leading industrial countries have 95% of all computers.

- The Third World, with nearly 80% of the world's population, has only 30% of the world's newspaper output.

- One Sunday *New York Times* reader consumes more newsprint than the average African uses in 1 year.

- The only Third World country that meets the United Nations Educational, Scientific and Cultural Organization's (UNESCO) standard for per-capita use of newspapers, radio, television, and cinema is Cuba, whose government the United States has made the greatest effort to overthrow (adapted and updated from Frederick, 1993, p. 75).

Third World leaders have often been incensed over Western press coverage of their countries that overemphasizes instability and chaos. This is often seen as reflecting a Western cavalier attitude toward non-Western societies and cultures and ignoring the West's own historical responsibility in bringing about the underdevelopment of the South. In the late 1970s and early 1980s, UNESCO, established in 1946, became the main forum for venting views on cultural imperialism, which eventually led to the U.S. withdrawal from that body under Reagan in 1985 and by Britain under Thatcher a year later. The Anglo-American alliance would brook no compromise of their cherished "free flow of information" doctrine, which declares, in effect, that governments have no right to restrict foreign media from accessing and transmitting news and other information content across national borders. Ironically, the only Third World state to join the Anglo-American protest in withdrawing from the UNESCO was Singapore, which has one of the most closed and restrictive state policies toward domestic and foreign media and which periodically prosecutes and shuts down foreign magazines and press that print unflattering stories about its government.

In response to the "Article 19" injunctions on freedom of expression, some Third World spokespersons, as a matter of national sovereignty, have expressed the need to limit the right of foreign journalists to uninhibited media practices;

others have asked that UNESCO assure reciprocity—the "right to communicate"—by helping Third World news agencies redress the one-way flow of information. In the late 1970s, UNESCO helped organize this debate, and in 1980 the agency published the most comprehensive study on the subject under the commission of Nobel and Lenin Prize winner Sean MacBride. The *MacBride Report* called on industrial nations to cooperate with and help finance the South's efforts to correct news and cultural flow imbalances, set up broadcast and other information systems, and improve the infrastructure for Third World telecommunications, communication satellites, and basic information services (see discussions of UNESCO and the *MacBride Report* in Frederick, 1993; MacBride, 1980; Preston, Herman, & Schiller, 1989; and Roach, 1993).

Even prosperous Canada, riven internally by the cultural and political tensions between its Anglophone and Francophone peoples that led to a near vote of secession in Quebec in 1995, also challenges the free-flow opinions of its southern neighbor, the United States. In November 1995, the public Canadian Broadcasting Corporation (CBC) announced that starting in fall 1996, it would cut all U.S.–produced programming, including *Fresh Prince of Bel Air, The Nanny,* and *Central Park West,* during prime time, an action similarly taken in Europe, where the European Community cultural ministers voted to reduce the number of U.S. imports carried within the continent. In Canada, where U.S. programs dominate three fourths of English–speaking household viewing time, over the air, on private stations and via cable, CBC expressed anxiety that Canadians are being "swallowed up" through commercial cultural submersion from the media superpower (Moore, 1995).

The issue of cultural imperialism is actually more complicated than control over the channels of communication. It also has to do with the preservation of cultural landmarks against the incursion of the "golden arches," or at least the right of small nations, both in their public and private domains, to protect themselves from having big and wealthy states chart their future cultural institutions. Already, in almost every Third World country, Western–style supermarkets and shopping plazas cater to foreign and local elites and offer things as diverse as Kellogg's Corn Flakes, Pizza Hut, and Nintendo. The depletion of foreign currency in bringing in these imported non-essential items is money not spent on basic needs like food production, health care, education, cleaning the environment, employment, and investment in local manufacturing.

About 80% of Coca–Cola's income is from outside the United States, and in many Third World countries it is substituting in children's diets for milk, essential vitamins, and protein. And cigarette advertising that is banned or restricted in U.S. media is ubiquitous in the Third World, where the Marlboro Man, now on billboards throughout China, is often more recognizable than the head of state. Mexican television, which is filled with advertising for Frito–Lay and other imported snack foods, is causing children to consume useless calories instead of more traditional and nutritious foods like tortillas and beans, but even these staples are increasingly being imported and becoming less affordable as the fields are appropriated to grow asparagus for export to the United States and Canada (Barnet & Cavanagh, 1994, pp. 246, 253; Pursell, 1995, p. 313). The North American Free Trade Agreement (NAFTA), passed by Congress in 1993, is likely to increase food and other imports, putting more Mexican farmers out of work and further eroding Mexico's economic and cultural sovereignty.

Another threat to Third World countries, regardless of ideological orientation, is the private Western communication satellite. Satellite spillover of American and European television programming into the Caribbean has long been an issue for governments and nationalists in the region concerned with defending local culture from the onslaught of *Dallas, Dynasty,* and MTV. This has more recently become controversial in Asia, where Star TV, a Hong Kong–based satellite system purchased by Rupert Murdoch in 1993, began spilling over American talk shows, CNN, Hong Kong news, the National Football League, a Japanese pornography channel, and other foreign fare to the People's Republic of China until that country agreed to accept its transmissions. Entrepreneurs view the overspill as part of the "free flow" of information, from which they expect windfall profits.

Those Chinese interested and with the means, especially in southern China, often spend their life savings to purchase and install rooftop dishes. The Chinese of Guangzhou (Canton) share the same language, Cantonese, with capitalist British Hong Kong (until 1997, when the colony was returned to China) and are otherwise the country's most Western–influenced and business-oriented people. Foreign investment in China has been largely concentrated in Canton, and the Cantonese have seen their region and language reinvigorated against attempts of the central government in Beijing to instill Mandarin as the national tongue and the nucleus of the national culture. Unable to stop the illegal influx of satellite dishes, the Communist Party will, in the meantime, have to contend with the likes of Bart Simpson.

Other countries, such as Saudi Arabia and Kuwait, two countries the United States defended in the 1991 Persian Gulf War, have banned satellite dishes in the name of defending Islam. Iran, the officially Islamic state with which the United States has had the most hostile relations since the fall of the shah and the seizure of American hostages in the late 1970s, initially tolerated and manufactured satellite dishes and received Rupert Murdoch's Star TV (which features *Dynasty, Wrestlemania, Simpsons,* NBA basketball, and an Asian version of MTV). In late 1994, the Iranian government issued a ban on satellite dishes (*New York Times* News Service, 1994).

Indonesia, a strong military ally of the United States in southeast Asia, has banned foreign distributors from directly importing or distributing foreign films, for which the United States retaliated with threats of trade sanctions (Bello,

1994, p. 81). And Singapore, though it separated itself from the Third World's attacks on cultural imperialism, also regards Western culture as a corruption of its supposedly Confucian heritage. Even though Singapore, an authoritarian and essentially single-party state, forbids home satellite dishes, the government has invited HBO and MTV to set up regional headquarters on the island (Shenon, 1994). Third World allies of the United States often make strange bedfellows. But elsewhere in Asia, some 38 countries are receiving Star TV's English–language, mostly U.S.–made programs in which more than 360 global advertisers, including Audi, Canon, Coca–Cola, Levi Strauss, MasterCard, Motorola, Nike, Panasonic, Pepsi–Cola, Reebok, Sony, Shell, Toshiba, and many other TNCs, are packaging seductive appeals for personal consumption (Schiller, 1996, p. 112).

The flow of Western information to the Third World is not inherently worthless or destabilizing. It is, however, when the flow is filled with toxic waste, such as recycled trashy television and film reruns of mordant violence and various shades of pornography or mindless consumerist advertising that soaks up the disposable income of the affluent (with kitschy imitations of "Western" goods for the poor). Intensely violent Japanese robot cartoons are packaged in special market appeals to children. Imported sensationalism (e.g., the trial of O. J. Simpson, other voyeuristic obsessions and public relations coups) are used to distract people from relevant economic, political, and social concerns. And the political biases of Western news agencies often insult the cultural values in Third World countries.

Many artistic and enlightening cultural products of the West that would resonate with Third World audiences tend not to be exported for lack of market power. These include the wide range of output from independent filmmakers, alternative presses, and small book publishers. The poor quality of most commercial American media becomes magnified in non-Western contexts even while their superior technical values often overwhelm their Third World media counterparts.

Although there are domestically owned media empires in Third World countries, Western commercial influence remains pervasive. The Samsung media and electronics conglomerate in South Korea is financially interlinked with the Corning Corporation (U.S.). In Hong Kong, Dow Jones (U.S.), which owns the *Asian Wall Street Journal,* also has large blocks of shares in the *South China Morning Post,* the *Far Eastern Economic Review,* and in Singapore, parts of the *Straits Times* and Times Business Publications. Singapore Press Holdings (SPH), in turn, which owns most of the Singapore mass media and parts of the *Asian Wall Street Journal,* also has holdings in various companies stretching from Borneo and New Caledonia to the United States and Britain and is in partnership with the Hachette Group, the biggest media owner and defense contractor in France. In support of government policies, SPH invokes threats of political repression to keep Singaporean journalists in line, at the same time that it has to put up with freer forms of expression in its foreign media shareholdings (Lent, 1993). Given the structure of corporate-military-government interlocks with mass media in the South, as in the West, it should not be surprising that the dominant information structures resist reporting news that points to corporate complicity in global issues of world hunger, political repression, and environmental devastation.

Another information flow issue upsets Third World countries, but it is not about cultural or news transfers. It is about information on physical geography that is gathered by remote sensing satellites, the French Satellite Pour l'Observation da la Terre (SPOT) and by the U.S. government's Landsat, now called EOSAT and under the private control of RCA and Hughes Aircraft. These satellites can photograph ground details of less than 10 miles in width and capture important information for Third World countries, such as impending weather changes or natural disasters. They also can be used to gather information on mineral, oil, water, crop patterns, or other crucial resources, giving those in possession of such data considerable bargaining leverage.

Third World countries won some degree of protection with the passage of the 1986 Principles Relating to Remote Sensing of the Earth from Outer Space treaty. In principle, countries whose land resources have been photographed by orbital satellites have certain rights of access to the data. Without available expertise, however, the photographs by themselves would not ensure internal sovereignty over the country's resources. Other sensitive issues involve satellite photography, including national security concerns such as in the observation of countries' defense systems (Mulgan, 1991, pp. 24, 227).

WHY IS THE "THIRD WORLD" THIRD?

It is important to establish what we mean by the term *Third World*. Although I cannot go into much detail here, history is the key to understanding the nature of the countries comprising the Third World (most of Africa, Asia, Latin America, the Caribbean, and the Pacific). Starting in the 1400s, the European powers began to explore and conquer non-European regions, establishing colonies and seeking gold, silver, spices, other forms of wealth and trade and often religious converts. Not all the consequences of colonialism were destructive, yet a record of brutality and tragedy, and its continuing legacy, is associated with the conquest. This legacy has fostered deep economic, political, cultural, technological, and informational inequalities in the world.

As West European countries embarked on global expansion and later modern industrialism, their Third World colonies remained economically stagnant even though the peoples of Egypt, China, Mexico, and other parts of the world had once been well ahead of the Europeans in many industrial arts and in scientific and commercial development. There is no single pattern or explanation for the economic underdevelopment of most of the Third World today, but under colonialism, internal development was subordinated to the needs of the colonizer. Cities were developed to service the export interests of the colonial power, land uses and food production were organized for their international exchange value more than for feeding people locally, transportation infrastructure was laid down along routes that tied mining and agricultural areas to the country's port facilities, and domestic administrative and business elites were either co-opted or coerced into submission. Uprisings, such as the Taiping Rebellion and Boxer Rebellion in China, the rebellions in British India (e.g., the Sepoy Mutiny), and the independence movements in South America revealed the resilient nature of oppressed peoples and the fragility of colonialism.

Nonetheless, the impulses of 19th-century European nationalism, industrialism, and chauvinism inspired a wave of new colonial ventures in Africa, Asia, the Pacific, and the Caribbean and deepening economic penetration, especially by the British, in the independent republics of Latin America. Cecil Rhodes (1853-1902), who endowed the famous Oxford University scholarship (originally available only to white students of northern European origins) and who pushed British expansion into South Africa, where he became prime minister of the Cape Colony, and into the country to the north that became known as Rhodesia, was one of the most forceful advocates for imperialist policy. His unequivocal defense of imperialism was that:

> We must find new lands from which we can easily obtain raw materials and at the same time exploit the cheap slave labour that is available from the natives of the colonies. The colonies would also provide a dumping ground for the surplus goods produced in our factories. (cited in Ponting, 1991, p. 222)

Foreign commerce dominated the relations between the colonial powers and the Third World according to a spatial and physical division of labor determined by the major powers. New technological implements from the Western world, including the railroad, the telegraph, and the telephone, were transferred to the remote outposts of capital to further the commercial exploitation of precious metals, other raw materials, goods, and services. As a result,

> many articles [in the colonized areas], *formerly cheap,* because unvendible to a great degree, such as fruit, wine, fish, deer, etc., became *dear* and were withdrawn from the consumption of the people, while, on the other hand, the *production itself,* I mean the special *sort of produce,* was changed according to its *greater or minor suitableness for exportation,* while formerly it was principally adapted to its consumption *in loco.* (Marx, in de la Haye, 1980, p. 152)

Thus, new means of transportation and communication helped solidify the control of foreign interests while straining the opportunities for independent existence in the areas under their imperial umbrella. And modern TNCs continue to exercise domination over the South through control of the sourcing and distribution of information.

Marx, in his first volume of *Capital,* written in 1867, found that "the cheapness of the articles produced by machinery, and the improved means of transport and communication furnish the weapons for conquering foreign markets" (Marx, 1967, p. 451). Telecommunications serves international capitalism not only in the transmission of influence over culture and popular consumption but more fundamentally by means of conquering space and time constraints in the creation of foreign industrial, labor, and commercial markets. Geographic distance is compressed through voice, image, and data linkages that add to the mobility of large businesses in the core industrial countries in their daily commerce with overseas subsidiaries. The result is greater decentralization of production with increased centralization of supervision.

By the time Third World countries achieved their political independence from their colonizers, in many cases only after continuous revolutionary activity, the terms of sovereignty were highly inequitable, greatly favoring the former rulers and their coteries of local administrators. Centuries of colonization left behind an agricultural system geared to servicing the industrial countries, extreme polarities in land ownership and class structure, a poorly developed education and technological infrastructure, often overdeveloped military forces beholden to the former colonizer for material support, an extremely weak industrial base, irrational national boundaries crafted by "divide and rule" strategies of the West, and shaky political institutions. The political, bureaucratic, and landed elites of the new nations often had more in common with the aspirations of their colonizers than with their own people, a social phenomenon elegantly described by Frantz Fanon (1967), Amilcar Cabral (1973), and Renato Constantino (1980).

Owning very few technological patents worldwide, the Third World controls less than 20% of them in their *own* countries. Third World outlays on science and technology represent a mere 3% of world expenditures, creating a heavy dependency on foreign technology and oil (including the OPEC countries, which rely on high petroleum prices for their stability) and a huge indebtedness to Western banks (about $1.5 trillion in 1995). The United States has used its sliding trade and dollar exchange value situation in the 1970s to push new lending on the South as a way of increasing demand for its own exports (Haynes, 1996, p. 78). The Third World is often characterized by high levels of dangerously infectious and water-borne diseases (e.g., cholera, diphtheria, malaria, schistosomiasis, tuberculosis, typhoid fever, dysentery, meningitis), malnutrition (some 40% of total inhabitants, higher rates among children in many countries), and infant mortality (Barnet & Cavanagh, 1994, p. 354; The Report of the South Commission, 1990).

It is often assumed that TNCs are the "engines of development" for Third World countries and that their direct investment creates job opportunities that would otherwise not be there. Mostly American, TNCs bring enormous resources to bear: The largest of them (1990 data) have more assets (General Motors, $173.3 billion; Ford, $160.9 billion; Exxon, $83.2 billion; IBM, $77.7 billion; and GE, $128.3 billion) than most Third World countries' gross domestic products (GDPs) (Egypt, $35 billion; Nigeria, $30 billion; Peru, $18.9 billion; Malaysia, $37.9 billion). It is also claimed that they bring foreign exchange, technology, skills, products, and a higher standard of living. Third World countries should therefore be grateful for whatever attention they receive from the TNCs.

Critics of TNCs argue, in contrast, not about the potential value of foreign investment but about the type, focus, and distribution of such investment. Transnationals, critics say, are not interested in development, only in maximizing profit opportunities and minimizing costs. And, in fact, in the Third World as a whole, TNCs have enjoyed hefty returns on their investments: in the 1975-1978 period, for example, an average of about 26%, compared with 12% at home (Spero, 1985, p. 276). Even though TNCs can present a threat to the authority of Third World governments, most, including those in nominally socialist countries, welcome their investments where there is government supervision and, in many cases, opportunities for secret partnerships and kickbacks.

The poorest countries of the world receive only a tiny fraction of private foreign investment, however, whereas just 14 countries get 61% of the total in the Third World. Five alone (Argentina, Brazil, Indonesia, Mexico, and Venezuela) pick up 40%. In Brazil, for example, foreigners in 1980 owned 25% of the country's industrial assets and 40% of all sales, while 10% of wealthiest Brazilians took in more than 50% of total household income (Bello & Rosenfeld, 1990, p. 37; Spero, 1985, pp. 270-271). Foreign investment has long held out the opportunity for domestic elites to play the role of imperialism's junior partners, which ultimately intensifies antagonistic relations between themselves and the masses of ordinary citizens. In most of these countries, decisions to expand foreign investment were made originally under repressive dictatorships or single-party governments that did not honor broad-based participation or political checks and balances.

Moreover, most of the positive claims made about the benefits of TNC investments do not hold up well under careful examination. The overwhelming proportion of TNC investments is funded by local banks or out of local currencies, not dollars, which is money not available to domestic businesses. New technology is not generally transferred to Third World host countries, but rather is used to generate further TNC earnings in the form of rents, royalties, and fees. Skill transfers are likewise very limited, as TNCs most often prefer to exploit the low wages of unskilled assembly and basic service workers (where automation has not already made them redundant), mostly drawn from the ranks of peasant women while retaining the far smaller number of skilled technical and management positions closer to headquarters.

In the transnational factories in the South, about 90% of workers are women, selected for their diligence, their often desparate acceptance of lower wages (in Mexico, $5 to $8 per day) than men, and their cultural vulnerability to patriarchy and the domination of male managers. Mexican wages, for example, are, on average, 12% of their U.S. counterparts. In Brazil, the average is 10%; and in China, which is the fastest-growing supplier of U.S. imports, 15% by 1995, the wage level is a mere 2%.

Even in the countries with higher wage levels, such as Singapore, South Korea, and Taiwan, workers receive only about one third or one fourth of their Western counterparts in the same jobs (compared with one tenth or one twentieth in the weaker industrializing economies of Asia and Latin America) and have been losing ground to business and professional income earners in their own societies. Other consequences of foreign investment include the inducement toward "brain drain": engineers, doctors, nurses, and managers attracted to foreign lifestyles, but without the means to attain them, leaving the country for Western shores. Workers who produce goods for TNCs (footwear, electronic instruments, automobiles, clothing, even locally produced fruits like mangoes, pineapples, or bananas) have to settle for poorer-grade products or none at all, while the "export quality" goods are shipped abroad for their dollar earnings.

THE PUSH FOR
EXPORT-ORIENTED INDUSTRIALIZATION

Beginning in the 1960s, the major industrial countries began to experience greater competition among themselves in their search for new investment opportunities and consumer markets. This was at a time when new technological breakthroughs were helping accelerate global capitalist expansion. Submarine cables carrying voice and message exchanges across oceans were rapidly proliferating (the first Pacific telegraphic cable having been laid in 1903 between San Francisco and the new U.S. colony, the Philippines, via another new territorial possession, Hawaii). Satellite communication was now available for point-to-multipoint transmissions. These technologies made it increasingly at-tractive for TNCs to move their factories, businesses, and jobs overseas, espe-cially to the Third World, where the cost of producing goods and services was much lower. With the conquest of time and space through communication technology, the world suddenly was seen as a more manageable marketplace—

for those corporations able to take advantage of it, a veritable borderless universe of profit incentives.

Many Third World countries at this point were having difficulty establishing markets for their own produced goods. Several, particularly in east and southeast Asia and in Latin America, began to take up the offer presented to them by the TNCs, the World Bank, the U.S. Agency for International Development, the U.S. Department of Commerce, and other powerful global banking and industrial institutions to help set up special economic zones that would produce goods for export to the industrial countries. These would become known as "free trade zones" or "export processing zones" (EPZs). In some cases, the commodities produced in these zones would be finished goods, such as clothing; in other cases, they would be partially finished goods or assembly operations, such as in automobiles and consumer electronics. The EPZs would house peasant workers drawn from nearby farming areas, where wage labor compensation was poor to non-existent.

In fact, Third World export industries are often based on the use of child labor. The U.S. Department of Labor reported that 19 Third World countries alone exploit at least 46 million children under severe working conditions for such goods as hand-stitched sport shoes and soccer balls in Indonesia and Pakistan, exported by U.S. companies to the United States and Europe. In the Philippines, some two thirds of all children, who should be in school, are instead in the workforce. In many low-income countries, children are often seen in the streets hawking cigarettes, flowers, fruits, or other small items or scavenging in garbage dumps for any food or items that can be resold or recycled, usually bringing home less than a dollar a day. Worse, millions of children are working as prostitutes, selling their flesh so that other members of their family can survive another day.

Globally, more than 100 million children are doing adult labor. Many of them help dig mines in Latin America and Africa; in Thailand, there are 5,000 "sweat-shops" where children earn 5 cents for sewing 200 buttons; in Nepal, there are about 5.7 million children making cotton clothes, many working 10 or more hours per day; in India, 17.5 million boys and girls between 5 and 15 years old produce rugs, brass vases, shoes, fireworks, and other goods for export; Africa employs one in every three children, many of them preteens who knot carpets (Morocco and Egypt), sew clothing (Lesotho), pan for gold (Zimbabwe and Ivory Coast), and mine diamonds (Ivory Coast) and chromium (Zimbabwe). The U.S.– owned *maquiladora* plants (rural export processing zones) in Mexico and Guatemala employ child labor with extremely low compensation or benefits. The value of these exports are worth tens of billions of dollars (Epstein, 1994).

Many of these EPZs were set up in remote parts of the country to try to attract rural labor, avoiding the problem of trying to invest in already overcrowded major cities. TNCs were willing to move to remote locations such as Bataan in the Philippines, Penang in Malaysia, Kaohsiung in South Korea, and Tijuana in Mexico because satellite or microwave links always kept them in touch with their branch offices in the capital cities of those countries and from there to their headquarters in New York, Tokyo, London, or elsewhere. They could, therefore, maintain centralized control of their "offshore" operations via advanced communication linkages and enjoy the benefits of lower production costs, especially labor. In Mexico and the Philippines, for example, television set components and assemblies are currently produced at about one tenth to one twentieth the labor cost in the United States. More than 90% of components for U.S.–produced color television sets are made in Mexico and East Asia (Bello, 1994, p. 89). With NAFTA and big foreign investors like Japan's Sony, Hitachi, and Matsushita and South Korea's Samsung moving to Tijuana, "the necessity of building TVs in the United States evaporated," and so did American jobs. TNC television set manufacturers substituted $9-an-hour American jobs with $50-a-week wages in Mexico (DePalma, 1996).

In the Philippines, large public and private debt was incurred to bring telecommunications into its export processing zones. Modern telecommunications permitted TNCs operating there to have clear channels to their subsidiaries and headquarters in other countries while the Filipino public absorbed the infrastructural costs and much of the risk. The large communication infrastructural development also allowed Philippine president Ferdinand Marcos (1965-86) to funnel millions of dollars in revenue as a silent partner in the communication enterprises and through TNC equipment supplier kickbacks into his Swiss bank accounts. By the end of his forced departure, the country was close to $30 billion in debt, the economy was in ruin, social conditions had seriously deteriorated, military abuses were rampant, and public access to telephone and telegraph was one of the worst in the world. For the transnational community, the country was one of the most profitable in which to invest, and Marcos' family and friends were secure tycoons in exile, soon to return home (Sussman, 1991).

This partnership of Third World plutocrats with TNCs and their Western state guarantors is certainly not confined to the Philippines. In Indonesia, a country of nearly 200 million people, the government has been under the military control and presidency of one leader, General Suharto, since his 1965 coup d'etat. Suharto's family and business associates have used state power to gain financial entry to virtually every transnational enterprise, including huge telecommunications contracts with Western TNCs such as AT&T. Elections are structurally rigged to guarantee electoral control by the military's Golkar Party, and despite the government's horrendous human rights record, repression of democratic parties, and mass slaughter of the Timorese, Indonesia ranks high among U.S. Asian allies and trade partners.

In 1976, the Indonesian government had a consortium of TNCs (Ford Aerospace, Hughes Aircraft, ITT, Siemens, and a few other OECD corporations) and the U.S. National Aeronautics and Space Administration (NASA) build and launch a communication satellite system. The users of its latest generation satellites continue to be mainly domestic and foreign corporations and governments, including HBO (Time Warner), ESPN (Capital Cities/ABC), Turner Broadcasting (Time Warner), and Star TV (Murdoch). Another study found that in nearby Papua, New Guinea, new modern telecommunications investments are being made, which favor an "export-oriented enclave economy" that is breaking down and integrating the domestic economy into the TNC-controlled world market, while only 0.6% of the people have telephones. The result is that rural economies have been severely disrupted and become dependencies of urban and foreign metropoles (Samarajiva & Shields, 1989, pp. 14, 17, 23). In Mexico, the national telephone monopoly has been turned over to a transnational consortium, led by Southwestern Bell (U.S.), while in Chile, the telephone network has been purchased by private Spanish interests.

The scientific management of production through the use of computers and telecommunications, recent innovations in flexible, small-batch production technology, and the broadening of the system of tariff preferences through the General Agreement on Tariffs and Trade (GATT) has also allowed corporations to take greater advantage of Third World (and First World) labor markets. *Flexible production* means that the adaptable character of computer-aided manufacturing permits relatively easy shifts in product design and assembly. This, combined with new strategies of "just-in-time" production, allows corporations to use smaller workforces in strategic locations (close to markets) and to cut costs in significant ways (e.g., inventory, scheduling, labor and management costs, equipment use). Much of this workforce is part-time, non-union, working at home, or abandoned altogether as capital moves overseas to take advantage of even lower labor costs.

Opportunities for investments in Asia and Latin America have made U.S.–based TNCs "footloose" in efforts to expand markets and reduce production costs. Microcomputers and telecommunications make it easier for financial, insurance, legal, and other firms, for example, to relocate data entry work, consumer credit reports, or even typing jobs to south and southeast Asia, the Caribbean, and even Ireland to exploit an extremely low-wage, largely female labor force, as American Airlines does in Barbados. An oversupply of underpaid engineers in countries like the Philippines, India, Mexico, and Russia allows electronics firms in the United States to employ a technical workforce in Asia, Latin America, or the former Soviet Union that works on chip design or software programming at a fraction of the U.S. cost. Satellites, oceanic cables, digital telephones, desktop computers, faxes, modems, and other modern communication equipment are at the disposal of TNCs to create this international division of labor.

An additional consequence of increased trade and investment in which telecommunications and digital information systems are intimately associated is the impact on the global environment. The industrial growth paradigm itself is, according to many experts, on a collision course with the earth's ecological integrity. In specific instances, especially in the Third World and some former European socialist countries, the integration (or competition during the era of eastern Europe and the Soviet Union) of less-developed economies within the world market has led to massive environmental plunder. The debt crisis of many Third World countries has induced their governments to find new sources of foreign exchange with which to begin to pay back interest due on loans incurred from Western banks during the 1970s and 1980s. In those years, European, Japanese, and U.S. banks were themselves desperate to unload the glut of dollars in their vaults that had been generated by the U.S. war in Indochina, the decline of dollar exchange rates, and the staggering oil price markups of 1973 and 1979.

Indonesia, the Philippines, and Brazil denuded a major portion of what were among the last major forest and wildlife reserves in the world. Mining for strategic mineral resources, on which the industrial countries are so dependent, and modern plantation agriculture, which is central to the global food production system, left toxic residues in their midst that filtered into aquifers, lakes, and rivers and seriously damaged or destroyed drinking water, the supply of fish (and protein), and other vital parts of the food chain. Soil erosion, siltation, desertification, and organically dead waterways, not to mention the greenhouse effects on the global atmosphere, result from indiscriminate tree cutting and pollution. Farming and fishing as sources of food, employment, and income are being devastated, forcing people to migrate, engage in the bleakest methods of survival, or end up as grim statistics on the disintegration of human existence. With the widening of transnationalization, the sustainability of all life forms has rapidly declined in the 20th century.

LIBERALIZE! DEREGULATE!
THERE'S NO NEED TO HAVE A STATE:
IDEOLOGICAL JINGLES OF THE 1990s

The growth of TNCs since the end of World War II and their insatiable need for new markets has eroded the autonomy of nation states in setting or pacing the course of social and economic change. TNCs themselves cannot be thought of as monolithic or having a singular agenda, but they do share the logic of integrating labor, technology, and other forms of capital on a global scale that maximizes productive efficiency and profit while reducing labor and other costs. Democratic–type states, loosely defined, must in some fashion, on the other hand, respond to broader social concerns than those of the TNCs, inasmuch as they represent multiclass interests. State power is, therefore, contested between those interests sympathetic and resistant to global market integration.

In the United States and several other advanced capitalist states, virtually all strategic industries (those that can make or break an economy) are already transnationalized, but in most Third World states and the former socialist industrial countries, transnationalization means foreign domination and is a more controversial matter (revitalizing the old communist parties in Russia and eastern Europe). The effects of socialist disintegration and capitalist integration of the world economy have been most severe in Russia, drawing that country into closer ties with China, a country that has taken more cautious steps in introducing private market forces and foreign investment.

Both China and Russia also have relatively independent foundations in electronics, mass communication, informatics, telecommunications, communication satellites, and other modern industries, even if not nearly as developed as in the West. As a result, they have stronger bargaining positions in trade and investment negotiations with TNCs in these sectors, compared with most less developed countries. In the Third World, with few exceptions, a self-reliant system of communications and information does not exist; most Third World countries get their news and data about the world and much of their internal cultural content from foreign transnational media and telecommunications corporations.

Led by the Reagan government, the United States in the 1980s pushed the deregulation of worldwide telecommunications, supposedly to promote greater domestic and international competition. TNCs, which constitute about 90% of all international telecommunications traffic, demanded lower information costs for their own continued expanded growth and productivity. Deregulation provided a rationale for separating local and long-distance telephone pricing, inducing greater product and service competition and encouraging fuller convergence of the computer and telecommunications industries. AT&T, which in the early 1980s was a virtual monopoly local telephone, domestic long-distance, and international service carrier, divested its regional operations following a 1982 federal "consent decree." But in the rush to deregulation in the United States and some other countries and in the price hikes for local telephone service that followed, as one communication scholar has noted, the "social well-being receives no ballots" (Schiller, 1981, p. 127).

In the early 1980s, the United States and Britain were both declining as net telecommunications exporters while Japan was rapidly advancing. Even rising Third World countries such as South Korea had "world class" corporate players in international telecommunications trade, such as Daewoo, Samsung and Goldstar (Mulgan, 1991, p. 152). Restricted to domestic markets before 1982 and since released through legislative action, AT&T has actively competed for sales and partnerships abroad. Meanwhile, despite the rhetoric of liberalization, U.S. telephone markets have been reagglomerating, leading to mergers in other communication areas, such as cable and cellular telephone. Almost unnoticed in the media and devoid of any real public debate was the 1994 merger of AT&T with the largest wireless telephone producers McCaw Cellular, a market expected to reach $16 billion by 1996 (Andrews, 1993).

The U.S. leadership in the deregulation of industry in the 1980s was part of the Reagan administration's shift to "supply side" doctrine, which essentially meant that public policy would favor the private corporate sector and move away from social equity issues. In telecommunications, the United States initiated the policy shift with the deregulation of the telephone industry, and Britain, under Prime Minister Margaret Thatcher, was not far behind in deregulating its own national telecommunications system. The director of the British Telecommunications corporation, J. A. White, in 1986 pressed Third World countries to abandon thoughts of a "New World Information and Communication Order" and to pursue the same direction:

> Developing countries must liberalize their markets in order to develop telecommunications systems and infrastructures. The only course is to hasten deregulation (of standards and functions) and to open up commercial competition, although such changes can lead to great confusion and turbulence in the markets and social structures. In any case, the trends are inevitable.
>
> The conflict is between free competition and national telecommunications systems. It is no longer sufficient to complain about the past, oppose those that have power and dream extravagant dreams. The initiative now depends on the United States, England and the industrialized countries. It is necessary to associate with those that can help with development. (cited in Gonzalez-Manet, 1992, p. 38)

It was not long before capitalists and their state backers in Third World countries began to follow the U.S. lead in privatizing and deregulating telecommunications and allowing foreign capital to enter formerly protected economic turf. Malaysia, Singapore, and Mexico were among those that began to privatize state-run telephone companies, and New Zealand, Australia, and the Philippines are among several countries that have allowed foreign equity participation. In the case of Malaysia, privatization of telecommunications was influenced by the World Bank's free trade-based lending policies. But just as important is the reality, described by one researcher, that the innovations embodied in "liberalization" were used as "tools (albeit very fashionable ones) taken up by political elites to further and to legitimate an historical process of wealth redistribution from the state [that previously owned the telecommunications system] to members of the elite" (in particular, the prime minister and his business associates in the ruling political party; Kennedy, 1989, p. 7).

Under pressure from the U.S.–controlled International Monetary Fund (IMF) for privatization and deregulation of debtor nations, Teléfonos de México

(Telmex) was sold off in late 1991 to private investors, producing windfall profits for stock investors on the Mexican and New York exchanges—and thousands of layoffs of telephone workers. Basic telephone rates rose immediately by 16%, with 4% increases for the next several months and another 15% increase by Christmas of the same year, yielding record profits but making service "one of the world's most expensive, and, paradoxically, most inefficient" (Ellison, 1991, p. 10A).

Telmex reveals the logic of transnational capitalism in the realm of telecommunications policy and planning. The biggest stockholder in Telmex was a consortium made up of a Mexican billionaire, Southwestern Bell International Holdings Corporation (U.S.), and France Cables et Radio, a France Telecom subsidiary. Reflecting the interests of its business user clientele and paralleling the pattern in the United States after the AT&T breakup, local calling rates have risen dramatically as long-distance and international charges have been reduced. The increase of domestic rates was intended to fund a $7.7 billion investment project (1992-94) for converting to digital switching systems and fiber optic oceanic cables, linking Mexico with the United States, the Caribbean, and Spain. Meanwhile, Mexico has one of the lowest telephone access rates in the world even though it is the 17th largest economy (Burns, 1990; Ellison, 1991).

Other Third World countries have followed the lead of the United States in pursuing new "value added networks" and services, such as in fiber optical transmission, teleports, paging services, and high-speed data retrieval and transfer services, most of which bypass the public switching systems and benefit large business users almost exclusively. Thus, public sector subsidies to private communication operators and their service users (through government and World Bank loans, government tax and tariff write-offs, legal monopolies, government diplomacy on behalf of telecommunications firms, government equipment and service purchases) have become a form of welfare for the elite. The assumption, where genuine theory is at work and not simply greed and collusion, is founded on notions of development organized from the top, otherwise known as "trickle-down economics." Squeezing the poorer classes with cuts in wages, benefits and social spending, the trickle has been more up than down.

Historically, colonial powers in most cases were able to capture local land-holding and commercial elites by giving them a minority share in the benefits of colonial trade and to isolate or eliminate organized elements that opposed foreign domination (e.g., nationalist cultural, religious, and business elements; socialists; communists). In this way, the static nature of the economy and class relations in Third World countries was preserved. Even after independence, economic dependency, or "neocolonialism," was perpetuated through foreign control of markets, science, technology, and information. A few countries, such as India after the British departure and China after the 1949 revolution, demonstrated a fierce determination to reestablish their independence and to do so, in part, by way of relatively self-reliant foundations in science and engineering.

Most Third World countries do not have such technical foundations. The World Bank, the IMF, and foreign government agencies continue to subsidize transnational domination in technology by persuading or forcing Third World states to induce TNC investments in their countries as a precondition for financial loans and other assistance. Tied loans often require Third World governments to cut assistance to education, cultural institutions, public health services, welfare recipients, farmers' associations, and other forms of social spending (to enable tax cuts and free up money for private-sector investment). The World Bank also insists on devaluation of the local currency (making foreign currency investment more valuable), an end to price controls (letting the market set prices), promotion of export development (favoring transnational enterprise over domestic manufacturing), and bringing in foreign experts for "national" economic decision making (thereby capturing the policy agenda into the future). In recent years, the World Bank has come under increasing criticism for lending policies that displaced Third World residents, destroyed their environments, and left more than a billion people in abject poverty (e.g., French, 1994).

The Association of Southeast Asian Nations (ASEAN) region is often associated with successful integration into the world economy and high economic growth rates. A more careful observation of those growth rates reveals very uneven patterns of class development and the continuing use of repressive measures against reformers, peasant organizations, and trade unionists as a way of eliminating democratic distribution demands on investment income. When computers began flowing into the region in the 1980s, it became obvious that social development was a low priority in the minds of state planners. According to one study of the region, only 3% of computers within ASEAN are used in such areas as agriculture, health, education, public works, water resources, public transportation, postal services, public information, population planning, rural and urban land development, or public utilities. The major users are the military, executive branches of government, TNCs, and major universities and research centers (Rahim & Pennings, cited in Lent, 1991, p. 187).

Although foreign technology and expertise under the right circumstances can contribute to improved social and economic conditions in Third World countries, abundant evidence suggests that no recipient country can rely on it, whether the state is socialist or capitalist. Nation states themselves embody deep social cleavages that do not tend toward the same ends. In the case of the United States, Britain, France, and the other economically and industrially developed states, the historical struggles of working people within political structures that permitted such struggle (even if often violent) can best explain the material and political gains they have achieved. Had Third World societies not been subordinated to colonial domination and political repression, this would be their experience as well. One cannot assume, therefore, that the most meticulously designed process of development, but one in which political participation is effectively denied to the poorer classes, can bring about genuine social change. Moreover, unlike the historical development of the now industrial countries, elite interests in Third World countries are too easily compromised by their subaltern (comprador) relationships with foreign mentors and partners, thereby undermining the interests of their fellow nationals. Broad-based political participation, rather than top-down decision making, is the best hope for social justice.

MORE NICs?

A group of political economists called the *dependencistas,* largely Latin American or writing about Latin America and whose work was particularly well known in the 1970s and into the 1980s, analyzed the world economy in terms of core, periphery, and semiperiphery. The *core* consists of national economies that have been and remain central in terms of industrial and technological development. *Periphery* denotes those Third World economies with a very weak or non-existent industrial base and that, through domination by the core, have been retained as suppliers of agricultural and mineral raw materials. The *semiperiphery* (NICs) are a handful of Third World economies (especially the Asian NICs, Singapore, Hong Kong, South Korea, and Taiwan) that have managed to achieve a second tier but increasingly important industrial base, with a relatively large manufacturing sector, such as in electronics or textiles, and a significant industrial working class. Mexico, Brazil, Argentina, and one or two other Latin American countries are said to loom as potential new members of the NIC group, but they remain burdened by huge national debt, deep polarizations of wealth and poverty, political instability, and in Mexico's case, a near economic collapse in 1994.

NICs are often held up to poorer countries as role models of non-revolutionary and capitalist economies. Not only are the human rights records of the NICs conveniently ignored, but little attention is paid to the heavily interventionist part the state plays in guiding their development, including, in many cases, ownership or control of major enterprises. Also, the "national" image of NIC corporations often disguises the degree of foreign participation or ownership. One successful industry in South Korea has been television manufacturing, but in fact, 85% of the value of South Korean color television components is made up of Japanese imports, and the South Korean Samsung corporation's VCR is actually a licensed product of Japan's consumer electronics giant Matsushita (which until 1995 also owned MCA studios in the U.S.; Universal Studios in Hollywood and Orlando; Panasonic, Quasar, Technics, JVC, and other communications industry holdings). The big South Korean trade surplus with the United

States is, in part, a hidden extension of the huge Japanese trade advantage with the United States (Bello & Rosenfeld, 1990, p. 2).

Among the NICs, Singapore is most often exemplified as the Third World success story. It is certainly a remarkable achievement that its people and government have established an important niche in the world economy and developed an enviable physical quality of life and an advanced system of education and technological research. With fewer than 3 million people living on a small island and with no rural population, however, this city state is hardly typical of the rest of the South. Singapore was founded by the British in 1819 as a special entrepot between Britain's South Asia and Chinese trade ports, and one main cargo in that trade was the opium harvested in India and imposed by the colonial power on a humiliated and unstable Chinese society. When Singapore became fully independent in 1965, it was already a well-manicured outpost for British and British-trained civil servants, a thriving trade port for Western exports and imports, a center of abundant merchandising skills, and a society with a degree of political stability unmatched in the region.

The relative cultural homogeneity (Chinese constituting almost 80% of the population) and the clever and headstrong leadership of Lee Kuan Yew, the first prime minister, who stepped down only in late 1990, were also factors in Singapore's ability to convert from a manufacturer and importer of cheap goods to a haven for export production, especially in electronics. Singapore has become the most concentrated enclave of TNCs in Asia, with recent massive investments from Japan, a country anxious to unload and invest its oversized savings pool. An American electronics company, Seagate, which manufacturers computer disk drives, is the biggest employer in Singapore. The island's prosperity is designed to rise or fall on its service relationships to TNCs and its dependence on continuing transnational investment (Bello & Rosenfeld, 1990, chap. 17).

Aside from questions about its sustainability, there is a dark side to Singapore's prosperity. Under Lee Kuan Yew and his hand-picked successor, Goh Chok Tong, the ruling People's Action Party (PAP) has been a political monopoly since the country's inception. It is a rigidly authoritarian state that interferes in every intimate aspect of its citizens' lives. The government conducts eugenics policies to ensure future demographic domination by elite, educated Chinese (rather than its Malayan, Indian, Pakistani, and other ethnic minorities), such as by sponsoring "love boat" cruises for selected invitees in support of its conjugal objectives. It also suppresses any media criticism from its own monopoly broadcast and newspaper outlets and the foreign press and has frequently closed down both when they contained objectionable journalistic observations. In 1993, the government brought charges against a Singaporean business newspaper and government employees for the publishing of government statistics that suggested a possible mild economic slowdown.

In the late 1980s, then Prime Minister Lee brought a personal libel suit against the Dow Jones–owned *Far Eastern Economic Review* in a Singaporean court and won a big monetary settlement, an opportunity since repeated against other Western press holdings. The prime minister also blamed Western media coverage of the ouster of Marcos in the Philippines in 1986 for inciting the Tiananmen Square uprising in China in 1989 (Erlanger, 1990). As protective as American media generally are toward the interests of big business, dictatorships do not trust their freewheeling tendencies. In Singapore, media are looked on mainly as organs of propaganda—to "educate" citizens in Orwellian repetition about the virtues of hard work and obedience to authority and to lecture them about the moral vices of spitting, littering, chewing gum, and urinating in the wrong places.

When an American economist teaching at the National University of Singapore published an article in the *International Herald Tribune* in 1994 in which he criticized unspecified Asian governments for using legal suits to suppress political dissent, he was soon visited and harassed by the Singapore police. He was forced to leave his post and the island for fear of arrest. A few months earlier, an American teenager studying in Singapore was imprisoned for 4 months, fined $2,000, and flogged ("caned") for allegedly spray-painting a few cars. The paint but not the scars were removable.

NICs like Singapore have concentrated on communication infrastructure as one central project intended to bring their economies in line with the world system of information, commodities, and financial trading and to serve as a means to lure TNCs into investment and trade relationships. If NICs or other Third World countries wish to become relatively independent of TNCs, information technology could potentially provide opportunities in that direction. By the late 1980s, the Asian NICs had created important niches in international markets for their low-end and medium-end production of electronics goods in the world market. At the same time, this tied their future development to the health of the Western economies and the stability of the U.S. dollar.

Only a few of the larger Third World countries (China, India, Brazil, Colombia, Mexico, Indonesia, an "Arabsat" consortium) and none of the Asian NICs have their own communication satellites, and the little research available does not indicate that these countries themselves are making much use of them for their own social and economic development. Rather, it appears that the heaviest users are TNCs, military services, foreign governments, and domestic television entertainment channels. The limited radio spectrum available for geostationary orbiting satellites has been put on a first-come, first-served basis, allocating almost all of the "parking spaces" to the United States, the other major industrial countries, and the former Soviet Union. The availability of communication technology does not, in any event, ensure that its uses will be focused on the social and economic development of the poor, and as discussed in Chapter 3, the pattern was little different during industrial-revolution-era America.

Communication technology also does not necessarily deliver domestic information access to those who most need it. The existence of Arabsat has not given women a greater voice in the public affairs of Arab nations. In few instances are computers used to distribute health information to regions of Third World countries with high rates of infant mortality and infectious diseases. In sub–Saharan Africa, where computers could be useful in delivering health services, they are almost unknown in all but 2 of the 49 countries. Third World countries are reluctant to use expensive satellite channels to deliver long-distance learning opportunities to remote provinces and villages. One country that has used foreign assistance to develop widespread health and education facilities and instructive media is Cuba, a country with few telephones and little state-of-the-art information technology.

Cuba's resilience is remarkable given the attempts of the Eisenhower and subsequent administrations to sink the revolutionary regime of Fidel Castro. The country has had to withstand a long-standing embargo on American travel, trade, investment, and communication; persistent and illegal penetration of its airspace; assassination attempts on its leaders; a barrage of hostile propaganda; and other attempts to overthrow its government and economic system. The United Fruit Company, long associated with support for "banana republics" (plantation-type economies guarded by military dictatorships) used its own extensive communication resources to illegally transmit radio transmissions into Cuba that were supposed to prepare Cubans for the ultimately disastrous Bay of Pigs invasion in 1961. Numerous other radio stations were set up, usually with CIA connivance, by Cuban exiles into the 1980s. At the same time, the United States is a signatory to several international organizations, including the United Nations Charter, and covenants founded on respect for the sovereignty and independence of all nation states and the principle of non-intervention in their internal affairs.

Although international agreements concerning the right of free expression raises high standards for governments all over the world, they have to be seen within the existing state and market practices that are often contradictory to those ideals. Many people believe that the lack of expressive freedom is peculiar to Third World societies, an assumption that often fails to take into account both the real limits on communication within First World countries and the ways the First World contributes to state repression elsewhere. The relative ease of communicating in the United States does not mean that articulate critical voices have the means to be heard, especially in the area of foreign policy. Purges of Asian experts within the State Department during the McCarthy era ensured that alternative views on the country's intervention in Vietnam would not be aired. The rest is history.

The Western powers also do not usually take human rights into account when they provide arms, police training, and other vital forms of support to governments that then turn their advantage against their own citizens. When Nelson Mandela was incarcerated by the white South African authorities in 1962, it was a CIA operative who turned over to the apartheid regime the critical intelligence that brought his arrest ("Southscan," 1990; "U.S. Embarrassed," 1987), while American computer corporations sold the government equipment to keep tabs on political dissidents. Before worldwide sanctions were imposed on South Africa in the 1980s, the United States was the regime's leading trade partner. When Angola and Mozambique were fighting for their independence against Portugal in the 1970s, the United States supplied the colonial power with the weapons used against the African revolutionaries. And when Ferdinand Marcos instituted martial law in the Philippines in 1972, the "hotline" telephone network he used to round up politicians, professors, journalists, and labor and student leaders was provided by the U.S. Agency for International Development.

Many countries would like to have the economic growth of NICs like South Korea. But there are not likely to be many new "South Koreas" because the circumstances of that country's material expansion are rather unique—in fact, largely a by-product of the Cold War. Its history since separation from Soviet–allied North Korea in 1945 (made permanent in 1948) brought a bloody war in the 1950s, a high level of Western investment intended to make it a showcase for capitalism (vs. socialist North Korea), an enormous inflow of U.S. military assistance, including the stationing of U.S. forces (now more than 50 years), and a long, only recently relaxed, police state system.

A positive aspect of its development was both its early commitment to land reform and the active involvement of the state in fostering a protective environment for domestic industry, policies that the landlord-dominated Philippines would be wise to emulate. South Korea's impressive growth statistics, however, hide low wages, extreme polarities in income distribution, and the continuing interference of government in efforts to bring about political democracy. In the media, the Kim Young Sam government has interfered with union activity and used the police to break up strikes in the press and television over efforts to elect management. At the same time, government courts in 1996 convicted the last two leaders of the dictatorship and many of the military generals of crimes against the people.

INFORMATION INFRASTRUCTURE
AND THE MAKING OF
A GLOBAL PROLETARIAT

The people in the Third World who make the commodities or provide the services that foreigners use are themselves the least likely to get access to the products of their labor. Workers make telephone sets, television components, computer circuits, audio- and videocassettes, as well as Levi's jeans, Nike running shoes, sports equipment, Barbie dolls, and clothes for Macy's and Bloomingdales, but with most working at or below the poverty line, they cannot share these goods with their own families. Poor peasant women make up the great majority of these workers. Their output is for export only.

Telecommunications has been essential to the development of export processing zones, where these commodities are produced, linking them to the port cities and their overseas headquarters. Meanwhile, poverty forces millions of Third World professionals, including engineers, doctors, and nurses unable to find rewarding work at home, to become the cheap imports of industrial countries. In the Philippines, the repatriation of dollars, some $6 billion annually from three to four million overseas Filipinos, is the largest single source of that country's foreign exchange, making people their number one export (Broad & Cavanagh, 1993, p. 14). As Marx wrote:

> Every development of new productive forces is at the same time a weapon against the workers. All improvements in the means of communication, for example, facilitate the competition of workers in different localities and turn local competition into national, etc. (cited in de la Haye, 1980, p. 52)

Transnational communication opened the Third World to deeper levels of economic, political, and cultural penetration and to vast areas of the world relatively untouched by Western capitalism. Economically developed and less-developed nations have adopted communication technologies in ways that improve some people's lives, but the opportunities have not been widely distributed much beyond privileged elite enclaves. The overwhelming majority of Third World citizens still have no telephones (with access rates in rural areas of 1 telephone per 1,000 population), no full-time electrical power, and no constant source of adequate food supply, much less advanced education or upward social mobility. In poor countries, the use of international circuits is dominated by the transnational enterprises of the OECD countries. During the last 30 years of the "information revolution" (starting with communication satellites), more war- and hunger-related deaths have occurred than in any previous human epoch (Lent, 1986, p. 14).

Moreover, despite political independence for Third World countries, the legacy of economic, cultural, and informational hegemony—structures of foreign control that undermine nominally independent political systems—remains. Many countries in the South that have had some success in building up an industrial base of their own are locked into a form of debt bondage resulting from their inability to pay back huge loans extended to them during a more promising phase in the world economy. With few exceptions, the Third World has not yet achieved the industrial or technological means to compete on a level playing field with the Western capitalist countries. The important social-psychological factor of national identity is undermined by the fact that cultural industries (e.g., broadcasting, film, theater, books, magazines, music, advertising) are permeated by Western, especially British and American, imprints.

India, which prides itself in having a relatively strong national and anti-imperialist political culture (having successfully fought an anticolonial war for independence in the 1940s) and having created a strong film industry (more than 800 films produced each year, more than a third of the world total), has experienced a new wave of cultural penetration since recently opening its economy to foreign investment. Both the English–language version and a dubbed Hindi version of *Jurassic Park* swamped Indian theaters in 1994, prompting Indian filmmakers to fear an influx of many more dubbed Hollywood films. Hollywood, of course, can produce more technically advanced and far better-funded films than the domestic variety, whose industry also has to contend with pirated foreign videos and satellite-delivered foreign television programs (Joshi, 1994). Before the 1980s, India was well known for its dedication to self-reliance, both in technical fields and in indigenous popular culture. Since the 1980s, however, that has changed as the government has moved toward more foreign investment and foreign advertising and a more urban-centered, elitist, and Western–style entertainment-oriented approach in media development (Pendakur, 1991).

Western news agencies are additional conduits for transmitting foreign viewpoints in Third World countries. Although Nigeria, by far the largest country in Africa, has been politically independent since 1960, its national news agency (NAN) is still heavily dependent on Western news content for its information. Of its foreign news reporting, a third comes from the British Reuters news agency, twice as much as the continental Pan–African News Agency, which gives the Europeans a great deal of authority in defining world events. Businesses in Nigeria and elsewhere in Africa prefer Reuters for on-line services, both news and financial reports. This preference greatly influences the economic outlooks of Nigerian business leaders and, ultimately, the type of capital that flows into the country and the continent, thereby tending to service the lifestyles of elite Reuters subscribers far more than the poor (Musa, 1990).

In Latin America, the picture has long been the same, although the U.S. news agencies, Associated Press and United Press International, have been the region's main foreign news sources. Latin Americans thus read about foreign events, even those occurring in neighboring countries, through the filtered lenses of U.S. publishers. Many stories that might be critical of U.S. corporations or the policies of the U.S. government or general news about the region may not reach the readers of Latin American newspapers because few domestic papers have the resources to support a network of foreign correspondents. Foreign commercial advertising encourages import consumption patterns that indulge the small elite fraction, thereby depleting the dollar reserves of national banks that would be available for investment in schools, public health, job creation, domestic cultural institutions, and other areas that serve the average citizen (Reyes Matta, 1976).

In explaining U.S. foreign policy behavior in the Third World, American news agencies can usually be relied on to take their own government's side. There are several explanations why this is the case. One is that media corporations, as described in Chapter 5, are large private entities with interlocking directorates and ownership in big industrial corporations and banks that have major investments in the Third World and therefore tend to share ideological prescriptions favoring private enterprise, "free trade," and unregulated media against nationalists who would prefer a more interventionist role for the state.

Another factor is that although major news agencies are capable of being quite resourceful in gathering overseas news by themselves, U.S. government sources (e.g., embassies, the CIA, the White House, cabinet-level departments, Congress) are important in feeding tips, stories, and contacts to the press on a regular basis. Without these sources, life would be much harder for ambitious journalists, their editors, and publishers. Third, most journalists and editors in the powerful media that operate overseas bureaus value their prestigious positions highly and have passed many gatekeepers' reliability checks along the way to the top. They will rarely jeopardize their careers by confronting the highest circles of state power and possibly end up unemployed or else writing obituary copy. Raymond Bonner, one of those exceptions at the *New York Times,* insisted on filing stories about the human rights abuses of the Salvadoran government, which the Reagan administration was supporting. Unwilling to accept his editors' censorship, Bonner was forced to leave the paper.

It is difficult enough doing local investigative reporting, let alone taking on foreign policy controversies, when, as is usually the case, there is general consensus about "national interest" objectives of the state, as defined by the two dominant political parties. The White House and the U.S. Information Agency (U.S.I.A.), which operates Voice of America and other overseas propaganda channels, can and do coerce news journalists into toeing the government line or holding back information from the public that would interfere with immediate official policy objectives. During the war in Vietnam, the U.S.I.A. frequently flew American journalists into the country at government expense as part of its news management function (U.S. Senate, 1966, pp. 4-6). The U.S.I.A. is an outgrowth of the World War II–era propaganda agency, Office of War Information. It runs U.S. Information Service branches throughout the world.

If these channels of influence are not sufficient, the CIA has long maintained "assets" among the American and foreign media, both Americans and local publishers, editors, and journalists on the payroll or in active cooperation with the agency. In 1977, Carl Bernstein (of Watergate fame) revealed in *Rolling*

Stone how the CIA maintained a global corral of mass media agents and staff "who pushed stories" or "sometimes blocked articles" at the behest of the agency (Lee & Solomon, 1990, p. 117). American television and print news organizations long gave credibility to a CIA–concocted story, without minimal standards of evidence, that claimed that the assassination attempt on Pope John Paul II by a Turkish neo-Fascist in 1981 was a plot hatched by the Soviet KGB and Bulgaria. Although the media eventually repudiated the story in a kind of news footnote, it revealed another case of media complicity in the manufacture of Cold War consciousness.

The CIA has had different methods of dealing with the U.S. media. CIA Director William Casey, who was a key conspirator in the secret and illegal operation to subvert Congress and the Constitution in the Iran–*Contra* episode, used the threat of prosecution to keep news organizations from releasing to the public what he regarded as sensitive information. His successor, William Webster, preferred the "good cop" approach of trading scoops in return for newspaper promises (including those from the *New York Times, Wall Street Journal,* and *Washington Post*) "to kill, alter or delay articles concerning CIA operations" (Kurkjian & McConnell, 1989, pp. 1, 5). The media critic Edward Herman (1992) sees U.S. media behavior involved in an Orwellian "doublespeak," an unselfconscious and internalized double standard of reporting that allows news organizations to defend repressive foreign governments allied with the United States and attack others, even if less repressive, that the government treats as "enemies" of the world capitalist order.

For jobs that the regular media could not handle, Casey's CIA had secret internal publications. One of them, the *Manual del Combatiente por la Libertad (Freedom Fighters' Manual),* was a "comic book" produced by the agency, which it circulated in the thousands by secret airdrop over Nicaragua. Paid for with U.S. tax dollars, the illustrated "comic" showed Nicaraguans how to sabotage their government and economy, with suggestions ranging from the almost innocuous (reporting late to work, losing tools, letting water faucets run), to the criminal (knocking out factory power, damaging state vehicles), to open rebellion (throwing Molotov cocktails at police stations). This was part of a larger extra–legal strategy to overthrow the elected socialist government, the Sandinistas. Other tactics included the mining of Nicaragua's commercial harbors and providing guerrilla training for a Honduras–based counterrevolutionary force, the *Contras,* which routinely carried out acts of terror against civilians inside the borders of Nicaragua.

The American business sector has its own information lines to and from the Third World. International communication "highways" have made it easy for transnational capital to bypass the barriers of time, space, and terrain and enlarge the scale, scope, and profitability of enterprise anywhere, anytime. Global networks of satellite and oceanic channels are for sale to the highest bidder.

Among the best clients are transnational financial institutions that use data entry workers and software writers and editors in the Caribbean, China, and south Asia to process and transmit information to headquarters at one-tenth to one-twentieth the pay scale of their U.S. counterparts. Not to be outdone, in the United States, female prison inmates in North Carolina are used to process hotel bookings from 800 numbers.

Telecommunications has made it possible for U.S. telephone companies to transfer toll-free call handling work to low-wage operators in Puerto Rico and to eliminate thousands of other jobs with voice recognition technology. In Taiwan, South Korea, the Philippines, Thailand, and China, workers churn out the repetitive graphic reproductions for the animated films of Hanna–Barbera, Warner Brothers, and Disney. Indian engineers, working at a fraction of the salary of their Western counterparts, write much of the software used by American computer firms, such as Texas Instruments. TNCs with the collaboration of transnational elites have organized the international division of labor on the basis of a Darwinian competition for personal and nation state survival.

The declining cost and improved transmission capacities of telecommunications has made it very efficient for TNCs to operate "offshore back offices" in Third World countries, but this requires the latter to provide modern and expensive infrastructure for that purpose. Banks, insurance companies, airlines, and hotels use workers, primarily women, in Jamaica to punch keystrokes in monotonous repetition at rates as low as $1 per 1,000, and in the Philippines and China, at 90 cents, and with few benefits. This relocation of work, according to a study by Mark Wilson, enables TNCs to employ non-unionized and large numbers of temporary and part-time employees and to segregate this section of the labor force from its northern coworkers. The major capital cost for the foreign employer is for the highly portable personal computers (Wilson, 1995, pp. 205, 211, 218).

Generally, the conditions of work in the South are far below standards established in Western countries. In Haiti, the basic wage in foreign corporations, more than 90% of which are American, is 13 cents an hour. Semiconductor workers at Intel, Motorola, and other U.S. companies in Malaysia, working with toxic substances under extremely stressful conditions, are prevented from organizing for better conditions, with the active intervention of the Malaysian state and the complicity of the U.S. government. In Mexico and many other countries, labor organizers live in fear of being arrested or liquidated by their foreign-backed governments. Newspaper and newsmagazine stringers in many Third World capitals are asked to provide Western news agencies and correspondents with the raw material that stereotypes their countries' proclivity to violence and disaster.

The United States also has deep divisions based on conspicuous class, gender, race, and ethnic differences in the hierarchy of power, including the communications and information industries. The U.S. Equal Employment Opportunity Commission (EEOC) found, for example, white men dominant in managerial and professional job categories in the computer and electronics companies, whereas the semiskilled labor jobs were largely filled by nonwhite women (cited in *Global Electronics,* 1993). Meanwhile, tens of thousands of jobs in communication-related industries have been relocated in recent years from the United States to Asia, Latin America, and the Caribbean. Manufacture of U.S. company-produced color television sets from the 1960s onward came to rely heavily on Mexican and Asian labor, from 23% of the value of components to more than 90% by the late 1970s, leading to massive American layoffs in the industry (Bello, 1994, p. 89). By the early 1980s, U.S. TNCs employed more than 10 million workers beyond its borders, including 65% of Kellogg's, 60% of Ford Motors and ITT, 55% of Colgate–Palmolive, and 50% of Goodyear and Citicorp, exerting downward pressure on remaining U.S. labor employment (Barry et al., 1984, p. 12).

The adoption of new communication technology also has many unforeseen consequences. The Communist Party–led New People's Army (NPA) in the Philippines has undergone a degree of "professionalization" following the release from prison of their leading cadres by the Aquino government in 1986. Computer technology became one of the modern weapons the NPA employed to keep files on party business and membership. By the early 1990s, two thirds of the executive committee of the party were arrested as a result of the government's capture of NPA computer diskettes. The most extensive research materials on the recent history of the party and NPA are now to be found inside the military's headquarters.

In El Salvador, the FMLN guerillas' "Radio Venceremos" (we shall win) transmitted nightly from the mountains from 1981 until the peace accord in 1992. It broadcast news of skirmishes with the U.S.–backed rightist military regime, details of atrocities committed by the government's army, including the first report of the El Mozote massacre of hundreds of civilians, revolutionary editorials, and folk songs. U.S.–supplied equipment helped the Salvadoran army locate and blow up the guerilla transmitter, but the rebels rebounded by cleverly running their signal along the barbed wire fences of the rich ranchers in Morazan province—poetically making the *hacendados* inadvertent participants in the struggle for land reform.

Following one staged retreat, the guerillas left behind a bomb packaged inside a radio transmitter, which U.S.–trained Lieutenant Colonel Domingo Monterrosa, the organizer of the El Mozote massacre, took as his victory trophy aboard a military helicopter. Once Monterrosa was airborne, rebel engineers on the ground remotely set off the device (Simon, 1995). The popularity of the rebel cause made it impossible for President Reagan, despite his counterrevolutionary fervor, to send U.S. troops to invade El Salvador. Given the U.S. history of intervention in the region, it was a restraint conceivable only in the context of the earlier Kennedy–Johnson–Nixon debacle in Indochina.

Despite the great income and technological inequalities between the indus-trialized countries and the Third World, the latter have been making headway in certain areas. Aware of the consequences of foreign domination of the media, many Third World countries have joined in regional and interregional news organizations, such as the News Agencies Pool of the Non–Aligned Countries, Inter–Press Service, the Caribbean News Agency, the Asia Pacific Broadcasting Union, the Organization of Asian News Agencies, the Asia–Pacific News Network, the Pan–African News Agency, the Union of African National Radio and Television, and others (Ansah, 1986, p. 68). Lack of funding is a major problem for these organizations. The economic recession in the 1980s made it even more difficult for them to survive, but they continue to work for more independent information structures.

An added problem is that the economic development and information access that does occur in most Third World countries is concentrated in urban business districts, worsening the income and information maldistributions between city and rural areas and alienating the Third World privileged classes from ordinary people. This also has led to migration and severe overpopulation of Third World cities that are not capable of providing services to the majority of new migrants, creating threats of epidemics and other crises of overcrowding. Lagos, a city that had 300,000 residents in the 1950s, had 8 million by 1990. Mexico City had 24 million inhabitants by 1995; Sao Paulo, almost 22 million; Seoul, 19 million; Bombay and Calcutta, 13 million; and Manila and Jakarta, 11 million. By the year 2000, these seven cities alone are expected to add 30 million people, a 30% urban population increase in just 5 years.

Absorption of Third World elites into the mainstream of transnational culture marginalizes the poor even further and leaves them few, often only the most extreme, alternatives, including war. Information technology and media do not create poverty, but their introduction into the mix of already severe class segregations are likely to make life worse for the majority while solidifying alliances among transnational elites. As one African writer noted, the "informa-tion revolution . . . [has] no geographical boundaries. It [does] . . . not stop at the equator. Its boundaries [are] . . . social" (Ansah, 1986, p. 69).

PART V

CONCLUSIONS

The concluding chapter sums up the main ideas of the book and considers areas for citizens to focus their attention in the coming new millennium. As powerful as transnational forces have become in the post-World War II era, there are many "cracks in the empire." Communications technology is the outgrowth and application of scientific research and discovery, but as we have seen, the science and technology community is not autonomous from the larger political and economic forces under which it is governed. The question turns on how to reformulate political ideas so that the ideals of democracy can be recaptured and communicated over the very channels that, in the past, prevented those ideals from blossoming. Every person made aware of the problems of the "information age" and of the alternatives is a new shoot of earthly possibilities.

8

Communication, Technology, and Politics

This land is your land, this land is my land.

—Woody Guthrie

. . . you take my life when you do take the means whereby I live.

—William Shakespeare

COMMUNICATION DISCOURSES:
REINSTATING THE "WHO"

If the dominant discourse on communication technology were to syntactically shift from the object-focus, "what" is changing, to the subject-focus, "who" is doing the changing, this would mean, first, identifying the individuals and their organizations involved in the design and production of communicating devices and media and the social marketing values they hold; knowing as much about corporate decision making as corporate executives know about consumer behavior. Second, it would emphasize the social needs in communication systems more than the needs of capital. And third, it would mean putting people first in the planning of communicative systems: Does the communication design maximize the capacity and opportunities for human development for the greatest number, as well as its poorest members? Does it encourage interactive, as opposed to top-down, participation? Is it accessible in terms of easy and inexpensive usage?

INFORMATION ACCESS AND
THE STRUGGLE FOR DEMOCRACY

The greatest victories in the building of democratic societies have been won through struggles. Shays' Rebellion was one important expression of resistance to tyranny that led to the Bill of Rights. But even the armed struggle of the Civil War was not sufficient to bring African Americans into full citizenship; that required continuous legal challenges, political mobilization, and civil disobedience before barriers based on race began to come down. Workers' efforts to organize or strike for better conditions and benefits and militant demands for greater social security were often met with business, state, and police violence before things got better. Harry Braverman (1974) noted that "capitalism creates a society in which no one is presumed to consult anything but self-interest" (p. 67). At the same time, increasing contact and communication between peoples of different nations and the protections of a democratic charter have widened the expectations, if not the fulfillment, of a more participatory discourse.

John Jay, president of the Continental Congress and the first chief justice of the U.S. Supreme Court, declared, "The people who own the country ought to govern it." James Madison wrote in 1822, however:

> A popular Government without popular information, or the means of acquiring it, is but a Prologue to a Farce or a Tragedy; or perhaps both. Knowledge will forever govern ignorance: And a people who mean to be their own Governors, must arm themselves with the power which knowledge gives. (cited in Katz, 1987, p. 2)

New communication channels offer conflicting opportunities for both broad delivery of useful information and more highly segregated systems of knowledge. The increasing polarization of education into privileged and neglected school systems is not a favorable indication of what may follow the introduction of a private sector-driven "information superhighway." Unbounded faith in marketplace approaches to resolving public policy disputes has only added to the problems. It has led to a higher rate of technology-driven unemployment: a dual economy of poorly paid low-end jobs on one end and a relatively small professional corps on the other. There has been a demise of public institutions, libraries, public schools, park services, funding for the arts, public broadcasting, and, in general, the philosophy of public service. A tabloid media-style system of selecting political representatives serves as the electoral process. And mass marketed culture-for-profit, as in cable TV, has been offered as a substitute for classical and critical studies of history, politics, aesthetics, and philosophy.

Modern communication technologies are embedded with a blueprint of modern social relationships, which, disconnected by space, are reunited in time. The commonplace sight of people in public spaces talking into portable telephones reveals how, in an atomistic world, people attempt to preserve some form of dyadic relations and how this, in turn, helps normalize the condition of social isolation. Tenaciously, commercial interests use such instruments to extend the reach of enterprise into areas once considered beyond possible penetration. Yuppies strutting in tranquil state parks talk over cellular telephones with their brokers, traveling sales representatives contact home office supervisors while 30,000 ft airborne, MBA students get stock market reports via satellite over their beeping wristwatches, subcontractors bring pagers to beach resorts, and restaurant patrons pause between courses to send e-mail and faxes.

In Cambridge, Massachusetts, in Portland, Oregon, and in cities in between, cafes provide desktop computers with communication software for companionship while patrons sip cappuccino. Everywhere one is reminded that time is money, and that leisure, like art and education, has no use beyond its exchange value. A utilitarian 24-hour commodities market culture calculates every moment "wasted" in lost opportunity costs and potential dollar earnings.

Within a society as influenced by mass media as the United States, the existence of 500 television channels plus thousands more of electronic bulletin boards and CD-ROMs would most likely induce greater passivity, a demise in critical thinking, banal commercial formulas for coping with crises, and more confusion and functional illiteracy among an already poorly educated populace. When Americans began to be wired through a widely diffused telephone system, public pressure was enough to force the dominant and regulated supplier and service operator, AT&T, to maintain a policy of affordable access to local lines. AT&T eventually had to allow people to purchase their own telephone sets, instead of leasing. Deregulation may very well have encouraged the creation of new equipment and services, but it has also segregated the market into customized user segments that make local telephone monopolies less attentive to universal service principles. As business user rates for long-distance telephone (using voice, video, data, or message services) decline, local residential rates (used primarily for voice) increase. If telephone charges were calculated on the basis of bits of information transmitted per second, the real advantage to corporations would be even more pronounced.

Mark Cooper, director of research at the Consumer Federation of America, contends that the government and business communities have not demonstrated a commitment to sharing the information superhighway with the general public. The prospect of yet more supermergers across telephone, cable, and television industries would give such conglomerates the option of catering only to high-density or the most profitable areas, ignoring the rural populace and inner-city poor. Government support for the superhighway, which Vice President Al Gore has actively promoted during the Clinton administration, may ultimately turn on the ability of huge communication empires to deliver movies on demand, rather than on the more useful but less glamorous possibilities, such as telemedicine (Hyatt, 1993). Federal, state, and local governments can be expected to pitch the educational potential of information networks with the same clichés that classroom computers were originally sold and with just as little public investment or study of its pedagogical returns. Many authors, from different ends of the political spectrum, have taken a kind of mystical leap of faith that a laissez-faire private market god, benevolent or Darwinian, will take care of everything.

A more skeptical but pragmatic view among activists for social change is that we have little choice but to fight for public access channels, lest the public sphere become even more marginalized or swallowed up by private corporate interests and culture. Some see electronic communication as a supplemental tool in restoring a sense of local community vitality; others passionately exude over the possibilities of a computer-based democracy. As Stanley Aronowitz and William DiFazio critically observe, "Many scholars, critics, and journalists have greeted the phenomenal dissemination of cybernetic technologies from the office and factory to the home and recreational sites as the precursor of a new era of community, of perfect communication and even of a new explosion of eros" (1994, p. 77). They regard the emerging "technoculture" as inherently authoritarian, but one that ultimately has to be viewed, not within a utopian or dystopian framework, but within concrete assessments of its impact on people's livelihoods and their social independence from corporate agendas of power.

Various social activists, recognizing that information technology is too powerful a tool to be left to corporate magnanimity, have taken the issue into their own hands and started up local projects to empower communities with computer access. Some cities, such as Santa Monica, Glendale, and Pasadena in California, and Seattle in Washington, have set up on-line systems for public use, although no spectacular breakthroughs in altering the imbalances in information access have yet been reported (e.g., see Guthrie & Dutton, 1992). Even where such services for the public do exist, it is generally not the case that most Americans are wired into the information complex. Most people continue to take their messages from worklife interchanges and from the snippets of news events and culture served by network television, celebrity-focused magazines, Hollywood films, and what John K. Galbraith called "a relentless propaganda" of ubiquitous commercial advertising (1967, p. 218).

As one writer wryly notes, "Lone, libertarian computer hackers, government officials, and corporate CEOs across the country share a vision. They see a national network of electronic connections revolutionizing society, its grand design unfolding naturally, as if proceeding from technological DNA" (Reed, 1995, p. 51). Although some may see "electronic democracy" as the wave of the future, others involved in struggles for popular and distributional justice see computers as only one of many tools. Information may indeed be an increasingly important asset in American society, but information by itself is not very valuable without the educational preparation on how to understand and apply it to real knowledge and real-life situations. As T. S. Eliot asked in Chorus from *The Rock,* "Where is the wisdom we have lost in knowledge? Where is the knowledge we have lost in information?"

In a society in which knowledge is given higher priority as a public good, the schooling system will have to receive a larger share of the national wealth. In fact, however, public education in the United States has been under continual assault, especially by neoconservatives who want to privatize it and eliminate the Department of Education and federal aid to public schools. But even high-quality and equitably distributed public education is only part of the solution to inequality, lack of economic opportunity, and social injustice. Unless jobs that provide adequate income and esteem are substituted for those taken away by technological innovation and footloose transnational corporations (TNCs), educational opportunity alone will not bring the allocative benefits with which it is usually associated. Genuine democracy must be present at the site of production itself. Americans also need to understand that a strong public sector and the right to organize in the workplace (as guaranteed in U.S. law and in Article 23 of the Universal Declaration of Human Rights) are among the most enlightened features of a democratic society.

HEGEMONIC ORDER VERSUS
INTERNATIONAL PEACE AND COOPERATION

The sites and managerial centers of private production, however, are the spaces most carefully protected by legal-political structures, police/intelligence surveillance, and corporate media ideology. Even though close to 60,000 deaths a year in the United States are related to industrial accidents and unhealthy workplace conditions, these tragic stories rarely make it to the nightly television news. In fact, hardly a newspaper in the country has a labor reporter, even as whole sections of the press, if not virtually all of it, are devoted to business news. The broad protections for capital and the legal constraints on worker rights, notwithstanding their occasional recourse to the picket-and-strike weapon, profoundly discourages any genuine bargaining relationship between the two. If the conditions for labor in the United States are bleak, they are still more brutal in other countries, including those where U.S. TNCs have set up "offshore" operations.

U.S. transnational executives may not be directly involved in the repression of labor union organizers in Third World countries, but they nonetheless do welcome having work environments free of their presence. It is not a big step for regimes seeking U.S. investment to create attractive conditions, which include the organization of a compliant workforce through whatever means necessary. It is well known, and was even revealed on one occasion by CBS's *60 Minutes,* that the U.S. government cooperates with governments and businesses in Central America in maintaining and passing along to potential U.S. investors blacklists on labor militants and organizers. In a kind of "don't ask, don't tell" relationship with Central American governments, TNC executives can then exploit indefensible Latino workers through miserable wages, benefits, and workplace conditions. It is difficult to imagine that they are ignorant of the behind-the-scenes arrests, tortures, and government-sponsored death squad murders of labor leaders that is so common in countries like Guatemala, Honduras, and El Salvador, especially with the exposure given by several prominent human rights monitoring organizations, such as Americas Watch and Amnesty International.

U.S. military intervention in the Third World, particularly in the Caribbean and Central America, can be thought of as a kind of overhead capital for protecting the interests of U.S. investors. Since the 19th century and especially after the building of the Panama Canal, the State Department has come to think of the Caribbean Basin as an American lake. Frequent military interventions in the region in the early 20th century furnished protection of the shipping lanes into the canal and suppressed local revolutionary movements, as in Woodrow Wilson's forays into Mexico, and U.S. Marine occupations in Nicaragua from 1912 to 1925, the Dominican Republic from 1915 to 1924, and Haiti from 1915 to 1934. With his incursions into Mexico, Wilson was determined to teach the Mexicans "to elect good men." Everywhere at the time, U.S. presidents were convinced of their own noble intentions, a sentiment echoed in the mass media of the day, and their best students in Latin America and the Caribbean were business and military leaders ready to cut a deal with their northern patrons.

Until the collapse of the Soviet Union, the U.S. military was spending $300 billion a year in defense contracts to help keep the world safe for American capitalism, and the CIA was given an estimated $3 billion (some say $30 billion) for related intelligence and sabotage missions. In April 1995, it was reported that the CIA was continuing covert "liaison" and financial support relationships with intelligence agencies and military forces in other countries, including Guatemala and Haiti. In the 1970s and 1980s, Saddam Hussein and Manuel Noriega had similar arrangements with the CIA that had been kept secret from President Clinton, the Congress, and most important, the American people. In 1994, it was reported that for many years the CIA had funded Japan's ruling

Liberal Democratic Party to undermine that country's left parties and protect the single-party rule that lasted 38 years (Weiner et al., 1994). In recent years, the CIA has shifted some of its emphasis to international commercial spying in support of U.S. overseas corporations and trade relations and using electronic espionage against economic rivals such as Japan (Risen, 1995).

Apart from the human capital drawn from the ranks of the American working class, U.S. corporations have an enormous advantage in wealth accumulation through the transnational system of information gathering and transmission. The global information tool allows TNCs to maintain overseas subsidiaries even in the remotest locations. This enables them to move jobs out of the United States, playing off one national workforce against another in an internationalized division of labor, set up foreign markets for its products, permeate Third World societies with Western commercial culture, move money around at will, manipulate foreign governments, and consolidate the power of transnational capital against smaller capital and working people. Without a working-class media (hundreds of pro-union newspapers existed in the 1950s; few have survived) or a labor party (unlike western Europe, although declining there too), American workers, now joined increasingly by sections of what was the middle class, still have a weak class or union identity. As a result, it is difficult for most Americans to make sense of the political economic changes that are occurring in their midst, to relate to working people of other countries (including northern Europe, where workers, on the whole, are much better off), or to figure out ways to mobilize in defense of their own interests.

The Cold War justifications for the militarization of the American economy and culture, which narrowed the range of political thought in the country, did not end with the collapse of the Soviet and east European communist parties. With relatively little provocation, the United States remains mobilized for nuclear and conventional wars around the world, even if wary of Vietnam–type engagements. The much-heralded post-Soviet "peace dividend" that was supposed to yield services for the lower and middle classes has yet to materialize, postponing investments in basic needs, such as job retraining, public infrastructural investment, national health care, assistance to public education, scholarships and the arts, a livable minimum wage, urban renewal, youth programs, support to AIDS research, and other programs that would alter the declining conditions of life for most Americans. The New Right in American politics is committed to trashing the public sector as "inefficient" and implies that those with misfortune have only their defective values or genes to blame. And the mass media both feed and profit from these stereotypes and scapegoats with sensational crime stories and courtroom dramas, outrageous talk shows, voyeuristic soap operas, dramatic superhero/supervillain television and film plots, and pure escapist fantasies. With few exceptions, the mass media degrade or ignore people's real experiences, failing to present either the economic conditions underlying the decline of America's social institutions and social behavior or the activists at the grassroots who are trying to make a difference.

COMMUNICATION STRUCTURES
AS IF PEOPLE MATTERED

After World War II, a segment of the intellectual and working class leadership in the United States, largely unknown outside their own inner circles, began to develop new public media dedicated to critical and sometimes radical critiques of modern U.S. capitalism and its structures of power. Among the early print media were the *Monthly Review* and the weekly *Guardian,* both of which were founded in lower Manhattan, as was the Communist Party's prewar *Daily Worker* (now *People's Weekly World*). Monthlies, such as the *Progressive* and *The Nation,* are considerably older (*The Nation,* starting in 1865, is the oldest U.S. weekly), with left-liberal, less revolutionary orientations, but known for their tradition of investigative journalism and crisp cultural critiques of conservatism and the arts. They are joined by other progressive periodicals (e.g., *In These Times, Dollars & Sense, Extra, Covert Action Information Bulletin, Lies of the Times*) with some of the most insightful analysis and demystification of American institutions but that have perennially limped along financially.

Publications like *Antipode* for radical geographers and *Science for the People,* a social-democratic and popular journal on science and society, have reached people who wish to be educated about social impacts of public policy, science, and technology. Unlike mainstream media, all of these publications have had difficulty attracting advertising or other forms of funding beyond supplemental reader contributions and some foundation support. The broadcast realm has seen Pacifica Radio and Alternative Radio, both socialist-oriented and usually transmitted over university or on a few public radio stations. National Public Radio (NPR) and the Public Broadcasting System (PBS) are both mixed government-, corporate-, and public-sponsored broadcast networks that stay close to the political center, with an occasionally more liberal analysis than found on commercial network news. By the 1990s, however, 70% of prime-time programming on PBS was funded in whole or in part by four major oil companies, prompting the epithet, "Petroleum Broadcasting System" (Parenti, 1995, p. 175).

Paper Tiger Television in New York City produces critical documentaries for sale to cable stations around the country that pick it up, but this, too, along with community cable channels, still reaches small urban audiences. Right-wing media that oppose social reforms, defend market capitalism, and boost imperial foreign policy interests generally get considerably more support, mainly from rich corporate patrons. Their strongest voices, including Rush Limbaugh, whose nightly television program had George Bush's former media advisor, Roger Ailes, as its executive producer, have made inroads on the centrist media. With his conservative broadsides carried to 20 million listeners on more than 600 radio stations across the country, Limbaugh is often credited with mobilizing public support for the Republicans, which captured the 1994 and 1996 House and Senate elections (though with only 38% and 49%, respectively, of the electorate voting).

Despite the current right-wing trend, the mediascape has greatly changed since the end of World War II. The struggles waged by African Americans for political enfranchisement and the enormous influence of their heritage on American culture as a whole has greatly opened up and altered the complex societal mosaic and ways of thinking about "Americanness." Yet, political equality in America is far from being achieved—in racial, ethnic, gender, sexual orientation, or class terms. Continuing inequities represent the primary focus of struggle in our time, particularly in the light of the backlash of reactionary anxieties that would turn back the clock on civil rights and economic democracy. Social equality, equity, and justice, combined with the struggle to preserve the earth's habitat, provide a set of goals by which to measure the value of American institutions, including those in media and communications.

There is no shortage of information to critically study the human and other elements of the living environment. Most of the established channels of information are private and faithful to principles of marketing and profit seeking, however, rather than forums for popular debate and critical analysis of society. Light-headed escapism or superficial docudrama in the mass media substitute for consciousness-raising, both in their "news" and entertainment formats. From a participatory standpoint, use of the term *interactive* television has come to mean little more than choosing from a menu of pay-per-view programs or having electronic home shopping channels.

Mass media hype the "information society," and decidedly there are more opportunities for gaining access to information than before, especially through electronic communications. But information is not necessarily education, and education is not necessarily socially useful knowledge, at least not knowledge of the sort that teaches people how to interpret and assess the human and ecological condition or that raises the standards of community well-being. Most will agree that the Internet is, for the most part, a reservoir of junk mail and low-grade information.

To the extent that more channels now exist for accumulating "facts," the mass media generally give little guidance as to their relative scientific status or how to fit them into larger composites of knowledge. One concerned academic notes that, in American educational institutions, students "are overwhelmed by the factuality of the observed world and have enormous difficulty making the jump to concepts which may controvert appearances" and that as "critical thinking is the fundamental precondition for an autonomous and self-motivated public or citizenry, its decline would threaten the future of democratic social, cultural, and political forms" (Aronowitz, 1987, p. 467).

Moreover, many of the dominant communication channels (television, films, video, advertising, public relations) consist largely of one-way flows of information that are visual in nature, including, increasingly, the Internet and print media, discouraging citizens from honing their interactive communication skills. With television, personality picture magazines, and pulp fiction mediating how people think about their world, community standards of literacy and speech could be expected to decline. Illiteracy climbed in the United States from 4.3% in 1930 to 8.2% in 1986, whereas newspaper readership declined from 32% to 26%, dipping most dramatically after the introduction of television. Average daily telephone conversations, in contrast, increased from 670 per 1,000 population to 6,900 (Schement & Curtis, 1995, pp. 50, 111), although the content of this use has not been carefully studied. These numbers do not provide precise causal indicators, but the weakening literary standards of Americans do correspond to the declining investment in the country's public education and educational opportunities.

As one study of global culture and communication notes:

> Video is so formidable a competitor for the attention of 9- and 10-year-olds that frequently teachers, especially in the demoralized classrooms of American inner cities, and stretched-out mothers trying to earn a living and keep house (and fathers when they are there) serve neither as effective communicators nor role models. True, the TV set, unlike the teacher, cannot command the child to sit down or even to watch, but for that very reason its influence may be greater. (Barnet & Cavanagh, 1994, p. 158)

Under the pervasive influence of TV culture, "children are being instructed by shadows on the screen on how to become precocious consumers and debtors" coming under the influence of "highly creative communicators with no interest at all in their education" (Barnet & Cavanagh, 1994, p. 158).

With the control of commercial and a theoretically "public" television system under private corporations, the television industry and its patrons benefit enormously from their ability to reach and treat nearly every household in America as a unit of consumption. Television organizes leisure time activity almost as conscientiously as factories organize labor time, one disciplining the production of goods, the other the function of consumption. Commercial television has little interest in edifying its audiences, because its agenda is to sell cereal, soft drinks, and sports cars, playing to gratification impulses, like sex and violence fantasies, and short attention spans in the hope of stimulating consumption behavior. Public television has yielded to increasing pressure to make itself more commercial so as not to "crowd out" airtime and audience attention, particularly as PBS viewers represent a higher-end income bracket and market for expensive consumer goods. Commercialism also tends to dull the political edge in programming, which is of import to status quo conservatives.

Alternative filmmakers, most often of documentaries with critical political, social, or environmental messages, like to meet with their audiences to both display their work and discuss its meaning, usually taking questions on content and technique. The sharing of knowledge and use of public feedback in the communication arts build bonds of interest and participation between artists and other citizens. In the early years of television, a few attempts were made to introduce large-screen community viewing, but this socially oriented approach had unfortunate associations with government propaganda, as in, for example, such efforts by Nazi Germany or the Stalinist-era Soviet Union's wired radio broadcasts. In some cities, like Austin, Boston, and Portland, Oregon, social activists have opened up the possibilities of more enlightened television entertainment through community cable access channels, inviting program production groups that are usually left out of commercial broadcasting's cast of characters—immigrants, African Americans, Latinos, and other inner-city cultures, political dissenters, radical feminists, radical artists, gays and lesbians, union leaders, and antipoverty activists, among others.

Much of the initial excitement echoed in the mid-1980s by educators, politicians, trade journals, and industry public relations officers about bringing American school children across a new threshold of creative and educational experience through the use of computers has since toned down. The actual results thus far have been mixed and, for the most part, disappointing. There are no problems in American education that computers alone are going to fix. To think so is to ignore the chasm of social inequality that exists in the United States, intensified by the uneven quality of education that children are being offered. Computers integrated within enriched learning environments, available at expensive private and elite community public schools, can contribute to a creative, skills-oriented, and stimulating curriculum, such as in learning math and science through electronic game-playing. But at the same time, as John Broughton has argued, computers embody their own boundaries, rules, logic, order of information processing, and other mechanical constructions for organizing data, which are not identical with human methods of learning, discovering, and problem solving (cited in Aronowitz & DiFazio, 1994, p. 135).

Nonetheless, children with learned and appropriate computer skills can, to a considerable extent, recast formal methods of pedagogy into self-initiated educational experiences. Such a climate of learning is not likely to be induced in schools that are too poor to attract the kind of teachers who can turn children on to education or where the encapsulating social environment is too turbulent to use education as a pathway to personal development. The inequality of access to computers roughly parallels the class inequality that exists in the country as a whole.

Many studies show that income and class inequality in the United States have significantly widened in the past two decades and, according to the Census Bureau, is the widest since World War II (Holmes, 1996a). Twenty years ago, the United States had the highest standard of living for working people; today, it is far behind leading industrial countries, 13th in terms of wages, paid vacation, health care, benefits, pensions, and educational opportunities. The income gap between factory workers and chief executive officers has grown to the extent of making the United States the most class-polarized country in the industrialized world. The wealthiest 1% of the U.S. population has more assets than the lower 90%. In recent years, the United States has moved toward rapidly becoming a two-class society: one wealthy and comfortable, and the other poor and struggling to stay afloat.

One result of this inequality is that children of the working classes, especially African American and Latino, receive little training in computer programming and are far less likely to own computers than are European Americans. This is largely because their schools are underfunded, their teachers undertrained, and their parents or guardians less likely to have professional occupations that require computers. Computers in poorer classrooms are instead used for passive, routine, and rote forms of learning that bore students to inertia, occasionally for electronic games, and most often just for typing. Lack of computer literacy follows these children through life, as they are far more likely than their more affluent counterparts to end up in mindless and poorly paid jobs where they may operate electronic cash registers or bar code readers or work as data entry clericals (Piller & Wieman, 1992). Good education for everyone starts with a strong and broad public commitment toward that objective.

Another communication technology with pedagogical potential is radio, but like television, its capture by private commercial interests has greatly limited its development and use as a vehicle for raising popular educational and cultural standards. With little supervision by government and a frequency allocation system that favors commercial broadcasting values, a brand of populism has taken over radio in the format of politically charged talk shows. These programs, such as the most famous one hosted by Rush Limbaugh, typically screen out or disconnect diverse points of view so that "interactive" radio becomes, in effect, a monologue of controlled, often hysterical, propaganda. The vast majority of talk show hosts in the United States are conservatives or ultraconservative who chant a continuous mantra of anti-government and antiliberal slogans. Unable to generate much commercial sponsorship, progressive radio programs are few, but some, such as Pacifica Radio, do circulate in large urban markets and university towns.

In Third World countries, radio is the most accessible communication medium, although it is usually government-licensed and carefully controlled. In Jamaica, where there are 45 hours weekly of talk shows, and in other parts of the Caribbean, interactive radio often serves as a lightning rod for popular dissent and complaints against public corruption, creating a great deal of concern among government officials. Talk radio feedback from listeners helped force a Jamaican minister of finance to resign in 1992 over a controversy involving favors given to a foreign oil company (Thompson, 1993). But in other countries, such exposés often result in retaliation and violence against media personalities. In Latin America, several community radio stations broadcast programs by and about women, helping women and their families organize within political systems that marginalize them to the farthest edges of citizenship and economic survival.

A democratic and egalitarian communication system would be one that promotes community initiative, equitable distribution of wealth, and public and private policy based on equal opportunity. It would also fight for safe working conditions, nondiscrimination in education, respect for diverse cultural expression, appreciation for the natural environment and sustainable ecology, and belief in public access to media and other societal institutions. An authoritarian system, in contrast, has an agenda, often covert, of highly concentrated, centralized, and inherited political and economic power, controlled public expression and participation, education that emphasizes deference to established authority, limited access to specialized knowledge, and a resource use strategy that permits the despoilment of the natural environment for the benefit of corporate interests. Since the late 1970s, the United States has drifted toward authoritarian politics, which also is reflected in the prevalent design and uses of its advanced technology. As Lenny Siegel has written:

> The downgrading of work is no accident. Nor is it the necessary result of the new technologies. Rather, employers have chosen to use the new computer technology to de-skill their work force. The knowledge and experience that it formerly took to run machine tools in auto plants and cash registers at fast-food restaurants have been programmed into the equipment.
>
> The clerk who greets one at a MacDonald's [sic] no longer needs to know how to figure change, memorize prices, or even write orders. The computerized cash register does it all. Some even substitute symbols—such as a drawing of hamburger—for words. Such employers can hire unskilled, non-English-speaking workers, and they can save money by paying extremely low wages. (cited in Bello, 1994, p. 93)

The defense of popular democratic interests requires the politicization of discourse (again, reinstating the "who") and the preservation of a public culture that fosters a relationship of citizens meeting in public places in search of a commonwealth of interests. In an age when commercialism has taken on a furious pace and corporate logos adorn everything from city plazas to Olympic athletes' sportswear to the cornerstones of state university research centers, there is a well-organized belief that nothing of intrinsic historical and cultural value is worth saving for its own sake. In the creed of "postmodernism," commercial interests exploit every cultural landmark and infuse them with "new meanings" in an effort to transform coherent symbolic artifacts into "individualized experiences" and in hopes of breathing new life into evanescent industrial capitalism. Essence does not exist, only the perceived image; social constructions of reality are reduced to the atomized construction of each mind's eye. Systemic understanding of institutions of power is eviscerated into anomic subjective images based on personal "interpretation," isolating the historical role of public and critical intellectuals and denying the existence of organized, narrow, material self-interests and the need for public vigilance.

In the "postmodern" interpretation of culture, the common experiences that mold social identity are reduced and "deconstructed" into separate, individualized meanings, rather than collective ones. This approach renders meaningless the ideas of class, gender, ethnicity, racism, political affiliation, or any other shared identity as objective explanations of consciousness, social status, and power—as if every person were totally autonomous, free of constraints and able to "choose" her or his lifestyle. As a construction of social reality in itself, postmodernism frees existing foundations of power and inequality from historical and political accountability and tends thereby to protect the conservative political order in society. Being rich or poor is little more than a state of mind.

The decline of public consciousness is substituted by technical rationality and corporate business culture where markets, not public terrain and egalitarian values, are what matter most. For marketers, the individual voice registers as a yes/no statistical unit of consumption. The social contract essentially is not between citizens and their elected representatives, but between buyers and sellers. This late-20th-century version of capitalism is particularly strong in Anglo-American society. Shopping malls are the "public parks" of affluent Californians. American college football bowl games now bear the names of corporate sponsors, such as the Tostitos Fiesta Bowl. Reebok effectively owns the athletic program at the University of Wisconsin. In England, the Tory government plans to sell off the names of underground (subway) stations to the highest bidder, so Piccadilly Circus may become General Motors Square or perhaps Matsushita Plaza. The flagpoles surrounding Buckingham Palace, usually used to fly the flags of nations, are to be rented out to big business to run their corporate logos. Everything has potential market value. Everything is for sale.

As corporations agglomerate at higher orders of scale, their sense of public responsibility and accountability seems to diminish. Television is a private corporate instrument for organizing communities of consumption. The quality of programming on American television becomes cruder with each quick series turnover, reaching deeper for the lowest common denominator of taste and the highest rate of profit. Massification of consumption is now more scientifically managed, custom designing programs and advertising to niche interests and generating more flotsam on a wider variety of topics. The leading architect of modern liberal, consumption-driven, economic theory, John Maynard Keynes, scornfully derided ordinary people as the "boorish proletariat." For a time, he moonlighted in England as the impresario of a combined theater and restaurant operation and studied how various types of entertainment offered on the stage corresponded with levels of food consumption. In the production of culture-for-profit, Keynes's dilettantish sideline has been greatly improved on by the more sophisticated corporate entertainment executives on both sides of the Atlantic.

Keynes, who died in 1946, did not live to see the postwar commercialization of culture in his native Britain and the United States or the gigantic institutions that would take over the transmission of mass communication, entertainment, and cultural values. He did not witness the late-20th-century standardization, control, and homogenization of culture and lived experience of most citizens, what one sociologist has called the "McDonaldization of society" (Ritzer, 1993). Often thought of as the standard-bearer of popular culture, Disneyland purports to offer an imaginative and creative flight into a world of fantasy, but its reality is quite the antithesis:

> Far from liberating the imagination, Disney succeeds mainly in confining it. Like the conveyor "cars" and "boats" that pull you along steel tracks through "Snow White" and "World of Motion" and the "Speedway" rides, Disney is a plodding, precise, computer controlled mechanism pulling an estimated 30 million visitors along the same calculated, unvarying, meticulously engineered entertainment experience. It occupies its customers without engaging them. It appeals to everybody while challenging nobody. (Bob Garfield, cited in Ritzer, 1993, p. 137)

The challenge in recapturing genuinely creative and more individual- and community-oriented forms of cultural expression requires that people avoid patronizing the purveyors of bottom-line mass entertainment and information, seek out alternative and progressive voices, public intellectuals, and media, and take initiatives to mobilize for humanistic public policies toward art and culture.

A public medium would emphasize restoring a sense of public community and public ethics, advancing democracy and democratic relationships in society and in the household, challenging obstacles and prejudices that inhibit the participation of those presently denied equal rights and opportunities, and working for the realization of a world actually governed by the people themselves and by their true representatives. The packaging of communication and information as a public good by private corporate interests should always be suspect, given that, as Mark Twain dryly noted, "virtue has never been as respectable as money."

ISSUES FOR THE 21ST CENTURY

The reduction of culture to values of private profit also extends to concerns about the broader natural environment. Since the publication of Rachel Carson's *Silent Spring* in the early 1960s, public awareness of the earth's ecology has ebbed and flowed, largely because of periodic media attention (and general inattention) or specific environmental crises, such as Love Canal, Three Mile Island, and the massive destruction of flora and fauna by chemical pesticides. Corporations have not missed the change in the public mood, and in recent years some of the worst offenders, including the oil and timber industries, have taken on the public relations persona of being environmentally sensitive and engaged in the "green marketing" of their goods and company practices. Although this in itself reflects the increased importance of the environment in the public mind, it does not address the true environmental costs of the U.S. industrial model of mass consumption.

Even if sponsorship could be found on commercial television for critical examination of the impact of uninhibited consumption on the environment, it would be difficult to discuss such a complex issue within formats that allot only a few minutes of programming between commercial breaks. Most regular television programming is set indoors, giving the natural world a kind of exotic status, and ironically, it is often only the outdoor advertising shots that give the viewer any glimpse of green space during the usual schedules of soap operas, talk shows, news programs, magazine formats, sitcoms, and staged entertainment, almost none of which bring up environmental issues. News reporting of the devastating consequences of polluted air and water resources and the production of waste is even more difficult when stations rely on sponsors that produce much of this debris. Thus, television's absorption with interiors (especially bedrooms!), with its production cost controls, its urban-centered indifference to the environment, and its fear of offending mass consumption-oriented sponsors, does not comport well with the need to reduce the commercial-industrial uses and destruction of air, water, and land resources and to recycle what has already been produced.

Western economies are so dependent on expanding the built environment that protection of wetlands, wildlife, air and water quality, the forests, the soil, and conservation areas has come to represent a threat to industrial production, jobs, income, and "standard of living." Cities, suburbs, and exurbs continue to penetrate remaining rural and undeveloped spaces. With such deep attachments to the materialist way of life and dependency on urban and industrial structures, it is hard for many Americans to imagine life measured by any other standard than "I am what I own."

Environmental disasters are waiting to happen. Yet, most governments seem unable to respond to the early warning signs before the onset of actual crises. At present, the pace of worldwide consumption, particularly in the Western countries, continues to grow—joined by the emerging entry of such mass markets as China and India, together having almost 40% of the world's population. At the same time, people in general are better informed than they were 50 years ago about the social and environmental consequences of industrial development. It is well worth asking, What impact will new communication and information technologies have on the world with respect to protecting the earth's fragile environment? How can they be used to help organize peaceful relations among peoples, provide them, especially the poor, with greater opportunities for global education, health care, food and medical services, useful employment, personal security, and a reasonably comfortable standard of living in harmony with other people and other life forms?

On the issue of world peace, the reduction of ideological and nuclear tensions between the United States and the USSR (now CIS) has not reduced the level of armed violence elsewhere. The spectacle of computer-based "smart warfare" orchestrated by the Department of Defense and American media in celebration of "intelligent" weapons systems during the Persian Gulf War created for many a false sense of achievement and national superiority. Target effectiveness was greatly exaggerated (ignoring the high rate of "collateral damage" and exaggerating the success of the Patriot missiles). More critically, over the long-term, the media failed to assess the actual human and economic costs of the war and contributed in obfuscating the reasons for the war in the first place.

Media images of techno-war helped sanitize death and destruction, making future interventions a relatively easy option for ambitious politicians interested in defending American honor, pleasing the military-industrial complex and associated vested interests, and preparing the public for the next election. Although global arms sales plunged with the end of the U.S.-USSR Cold War ($67.9 billion in 1988 to $31.9 billion in 1993), U.S. arms sales rapidly escalated ($10.1 billion in 1989 to $22.3 billion in 1993), reaching 70% of the world total (Burns, 1995).

Most media reporting on the future of communications accept at face value industry claims about unending global technological progress, the far-sighted social responsibility of corporate planners, and the need for a deregulated workplace. This view unites futurologist Alvin Toffler with liberal and conservative political leaders Al Gore and Newt Gingrich. Given such self-assured apolitical, ahistorical beliefs, we might reasonably expect peasants in Nepal and Tanzania to be soon surfing the Internet. But the unfortunate reality is that, in the United States, more so in the Third World, the weakening foundations of economic democracy (e.g., poverty growing 17% between 1973 and 1988) are better indicators of technological opportunity than engineering triumphs in the compression of transmission frequencies. Even if intercontinental mobile telephone or television is already an option for some, under the present political system it will never become a practical choice for most people. It would be more reasonable to expect that in the rush to seize the electronic frontier, ordinary services, such as plain old residential telephone (voice), will rise in price and decline in relative quality.

Even the guru of "third wave" information technology, Toffler, is not entirely upbeat about the future of communications technology. Among other scenarios, he sees the possibility of organized crime syndicates using computers to shut down the world's banking system, Third World dictators transmitting phony incidents on satellite television to justify aggression, and newer applications of global positioning satellites, as were employed in the 1991 Persian Gulf invasion, to pinpoint military installations in the desert. According to one former ranking U.S. intelligence official, a well-trained computer bandit could shut down the country's automated teller machines, the Federal Reserve system, Wall Street, and the major hospital and business computer operations (Elias, 1993).

It should not be unexpected if a few skilled hackers become electronic Robin Hoods, pulling off illegal electronic bank account and credit transfers or scotching computerized debt records for the benefit of community groups or needy individuals—especially if the economy continues its downward trajectory. Over the past decade, official U.S. unemployment rates have averaged well above the 6% range (in the OECD countries as a whole, 35 million were unemployed in 1995). Counting the partially employed, undocumented workers and those who have stopped looking for work (all of whom are not included in the official numbers), the rate is actually much higher (about 12% in 1994). And the official tally of poverty at 15.1% (1993) of the population, tragic as that is in a country as affluent as the United States, is also too conservative an estimate (calculated at $14,400 total income for a family of four in 1993). It is not too hard to imagine that a new generation of hackers, either humanistically or selfishly inspired, could incite government justifications for the suspension of broad civil liberties in the name of a "national emergency"; there is already wide use of censorship and surveillance on the Internet.

Certainly other alternatives in liberating information are available for the benefit of the disenfranchised, but the more democratic options require the commitment and financial support of the public sector. The Internet and its offspring, the World Wide Web, already provide a multitude of accessible networks and individual accounts for collecting information about almost every subject imaginable, including subjects only the most ardent libertarians can defend. With a mass of users now in place and a new wave of mergers on the horizon, however, it can be expected that tiers of access and pricing will be set up, putting higher-quality services on a premium fee basis, flooding lower-cost/lower-quality basic services with commercial advertising (with faster, congestion-based price increases and slower up/down linking), and forcing out low-income/low-consumption households.

Once cable and telephone companies smell earnings opportunities, separate Internet accounts could cost individual users far more than they ever imagined, with private service providers enticing them to purchase multimedia (Internet/cable/telephone) packages. And regardless of cost, millions of people with time on their hands will have to have their on-line fixes. Meanwhile, the withdrawal of National Science Foundation support for the Internet, starting in 1995, could find academic users surfing on their own or dependent on cost pickup by their financially strapped universities, the latter option a perquisite that eventually could be tightly restricted or, like health benefits, put on a copayment basis.

If metering of cyberspace has become the standard mode of access (e.g., America Online, CompuServe, Prodigy), this is also true for other electronic communications. The "information revolution" in television looks more like a counterrevolution from the perspective that the "free" television programming in its early years (e.g., sports, theater productions, movies, public affairs) has shifted to a pay-as-you-go basis, either through cable packages or pay-per-view options. In the future, cable may allow viewers to unbundle their subscription packages by selecting only preferred channels, but with a captive market, the monopoly operator could then shift to season pass "discount" rates while raising monthly and pay-per-view rates, ultimately increasing yearly use cost.

The hucksters of the "information age" do not remind us very often that the business of business is more business—and more profit—and that deregulation means more concentration, monopolization, and vertical integration in the industry. Empowered wealthy conservative politicians disinterested in social democracy and putting all their faith (and their chips) in the market may choose to wipe out what little "public" remains in public television—or marginalize it to the status presently occupied by extremely low budget public access cable stations. Recall Huxley's *Brave New World*, in which the "savages" were banished beyond the pale of "civilization."

Self-proclaimed "cybernauts," many without any coherent political ideology, often ignore the decline of worklife and citizenship and the governmental and commercial invasion of privacy. Lack of social awareness is centered in the industrialization and privatization of communications and, given the present state of U.S. and world politics, could get worse into the 21st century. Unlike the promises in the late 1970s of a cybernetic world in which work would be unnecessary because of the surplus created through automation and robotization, people are, in fact, increasingly desperate for work, *any* work. Rather than working fewer hours, most people are working *more* hours just to remain in place financially.

Liberalization in trade and investment rules under the GATT and NAFTA agreements will move more jobs abroad, especially information-processing type functions, that can easily be managed via telecommunications in transnational back offices located in the poorest, lowest-wage Third World countries. From early 1994 to early 1995, the United States went from a $1.3 billion trade surplus with Mexico, one of its two NAFTA partners, to an $857 million trade deficit, with the loss of more than 60,000 American jobs (*Public Citizen*, 1995). Neither of the two dominant political parties in the United States appears capable of responding to the labor crisis, as both are committed to the religion of "free trade," fantasies of retraining workers in once-skilled labor jobs, and the privileged status of TNCs. Conservative governments in the leading industrial societies, stripping away welfare legislation and the last remaining defenses of labor unions, could well bring on the social conditions that Marx foresaw as the setting for proletarian revolution.

Marx's proletariat was an industrial factory workforce. Today, the factories of the world still include cold or overheated, unventilated, dirty, dark, 19th-century-type workplaces that Marx described, especially in some of the remoter enclaves of production (and the garment-making sweatshops in California and New York). But they also include the better-lit, air-conditioned offices of the modern era, where telephone operators, data entry workers, clerical typists, fast-food employees, department store and supermarket checkout clerks, and electronics assemblers earn their daily bread.

The cyber-community that some see as the "global village" of the future has little meaning for the poorly paid classes, who are unlikely to interact within the better-educated and more mobile, affluent, and cosmopolitan networks communicating in cyberspace. Today's virtual communities are no substitute for real communities, even if much of the popular discourse about cyberspace would have us believe otherwise, decontextualizing, as it does, the meaning of communication to a limited set of written, sometimes visual, exchanges. Much of what constitutes communication is subliminal; electronically mediated discourse habituates people to a different form of depersonalized exchange that causes them to lose track of the art and ethos of face-to-face interpersonal communion.

The most frequently heard concern about electronic communication narrowly focuses on free speech protection of on-line computer users—the defense of individual expression against the encroachment of government and private censorship. Although this is certainly an important right and issue, it often overlooks the larger context of First and Fourth Amendment rights that have long been endangered. Electronic eavesdropping and record keeping by businesses and by bureaucratic, police, and intelligence functionaries represent real threats to the cherished liberties of free citizens. But, as discussed in earlier chapters, the protection of privacy and the sanctity of the individual, going back to the witch trials in Salem, has never been quite as honored an American tradition as is so often claimed in conventional courses on American government and history.

Excited clichés of "patriots," like the U.S. Air Force commander who believed that he had to destroy Vietnamese ancestral villages to "save" the people from socialism, puts violent chauvinism ahead of cultural sovereignty. The memorable phrase of Patrick Henry, "Give me liberty or give me death," was spoken by a wealthy land speculator who kept slaves. The government surveillance of unarmed and nonviolent political dissident groups continues to the present.

Illegal surveillance and invasion of legitimate political organizations by the FBI, CIA, National Security Agency, and various police instrumentalities throughout the 20th century and the easing of restrictions on police information-gathering techniques suggest that the original constitutional protections against illegal forms of search and seizure have been greatly narrowed in meaning. The trespass of household privacy by telemarketing, mailing lists, consumer data-bases, caller-ID services, credit bureaus, computerized dialing, required blood and urine tests in athletics and many occupations, and the use of employee polygraphs by many businesses have become normal. So too have the sharing of medical records between health care centers and insurance companies, government interagency record swapping of citizen data, and the selling of driver license data by state motor vehicle registration agencies. These are but some forms of record searching, keeping, and sharing that represent grave threats to personal sanctity and security. Civil liberties attorneys will have a full docket of cases on privacy and surveillance law well into the next century.

The 1980s themes celebrating the arrival of cybernetic freedom and electronic democracy, with stackloads of literature echoing such rhetoric, has run head-first into the political economic realities of unregulated capitalism in the 1990s. Not only is the Internet becoming as commercial as radio was by the 1930s, but the new versions of the World Wide Web routinely provide intrusively penetrating biodata banks on private- and public-sector employees. Many universities now feature "Web pages" of faculty and staff, with photographs and résumés open to anyone curious to know. This means that no longer does the Internet protect or respect the privacy of on-line users (or even people who choose not to use the network), raising the specter of jobs, tenure, grants, promotions, awards, and so on being anonymously judged on such irrelevant criteria as appearance, gender, age, ethnicity, ideology, and other screens, thereby reinforcing historical patterns of social and political discrimination.

A neologism with a common derivative of privacy is privatization, which in contemporary parlance practically means organizing "private-sector accumulation with public-sector funds" (Aronowitz & DeFazio, 1994, p. 129). For example, Senator Jesse Helms (R–North Carolina) would eliminate federal funding of AIDS research but preserve government subsidies for tobacco growers in his home state. Both Republicans and Democrats have pushed for less government regulation of business, while supporting multi-billion-dollar federal aid to mismanaged private banks, automobile manufacturers, and huge defense industries; subsidies to mining, timber-cutting, grazing, and irrigation investments; and tax deductions for home mortgage interest payments, retirement plans, and health insurance premiums paid by corporations (Folbre, 1995). Affirmative action, in contrast, is seen by conservatives as inconsistent with a free society even though race, gender, and class privilege has always been relevant to economic advantage. A kind of affirmative action for the well-off can be found in the original state and federal exclusions that permitted only white, male landholders to vote, in the confiscation of Native American lands for their transfer to white homesteaders, and in the restrictive entry to universities, country clubs, and many forms of work against women and people of color.

In communications policy, "affirmative action" has meant an FCC policy of turning over control of the airwaves to the corporate elite for commercial gain, rather than of treating the spectrum as a public resource. Broadcasting, cable, and computer communications are far too valuable as educational and informational instruments to be left to the capriciousness of the marketplace, and yet that is where they are essentially governed. On the basis of the general success in delivering telephone, electric, gas, and water supply, broadcast, cable, and computer communications could be run as government-regulated public utilities, ensuring both transmission and reception opportunities by public institutions (e.g., schools, libraries, hospitals) or even, as in Italy, allocated among different political parties. Another option would be to have broadcasting and cable TV, or at least a significant quota of broadcasting licenses (including license requirements for cable operators), directly run out of public universities (as is already the case for some local PBS broadcasting affiliates). In western Europe, the shift to privatization and deregulation has allowed transnational media conglomerates to swamp national broadcasters and undermine local cultural traditions, especially in smaller countries like the Netherlands.

Ultimately, it is misguided to look for policy solutions, whether market- or regulation-driven, that do not include changes in the political structure itself. In recent years, politics in the United States has meant using experts, computer technology, and the mass media to engineer electoral victories and little about representing popular interests. Political action committees and their corporate interest group sponsors get to vote early in selecting political candidates, well before the first "election day" ballot is ever cast. This has to change before meaningful political debate and reform can take place that speaks to a future that most Americans really desire, including a design for communication and information structures as if people mattered. In a more democratic discursive and decision-making environment where previously unheard voices are given the podium and airtime, there is no predicting what imaginative and practical ideas would take shape.

Until a critical mass of Americans comes to actively reject the present electoral charade (passively most already do, judging by voting turnouts) and direct their sense of betrayal into support for a new political formation or party founded and mobilized on the basis of diverse grassroots interests, public communications will continue to reflect the needs of competitive market monopolies. Although interactive computers have their uses as conduits of information, they are also a great diversion from the everyday world of three-dimensional human encounters, which rely on physical and social context and the full panoply of sensory indicators: the smile; the look of concern and trust; intonation; the appearance of disappointment, grief, or anger; the gesture of assurance; the handshake or the nod; eye contact; the touch; and other communicated cues that machines cannot reproduce.

Communication technologies are extensions of opportunity within the rules of the political economy that control the allocation of resources in society. It is naive or worse to believe that computers, without reference to the larger context of power, can bring about new forms of economic, social, or political relationships. Such anthropomorphic notions deceptively ignore the actual actors involved in designing and implementing the information architecture and infrastructure of a transnationalized capitalist economy like the United States. Most of these front row actors, Fortune 500 directors, transnational corporate planners, millionaire communication enterprise executives, bankers, electronics and software engineers, military brass, arms industry contractors, and others have privileged lives far removed from the concerns of ordinary working people, and their interests do not often coincide.

From the 1930s to the 1960s, the Democratic Party carried forward many of the New Deal policies of Franklin D. Roosevelt and represented a multiclass coalition of interests that included labor unions and blue-collar workers, African Americans, Jews, liberals, sections of big capital, the southern states, and most big-city voters, The decline of liberalism, though never an overwhelming political force in the Democratic Party (especially the southern wing) or in the

United States, represents a major demographic shift in the country, as well as the failure of Democrats to consistently support ethnic minority and working-class concerns.

Starting with Nixon in the 1968 election, conservatives and Republicans actively marshaled political media, public relations, and advertising resources for their electoral campaigns. The most successful manipulation of national public images started in the Reagan years. A coalition of well-funded right-wing Republicans, Christian conservatives, anti-abortion activists, the National Rifle Association, racist and militarist fringe groups, and major segments of corporate capital have made a well-organized attempt to destroy the vestiges of institutional liberalism in American society. In the early 1990s, one of their coordinating bodies was the secretive Council for National Policy, which included Senator Jesse Helms, 1994 Senate candidate Oliver North, and House Representative Robert Dornan (R–California).

Indispensable to any political campaign is a media strategy to win the minds and hearts of the voting public. Unable to finance television advertising, minority party candidates receive virtually no media attention (with the rare exception of wealthy or corporate-financed candidates like Ross Perot or John Anderson) and therefore cannot effectively get their messages across to a nation riveted to the tube for its news and information. In 1992, Bush and Clinton together received more than $166 million in campaign funds, about two thirds of which went to broadcast advertising (Ross, 1993, p. 325). Major media have replaced the parties as the control center of U.S. elections. But more than that, the corporate media depoliticize politics by stripping elections of their real content and meaning of the issues and replacing informative discussion with the language of commercial television: good images, good sound bites, sideshow dramas, and emotional rhetoric that flies in the face of fact or reason. Marc Cooper comments that television "erases literacy, reduces the level of public discourse, atomizes all sense of a cohesive society . . . rewards emotion over thought, substitutes spectacle for analysis, symbols for ideas: it celebrates the market as a proxy for democracy and ultimately and inexorably turns citizens into mere consumers" (cited in Ross, 1993, p. 384).

After the election is over, who keeps track of what was true and what was not? Not network television owners and executives; laughing their way to the bank, they are too satisfied to notice. The FCC has taken the public domain of the airwaves and effectively handed it to corporate entities as a license to print money—and ultimately print ballots. Reforms that would give equal time to all electoral candidates, require public and limited funding as the sole source of financing, and greatly reduce the length of the campaign, would start to bring politics closer to the issues and the people. Democrats more than Republicans may want to put greater limits on campaign spending, but not to the extent of opening the political system to more than two voices and certainly not to those not certifiably establishment.

Can any instrument of electronic communication be left to the vox populi? The institutional managers of information would have us believe not. As a people, Americans are organized into work and communication spaces in which they can be told what they need and, for the right price, be given what they want. In this information age, there is a critical need to demystify the political economy of communications. The English novelist George Eliot (née Mary Ann Evans) made the point more than 100 years ago in *Adam Bede:* "An election is coming. Universal peace is declared, and the foxes have a sincere interest in prolonging the lives of the poultry."

References

Allen, B., & Closson, M. (1994, Spring). "Arms R Us." *Positive Alternatives,* p. 1.

Altschull, J. H. (1995). *Agents of power: The media and public policy.* White Plains, NY: Longman.

American Friends Service Committee. (1979). *The police threat to political liberty.* Philadelphia: Author.

Andrews, E. (1993, August 8). AT&T reaches out (and grabs everyone). *New York Times,* pp. F1, F8.

Andrews, E. (1994a, February 11). AT&T will cut 15,000 jobs to reduce costs. *New York Times,* pp. D1, D14.

Andrews, E. (1994b, February 12). White House seeking software to aid in wiretaps. *New York Times,* pp. 1, 44.

Ansah, P. A. V. (1986). The struggle for rights and values in communication. In M. Traber (Ed.), *The myth of the information revolution: Social and ethical implications of communication technology* (pp. 64-83). Newbury Park, CA: Sage.

Aptheker, H. (1976). *Early years of the republic: From the end of the Revolution to the first administration of Washington (1783-1793).* New York: International Publishers.

Aronowitz, S. (1987). Mass culture and the eclipse of reason: The implications for pedagogy. In D. Lazere (Ed.), *American media and mass culture: Left perspectives* (pp. 465-471). Berkeley: University of California Press.

Aronowitz, S., & DiFazio, W. (1994). *The jobless future: Sci-tech and the dogma of work.* Minneapolis: University of Minnesota Press.

Associated Press. (1988, February 3). Computers fueled stock crash—SEC. *Boston Globe,* p. 27.

Associated Press. (1996, December 8). Nixon urged IRS audits of Jewish campaign donors. *The Oregonian,* p. A26.

Athanasiou, T. (1985). Artificial intelligence: Cleverly disguised politics. In T. Solomonides & L. Levidow (Eds.), *Compulsive technology: Computers as culture* (pp. 13-35). London: Free Association Books.

288

Bagdikian, B. H. (1987, September). The media brokers: Concentration and ownership of the press. *Multinational Monitor,* pp. 7-12.

Bagdikian, B. H. (1989, June 12). The lords of the global village. *The Nation,* pp. 805-820.

Bagdikian B. H. (1992). *The media monopoly* (4th ed.). Boston: Beacon.

Balabanian, N. (1980, March/April). Presumed neutrality of technology. *Society,* pp. 7-14.

Baldasty, G. J. (1993). The rise of news as a commodity: Business imperatives and the press in the nineteenth century. In W. S. Solomon & R. W. McChesney (Eds.), *Ruthless criticism: New perspectives in U.S. communication history* (pp. 98-121). Minneapolis: University of Minnesota Press.

Bamford, J. (1982). *The puzzle palace: A report on NSA, America's most secret agency.* Boston: Houghton Mifflin.

Baran, P. (1957). *The political economy of growth.* New York: Monthly Review Press.

Barlow, M. H., Barlow, D. E., & Chiricos, T. G. (1995). Economic conditions and ideologies of crime in the media: A content analysis of crime news. *Crime & Delinquency, 41*(1), 3-19.

Barnaby, F., & Williams, I. (1983, November). Heavenly watchdogs. *South,* p. 13.

Barnet, R. J. (1990). *The rockets' red glare: War, politics, and the American presidency.* New York: Simon & Schuster.

Barnet, R. J., & Cavanagh, J. (1994). *Global dreams: Imperial corporations and the new world order.* New York: Simon & Schuster.

Barnet, R. J., & Müller, R. E. (1974). *Global reach: The power of the multinational corporations.* New York: Simon & Schuster.

Barnouw, E. (1966). *A tower in babel: A history of broadcasting in the United States: Vol. 1. To 1933.* New York: Oxford University Press.

Barnouw, E. (1968). *The golden web: A history of broadcasting in the United States: Vol. 2. 1933-1953.* New York: Oxford University Press.

Barnouw, E. (1975). *Tube of plenty: The evolution of American television.* New York: Oxford University Press.

Barry, D. et al. (1984). *Labor confronts the transnationals.* New York: International Publishers.

Baudot, B. S. (1989). *International advertising handbook.* Lexington, MA: Lexington.

Baudrillard, J. (1986). Requiem for the media. In J. G. Hanhardt (Ed.), *Video culture: A critical investigation* (pp. 124-143). Rochester, NY: Visual Studies Workshop Press.

Bell, D. (1960). *The end of ideology.* New York: Free Press.

Bell, D. (1979, May/June). Communications technology: For better or for worse. *Harvard Business Review,* pp. 20-42.

Bellant, R. (1991). *Old Nazis, the new Right, and the Republican Party: Domestic fascist networks and their effect on U.S. cold war politics.* Boston: South End.

Bello, W. (1994). *Dark victory: The United States, structural adjustment, and global poverty.* London: Pluto.

Bello, W., & Rosenfeld, S. (1990). *Dragons in distress: Asia's miracle economies in crisis.* San Francisco: Institute for Food and Development Policy.

Beniger, J. R. (1986). *The control revolution: Technological and economic origins of the information society.* Cambridge, MA: Harvard University Press.

Benjamin, W. (1969). *Illuminations* (H. Zohn, Trans.). New York: Schocken.

Bennett, W. L. (1996). *The governing crisis: Media, money, and marketing in American elections* (2nd ed.). New York: St. Martin's.

Berke, R. (1994, January 30). GOP TV: New image in appeal to voters. *New York Times,* p. 20.

Bernal, J. D. (1971a). *Science in history: Vol. 2. The scientific and industrial revolutions.* Cambridge: MIT Press.

Bernal, J. D. (1971b). *Science in history: Vol. 3. Science and industry in the nineteenth century.* Cambridge: MIT Press.

Bernhard, N. E. (1993). Ready, willing, able: Network television news and the federal government, 1948-1953. In W. S. Solomon & R. W. McChesney (Eds.), *Ruthless criticism: New perspectives in U.S. communication history* (pp. 291-312). Minneapolis: University of Minnesota Press.

Berry, J. et al. (1994, September 5). Database marketing. *Business Week,* pp. 56-62.

Binder, D. (1994, August 28). Shortwave radio: More preachers, less propaganda. *New York Times,* p. E5.

Bissio, R. R. (1990). Ten myths on democracy and communication. *Kasarinlan* [Manila], *5*(3), 83-86.

Blanc, G. (1985). Beware of the information age. *Development, 1,* 78-79.

Bleifuss, J. (1994, January/February). Flack attack. *Utne Reader,* pp. 72-77.

Boafo, S. T. K. (1991). Communication technology and dependent development in sub-Saharan Africa. In G. Sussman & J. A. Lent (Eds.), *Transnational communications: Wiring the Third World* (pp. 103-124). Newbury Park, CA: Sage.

Bolter, J. D. (1984). *Turing's man: Western culture in the computer age.* Chapel Hill: University of North Carolina.

Bolter, W. G., McConnaughey, J. W., & Kelsey, F. J. (1990). *Telecommunications policy for the 1990s and beyond.* Armonk, NY: M. E. Sharpe.

Bolton, B. (1993). Negotiating structural and technological change in the telecommunications services in the United States. In B. Boltan et al. *Telecommunications services: Negotiating structural and technological change* (pp. 123-143). Geneva, Switzerland: International Labour Office.

Boorstin, D. J. (1978). *The republic of technology: Reflections on our future community.* New York: Harper & Row.

Bottomore, T. et al. (1983). *A dictionary of Marxist thought.* Cambridge, MA: Harvard University.

Boyer, R. O., & Morais, H. M. (1955). *Labor's untold story.* New York: United Electrical, Radio & Machine Workers of America.

Branscomb, A. W. (1994). *Who owns information? From privacy to public access.* New York: Basic Books.

Braverman, H. (1974). *Labor and monopoly capital: The degradation of work in the twentieth century.* New York: Monthly Review Press.

Broad, R., & Cavanagh, J. (1993). *Plundering paradise: The struggle for the environment in the Philippines.* Berkeley: University of California Press.

Brock, G. W. (1981). *The telecommunications industry: The dynamics of market share.* Cambridge, MA: Harvard University.

Brooks, J. (1976). *Telephone: The first hundred years.* New York: Harper & Row.

Brown, L. R. et al. (1992). *State of the world, 1992.* New York: Norton.

Bunce, R. (1976). *Television in the corporate interest.* New York: Praeger.

Burns, C. (1995, June 14). Global arms market shrinks, pushing exporting nations to new salesmanship tactics. *The Oregonian,* p. A14.

Burns, E. (1990, August 29). Mexico firm tries to hang up on operators. *The Guardian* [New York], p. 16.

Cabral, A. (1973). *A return to the source.* New York: Monthly Review Press.

Canham-Clyne, J. (1995, July/August). Message to Congress: Cut corporate welfare, not Medicare. *Public Citizen,* pp. 1, 9-11.

Carey, A. (1987). Reshaping the truth: Pragmatists and propaganda in America. In D. Lazere (Ed.), *American media and mass culture: Left perspectives* (pp. 34-42). Berkeley: University of California Press.

Carman, H. J., Syrett, H. C., & Wishy, B. W. (1967). *A history of the American people: Vol. 2. Since 1965.* New York: Knopf.

Carnoy, M. (1993). Multinationals in a changing world economy: Whither the nation-state? In M. Carnoy et al., *The new global economy in the information age: Reflections on our changing world* (pp. 45-96). University Park: Pennsylvania State University Press.

Carter, B. (1989, June 1). TV viewers beware: Nielsen may be watching. *New York Times,* p. 1.

Carter, B. (1994, November 30). NBC is challenging Fox TV on foreign ownership rules. *New York Times,* pp. A1, C6.

Cassidy, P. (1995, Spring). Silent coup in cyberspace. *Covert Action,* pp. 54-60.

Caute, D. (1978). *The great fear: The anti-Communist purge under Truman and Eisenhower.* New York: Simon & Schuster.

Chaliand, G. (1989). *Revolution in the Third World* (Rev. ed.; D. Johnstone & T. Berrett, Trans.). New York: Penguin.

Chapman, G., & Rotenberg, M. (1993, Summer). The national information infrastructure: A public interest opportunity. *CPSR Newsletter* (Publication of Computer Professionals for Social Responsibility), pp. 1-5, 16-17.

Chomsky, N. (1989). *Necessary illusions: Thought control in democratic societies.* Boston: South End.

Chomsky, N. (1991). *Media control: The spectacular achievements of propaganda* (Speech printed in the *Open Magazine Pamphlet Series,* No. 10). Westfield, NJ: Open Magazine.

Churchill, W., & Vander Wall, J. (1988). *Agents of repression: FBI attacks on the Black Panthers and the American Indian movement.* Boston: South End.

Cirino, R. (1972). *Don't blame the people.* New York: Random House.

Coakley, T., & Murphy, S. P. (1994, May 28). Dunkin' Donuts to stop recording. *Boston Globe,* pp. 1, 6.

Cohen, J., & Solomon, N. (1992, April 15). Media beat. *The Guardian* [New York], p. 19.

Cohen, J., & Solomon, N. (1993). *Adventures in medialand: Behind the news, behind the pundits.* Monroe, ME: Common Courage Press.

Cohen, R. (1987). *The new helots: Migrants in the international division of labor.* Avebury, UK: Aldershot.

Cohen, S. S. (1993). Geo-economics: Lessons from America's mistakes. In M. Carnoy et al., *The new global economy in the information age: Reflections on our changing world* (pp. 97-147). University Park: Pennsylvania State University Press.

Constable, P., & Valenzuela, A. (1991). *A nation of enemies: Chile under Pinochet.* New York: Norton.

Constantino, R. (1980). *Identity and consciousness: The Philippine experience.* Quezon City, Philippines: Malaya Books.

Cornford, J., & Robins, K. (in press). Beyond the last bastion: Industrial restructuring and the labor force in the British television industry. In G. Sussman & J. A. Lent, *Global productions: Labor in the making of the "information society."* Cresskill, NJ: Hampton.

Cowell, A. (1994, March 24). Italy's knight in Teflon: Tycoon repulses charges. *New York Times,* pp. A1, A12.

Danielian, N. R. (1939). *AT&T: The story of industrial conquest.* New York: Vanguard.

de la Haye, Y. (1980). *Marx and Engels on the means of communication.* New York: International General.

Demac, D. (1988). *Liberty denied: The current rise of censorship in America.* New York: PEN.

Dennis, E. (1991). The military and the media: The Gulf conflict in historical perspective. In C. LaMay, M. FitzSimon, & J. Sahadi (Eds.), *The media at war: The press and the Persian Gulf conflict.* New York: Gannett Foundation.

DePalma, A. (1996, June 2). Tijuana booms as Asian companies build close to U.S. market. *San Jose Mercury News,* pp. 1E, 4E.

Diamond, E., & Bates, S. (1992). *The spot: The rise of political advertising on television* (3rd ed.). Cambridge: MIT Press.

Dietrich, W. (1992). *The final forest: The battle for the last great trees of the Pacific Northwest.* New York: Penguin.

Disney drops its efforts to build theme park in northern Virginia. (1994, September 29). *New York Times,* pp. A1, A8.

Dizard, W. (1982). *The coming information age.* White Plains, NY: Longman.

Documents show FBI kept files on leading U.S. writers. (1987, September 30). *Washington Post,* pp. A1, A7.

Domhoff, G. W. (1983). *Who rules America now? A view for the '80s.* New York: Simon & Schuster.

Dore, E. (1992, July-September). Debt and ecological disaster in Latin America. *Race and Class,* pp. 73-87.

Dorfman, A. (1987). The infantalizing of culture. In D. Lazere (Ed.), *American media and mass culture: Left perspectives* (pp. 145-153). Berkeley: University of California Press.

Dowd, M. (1994, January 29). Selling chips? Or is it Quayle? It's all a blur. *New York Times,* p. 6.

Downing, J. (1990, May). The political economy of U.S. television. *Monthly Review,* pp. 30-41.

Dreyfus, H. L. (1992). *What computers still can't do: A critique of artificial reason.* Cambridge: MIT Press.

Driscoll, D-M., & Goldberg, C. R. (1994, January 9). Women, jobs, and the press. *New York Times,* p. F13.

Drucker, P. F. (1966, Spring). The first technological revolution and its lessons. *Technology and Culture, 7*(2), 143-151.

Du Boff, R. B. (1984, October). The telegraph in nineteenth-century America: Technology and monopoly. *Comparative Studies in Society and History,* pp. 571-586.

Du Boff, R. B. (1989). *Accumulation and power: An economic history of the United States.* Armonk, NY: M. E. Sharpe.

Dye, T. R., Zeigler, H., & Lichter, S. R. (1992). *American politics in the media age* (4th ed.). Pacific Grove, CA: Brooks/Cole.

Ebenstein, W., & Ebenstein, A. O. (1992). *Introduction to political thinkers.* Orlando, FL: Harcourt Brace.

Editorial. (1992, February 2). *New York Times,* p. A22.

Elias, T. (1993, December 31). Toffler foresees rise of computer terrorism. *San Jose Mercury News,* p. G2.

Ellison, K. (1991, December 24). Privatization costs Mexicans more for bad phone service. *San Jose Mercury News,* pp. 1A, 10A.

Enzensberger, H. M. (1974). *The consciousness industry: On literature, politics, and the media.* New York: Seabury.

Epstein, E. J. (1974). *News from nowhere: Television and the news.* New York: Vintage.

Epstein, K. (1994, October 2). Children in sweatshops produce goods for U.S. use. *The Oregonian,* p. A5.

Erlanger, S. (1990, October 16). Singaporean cites siren song of TV. *New York Times,* p. A9.

Estabrooks, M. (1988). *Programmed capitalism: A computer-mediated global society.* Armonk, NY: M. E. Sharpe.

Ewen, S. (1976). *Captains of consciousness: Advertising and the social roots of the consumer culture.* New York: McGraw-Hill.

Ewen, S., & Ewen, E. (1982). *Channels of desire: Mass images and the shaping of American consciousness.* New York: McGraw-Hill.

Fabrikant, G. (1994, September 1). Media giants said to be negotiating for TV networks. *New York Times,* pp. A1, C16.

Fallows, J. (1994, March 24). The computer wars. *New York Review,* pp. 34-41.

Fanon, F. (1967). *Black skins, white masks.* New York: Grove.

Feder, B. J. (1991, November 3). Toward defining free speech in the computer age. *New York Times,* p. E5.

Fischer, C. S. (1992). *America calling: A social history of the telephone to 1940.* Berkeley: University of California Press.

Folbre, N. (1995). *The new field guide to the American economy.* New York: New Press.

Foucault, M. (1979). *Discipline and punish: The birth of the prison.* New York: Vintage.

Forester, T. (1987). *High-tech society: The story of the information technology revolution.* Cambridge: MIT Press.

Fox, E. (1988). Media policies in Latin America: An overview. In E. Fox (Ed.), *Media and politics in Latin America: The struggle for democracy* (pp. 6-35). Newbury Park, CA: Sage.

Fox, R. G. (1993, November/December). An update on S.W.I.F.T. *World of Banking,* pp. 25-26.

Frank, A. G. (1979). *Dependent accumulation and underdevelopment.* New York: Monthly Review.

Franklin J. H., & Moss, A. A., Jr. (1988). *From slavery to freedom: A history of Negro Americans* (6th ed.). New York: McGraw-Hill.

Frederick, H. H. (1993). *Global communication and international relations.* Belmont, CA: Wadsworth.

Frederick, H. H. (1994). Computer networks and the emergence of global civil society. In L. M. Harasim (Ed.), *Global networks: Computers and international communication* (pp. 283-295). Cambridge: MIT Press.

Frederix, P. (1959). Agence havas. In A. Mattelart & S. Sieglaub (Eds.). (1979). *Communication and class struggle: Capitalism, imperialism* (pp. 225-227). New York: International General.

French, H. F. (1994, July/August). The World Bank now fifty, but how fit? *World Watch,* pp. 10-18.

Fukuyama, F. (1989, Summer). The end of history. *National Interest,* pp. 13-18.

Gabriel, T. (1995, January 14). Ex-prisoner now called virtual hero by friends. *New York Times,* p. 12.

Galambos, L., & Pratt, J. (1988). *The rise of the corporate commonwealth: United States business and public policy in the 20th century.* New York: Basic Books.

Galbraith, J. K. (1967). *The new industrial state.* Boston: Houghton Mifflin.

Gandy, O. H., Jr. (1989, Summer). The surveillance society: Information technology and bureaucratic social control. *Journal of Communication,* 61-76.

Gandy, O. H., Jr. (1993). *The panoptic sort: A political economy of personal information.* Boulder, CO: Westview.

Garnham, N. (1986). The media and the public sphere. In P. Golding, G. Murdock, & P. Schlesinger (Eds.), *Communicating politics.* Leicester, UK: Leicester University Press.

Garnham, N. (1990). *Capitalism and communication: Global culture and the economics of information.* Newbury Park, CA: Sage.

Garson, B. (1988). *The electronic sweatshop: How computers are transforming the office of the future into the factory of the past.* New York: Simon & Schuster.

Gibson, J. W. (1988). *The perfect war: The war we couldn't lose and how we did.* New York: Vintage.

Ginger, R. (1975). *The age of excess: The U.S. from 1877 to 1914.* New York: Macmillan.

Ginsberg, B. (1986). *The captive public: How mass opinion promotes state power.* New York: Basic Books.

Gitlin, T. (1985). *Inside prime time.* New York: Pantheon.

Glick, B. (1989). *War at home.* Boston: South End.

Global Electronics. (1993, September). p. 1.

Goldberg, D. (1986, October). Captain Midnight, HBO, and World War III. *Mother Jones,* 26-29+.

Gonzalez-Manet, E. (1992). *Informatics and society: The new challenges* (L. Alexandre, Trans.). Norwood, NJ: Ablex.

Goulden, J. C. (1970). *Monopoly.* New York: Pocket.

Gouldner, A. W. (1976). *The dialectic of ideology and technology: The origins, grammar, and future of ideology.* New York: Oxford University Press.

Graham, S., & Marvin, S. (1996). *Telecommunications and the city: Electronic spaces, urban places.* London: Routledge.

Greenberg, E. S. (1989). *The American political system: A radical approach* (5th ed.). Glenview, IL: Scott, Foresman.

Greider, W. (1992). *Who will tell the people: The betrayal of American democracy.* New York: Simon & Schuster.

Guimaraes, C., & Amaral, R. (1988). Brazilian television: A rapid conversion to the new order. In E. Fox (Ed.), *Media and politics in Latin America: The struggle for democracy* (pp. 125-137). Newbury Park, CA: Sage.

Guthrie, K. K., & Dutton, W. H. (1992). The politics of citizen access technology: The development of public information utilities in four cities. *Policy Studies Journal, 20*(4), 574-597.

Hagstrom, J. (1992). *Political consulting: A guide for reporters and citizens.* New York: Columbia University, Freedom Forum Media Studies Center.

Hallin, D. C. (1989). *The uncensored war: The media and Vietnam.* Berkeley: University of California Press.

Hamelink, C. J. (1994a). *The politics of world communication.* Thousand Oaks, CA: Sage.

Hamelink, C. J. (1994b). *Trends in world communication: On disempowerment and self-government.* Penang, Malaysia: Southbound.

Hamelink, C. J. (1995). Information imbalance across the globe. In J. Downing, A. Mohammadi, & A. Sreberny-Mohammadi (Eds.), *Questioning the media: A critical introduction* (2nd ed., pp. 293-307). Thousand Oaks, CA: Sage.

Hartigan, P. (1992, March 4). Targeting PBS. *Boston Globe,* pp. 1, 14.

Harvey, D. (1989). *The condition of postmodernity.* Oxford, UK: Basil Blackwell.

Hayes, D. (1989). *Behind the silicon curtain: The seductions of work in a lonely era.* Boston: South End.

Haynes, J. (1996). *Third World politics: A concise introduction.* Oxford, UK: Blackwell.

Headrick, D. R. (1981). *The tools of empire: Technology and European imperialism in the nineteenth century.* New York: Oxford University Press.

Headrick, D. R. (1991). *The invisible weapon: Telecommunications and international politics, 1851-1945.* New York: Oxford University Press.

Heilbroner, R. L. (1992). *The world philosophers: The lives, times, and ideas of the great economic thinkers* (6th ed.). New York: Simon & Schuster.

Herman, E. S. (1990, May). The deepening market in the West: Part 3. Commercial broadcasting on the march. *Z Magazine,* pp. 63-66.

Herman, E. S. (1992). *Beyond hypocrisy: Decoding the news in an age of propaganda.* Boston: South End.

Herman, E. S. (1995a). The externalities effects of commercial and public broadcasting. In K. Nordenstreng & H. I. Schiller (Eds.), *Beyond national sovereignty: International communication in the 1990s* (pp. 85-115). Norwood, NJ: Ablex.

Herman, E. S. (1995b). Media in the U.S. political economy. In J. Downing, A. Mohammadi, & A. Sreberny-Mohammadi (Eds.), *Questioning the media: A critical introduction* (2nd ed., pp. 77-93). Thousand Oaks, CA: Sage.

Herman, E. S., & Chomsky, N. (1988). *Manufacturing consent: The political economy of the mass media.* New York: Pantheon.

Hirota, P. (1983, March/April). New police technologies. *Science for the People,* pp. 25-29.

Hoare, Q., & Smith, G. N. (Eds. and Trans.). (1971). *Selections from the prison notebooks of Antonio Gramsci.* New York: International Publishers.

Hohler, B. (1994, May 12). For Disney, Civil War battle. *Boston Globe,* p. 3.

Holmes, S. A. (1996a, June 20). Gap between rich, poor widens in America. *The Oregonian,* p. A17.

Holmes, S. A. (1996b, June 20). Income disparity between poorest and richest rises. *New York Times,* p. A1.

Howard, R. (1986). *Brave new workplace.* New York: Penguin.

Hoynes, W., & Croteau, D. (1990, Winter). All the usual suspects: MacNeil/Lehrer and Nightline. *Extra!* [Special issue].

Huberman, L. (1970). *We, the people.* New York: Monthly Review.

Hunt, E. K., & Sherman, H. J. (1990). *Economics: An introduction to traditional and radical views* (6th ed.). New York: Harper & Row.

Hyatt, J. (1993, November 28). The information underclass. *Boston Globe,* pp. 81, 83.

Inose, H., & Pierce, J. R. (1984). *Information technology and civilization.* New York: Freeman.

Internet ventures into world of computer capitalism. (1994, September 11). *The Oregonian,* p. K5.

Isaac, K. (1992). *Civics for democracy: A journey for teachers and students.* Washington, DC: Essential Books.

Janofsky, M. (1994a, September 23). Disney's Virginia project gets first two approvals. *New York Times,* p. A8.

Janofsky, M. (1994b, March 10). U.S. judge frees *Daily News* from Maxwell case claim. *New York Times,* pp. D1, D20.

Johnson, R. C. (1993). Science, technology, and black community development. In A. H. Teichert (Ed.), *Technology and the future* (6th ed., pp. 265-282). New York: St. Martin's.

Joll, J. (1978). *Antonio Gramsci.* New York: Penguin.

Josephson, M. (1962). *The robber barons.* Orlando, FL: Harcourt Brace.

Joshi, V. (1994, May 31). "Jurassic Park" takes a bite out of India. *Boston Globe,* p. 64.

Kadi, M. (1995, March-April). Welcome to Cyberia. *Utne Reader,* pp. 57-59.

Katz, S. L. (1987). *Government secrecy: Decisions without democracy.* Washington, DC: People for the American Way.

Katznelson, I., & Kesselman, M. (1987). *The politics of power: A critical introduction to American government* (3rd ed.). Orlando, FL: Harcourt Brace.

Kellner, D. (1990). *Television and the crisis of democracy.* Boulder, CO: Westview.

Kellner, D. (1992). *The Persian Gulf TV war.* Boulder, CO: Westview.

Kennedy, L. B. (1989, October). *Privatization as a method of wealth redistribution: The case of Malaysian Telecoms.* Paper presented at the Seventh International Conference on Culture and Communication, Temple University, Philadelphia.

Kerr, P. (1990, September 25). Spying by computer: Is it a Trentongate? *New York Times,* p. B1.

Kolko, G. (1988). *Confronting the Third World: United States foreign policy, 1945-1980.* New York: Pantheon.

Kotz, D. M. (1978). *Bank control of large corporations in the United States.* Berkeley: University of California Press.

Kramer, F. (1996, April 17). Media giants about to have growth spurt. *The Columbian* [Vancouver, WA], p. A1.

Kunzle, D. (1975). Introduction to the English edition. In A. Dorfman & A. Mattelart, *How to read Donald Duck: Imperialist ideology in the Disney comic* (pp. 11-21, D. Kunzle, Trans.). New York: International General.

Kurkjian, S., & McConnell, J. (1989, August 22). Restraining the media at the CIA. *Boston Globe,* pp. 1, 5.

LaFeber, W., Polenberg, R., & Woloch, N. (1986). *The American century: A history of the United States since the 1890s* (3rd ed.). New York: Knopf.

Lairson, T. D., & Skidmore, D. (1993). *International political economy: The struggle for power and wealth.* Orlando, FL: Harcourt Brace.

Lander, M. (1996, January 23). AT&T enters TV business via satellite broadcasting. *New York Times,* pp. C1, C20.

Landler, M., & Grover, R. (1993, August 2). Is Disney ready to go to school? *Business Week,* pp. 22-23.

Lappe, F. M., & Collins, J. (1978). *Food first: Beyond the myth of scarcity.* New York: Ballantine.

LaQuey, T. (1993). *The Internet companion: A beginner's guide to global networking.* Reading, MA: Addison-Wesley.

Lazere, D. (Ed.). (1987). *American media and mass culture: Left perspectives.* Berkeley: University of California Press.

Lee, M. A., & Solomon, N. (1990). *Unreliable sources: A guide to detecting bias in news media.* New York: Lyle Stuart.

Left Business Observer. (1989, January 17). pp. 4-5.

Lent, J. A. (1986). *Devcom: A view from North America.* Unpublished manuscript.

Lent, J. A. (1991). Telematics in Malaysia: Room at the top for a selected few. In G. Sussman & J. A. Lent (Eds.), *Transnational communications: Wiring the Third World* (pp. 165-199). Newbury Park, CA: Sage.

Lent, J. A. (1993). The world's media titans. *Jurnal Komunikasi* [Malaysia], *9,* 91-100.

Lerner, D. (1963). Toward a communication theory of modernization. In L. Pye (Ed.), *Communications and political development* (pp. 327-350). Princeton, NJ: Princeton University Press.

Lerner, D. (1969). Managing communication for modernization: A developmental construct. In A. A. Rogow (Ed.), *Politics, personality, and social science in the twentieth century: Essays in honor of Harold D. Lasswell* (pp. 171-196). Chicago: University of Chicago.

Lewis, T. (1991). *Empire of the air: The men who made radio.* New York: HarperCollins.

Lyon, D. (1988). *The information society: Issues and illusions.* Cambridge, MA: Polity.

Lyon, D. (1994). *The electronic eye: The rise of surveillance society.* Minneapolis: University of Minnesota Press.

MacArthur, J. R. (1993). *Second front: Censorship and propaganda in the Gulf War.* Berkeley: University of California.

MacBride, S. (1980). *Many voices, one world: Communication and society, today and tomorrow.* New York: Unipub.

Mander, J. (1991). *In the absence of the sacred: The failure of technology and the survival of the Indian nations.* San Francisco: Sierra Club Books.

Mao, Z. (1968). *Four essays on philosophy.* Beijing: Foreign Language Press.

Marcuse, H. (1964). *One-dimensional man: Studies in the ideology of advanced industrial society.* Boston: Beacon.

Markoff, J. (1994a, March 13). The rise and swift fall of cyber literacy. *New York Times,* pp. E1, E5.

Markoff, J. (1994b, April 13). Commerce comes to the Internet. *New York Times,* p. D5.

Marx, K. (1967). *Capital: A critical analysis of capitalist production* (Vol. 1, F. Engels, Ed.). New York: International Publishers.

Marx, K., & Engels, F. (1970). *The German ideology* (C. J. Arthur, Ed.). London: Lawrence & Wishart.

Mattelart, A. (1978, Winter). The nature of a communications practice in a dependent society. *Latin American Perspectives,* 13-34.

Mattelart, A. (1991). *Advertising international: The privatization of public space* (M. Chanan, Trans.). New York: Routledge.

Mattelart, A. (1994). *Mapping world communication: War, progress, culture* (S. Emanuel & J. A. Cohen, Trans.). Minneapolis: University of Minnesota Press.

Mazzocco, D. W. (1994). *Networks of power: Corporate TV's threat to democracy.* Boston: South End.

McChesney, R. W. (1993). Conflict, not consensus: The debate over broadcast communication policy, 1930-1935. In W. S. Solomon & R. W. McChesney (Eds.), *Ruthless criticism: New perspectives in U.S. communication history* (pp. 222-258). Minneapolis: University of Minnesota Press.

McChesney, R. W. (1994). *Telecommunications, mass media, and democracy: The battle for the control of U.S. broadcasting, 1928-1935.* New York: Oxford University Press.

McClelland, D. (1961). *The achieving society.* New York: Van Nostrand.

McCormick, T. J. (1989). *America's half-century: United States foreign policy in the cold war.* Baltimore: Johns Hopkins University Press.

McDermott, J. (1994, November 14). And the poor get poorer. *The Nation,* pp. 576-580.

McDougall, C. (1996, February 18). Kasparov beats Deep Blue to take historic match. *The Oregonian,* p. A16.

McGinness, J. (1993). Politics as a con game. In A. Alexander & J. Hanson (Eds.), *Taking sides: Clashing views on controversial issues in mass media and society* (2nd ed., pp. 162-167). Guilford, CT: Dushkin.

McLuhan, M. (1964). *Understanding media: The extensions of man.* New York: New American Library.

McPhail, T. (1993). *Television as an extension of the nation state: CNNI and the Americanization of broadcasting.* Unpublished manuscript.

Meehan, D. M. (1988). The strong-soft woman: Manifestations of the androgyne in popular media. In S. Oscamp (Ed.), *Television as a social issue* (pp. 103-112). Newbury Park, CA: Sage.

Meehan, E. R., Mosco, V., & Wasko, J. (1993, Autumn). Rethinking political economy: Change and continuity. *Journal of Communication,* 105-116.

Melman, S. (1985). *The permanent war economy: American capitalism in decline.* New York: Simon & Schuster.

Mercury News Wire Service. (1993, December 22). Gore outlines info highway. *San Jose Mercury News,* pp. 1A, 6A.

Miller, M. C. (1992, June 24). Operation Desert Sham. *New York Times,* p. A21.

Miller, M. C., & Biden, J. J. (1996, June 3). The national entertainment state. *The Nation,* pp. 23-28.

Minnick, W. (1995, Spring). Trawling the Internet. *Covert Action,* 61-62.

Mokhiber, R. (1996, December). The ten worst corporations in 1996. *Multinational Monitor,* 7-15.

Moldea, D. E. (1987). *Dark victory: Ronald Reagan, MCA, and the mob.* New York: Penguin.

Moore, B., Jr. (1966). *The social origins of dictatorship and democracy: Lord and peasant in the making of the modern world.* Boston: Beacon.

Moore, D. (1995, November 24-26). Canadian TV bars U.S. shows from prime time. *USA Today,* p. 1A.

Mosca, G. (1939). *The ruling class.* New York: McGraw-Hill.

Mosco, V. (1982). *Pushbutton fantasies: Critical perspectives on videotex and information technology.* Norwood, NJ: Ablex.

Mosco, V. (1988). Introduction: Information in the pay-per society. In V. Mosco & J. Wasko (Eds.), *The political economy of information* (pp. 3-26). Madison: University of Wisconsin Press.

Mosco, V. (1993). Communication and information technology for war and peace. In C. Roach (Ed.), *Communication and culture in war and peace* (pp. 41-70). Newbury Park, CA: Sage.

Mulgan, G. J. (1991). *Communication and control: Networks and the new economies of communication.* New York: Guilford.

Musa, M. (1990). News agencies, transnationalization, and the new order. *Media, Culture & Society, 12,* 325-342.

Myerson, A. R. (1994, March 29). For Alamo's defenders, new assaults to repel. *New York Times,* p. A18.

Nader, R. (1965). *Unsafe at any speed: The designed-in dangers of the American automobile.* New York: Grossman.

Naisbitt, J. (1984). *Megatrends.* New York: Warner.

Nash, J. (1989). Community and corporations in the restructuring of industry. In M. P. Smith & J. R. Feagin, *The capitalist city: Global restructuring and community politics* (pp. 275-296). Cambridge, MA: Blackwell.

Naureckas, J. et al. (1991, May). Gulf War coverage: The worst censorship was at home. *Extra!* [Special issue].

Neier, A. (1981). Surveillance as censorship. In A. Janowitz & N. J. Peters (Eds.), *The campaign against the underground press* (pp. 9-17). San Francisco: City Lights.

Nelkin, D. (1991). The political impact of technical expertise. In W. B. Thompson (Ed.), *Controlling technology: Contemporary issues* (pp. 275-290). Buffalo, NY: Prometheus.

Nelson, J. (1989). *Sultans of sleeze: Public relations and the media.* Monroe, ME: Common Courage.

Ness, E. (1994, December). BigBrother@cyberspace. *The Progressive,* 22-27.

Neuffler, E. (1994, June 12). Neo-Nazis spreading hate with high-tech. *Boston Globe,* pp. 1, 26.

New York Times News Service. (1994, December 27). Iran pulls the plug on satellite dishes. *San Jose Mercury News,* p. A8.

Noble, D. F. (1979). *America by design: Science, technology, and the rise of corporate capitalism.* New York: Oxford University Press.

Noble, D. F. (1986, January/February). Automation madness: Progress without people. *Science for the People,* 22-26.

Norris, F. (1995, September 21). AT&T shareholders will get stock in three companies. *New York Times.*

Nossiter, B. (1987). *The global struggle for more.* New York: Harper & Row.

Office of Technology Assessment (OTA). (1985). *Electronic surveillance and civil liberties.* Washington, DC: Government Printing Office.

Office of Technology Assessment (OTA). (1990). *Critical connections: Communication for the future.* Washington, DC: Government Printing Office.

Oliveira, O. S. (1993). Brazilian soaps outshine Hollywood: Is cultural imperialism fading out? In K. Nordenstreng & H. I. Schiller (Eds.), *Beyond national sovereignty: International communication in the 1990s* (pp. 116-131). Norwood, NJ: Ablex.

Oslin, G. P. (1992). *The story of telecommunications.* Macon, GA: Mercer University Press.

Pacey, A. (1991). Technology: Practice and culture. In W. B. Thompson (Ed.), *Controlling technology: Contemporary issues* (pp. 65-75). Buffalo, NY: Prometheus.

Parenti, M. (1977). *Democracy for the few* (2nd ed.). New York: St. Martin's.

Parenti, M. (1992). *Make-believe media: The politics of entertainment.* New York: St. Martin's.

Parenti, M. (1993). *Inventing reality: The politics of the mass media.* New York: St. Martin's.

Parenti, M. (1995). *Democracy for the few* (6th ed.). New York: St. Martin's.

Parker, J. (1995, September 5). China software pirates do roaring trade. *The Star* [Kuala Lumpur], p. 5.

Paterson, T. G. (Ed.). (1989). *Major problems in American foreign policy: Vol 2. Since 1914* (3rd ed.). Lexington, MA: D. C. Heath.

Paterson, T. G., Clifford, J. G., & Hagan, K. J. (1991). *American foreign policy: A history/1900 to present* (3rd ed., rev.). Lexington, MA: D. C. Heath.

Pear, R. (1994, October 12). U.S. is said to waste billions on computers. *New York Times,* p. A10.

Peirce, N. (1994, October 17). Old way of doing business defeated along with Disney. *The Oregonian,* p. B7.

Pendakur, M. (1991). A political economy of television: State, class, and corporate confluence in India. In G. Sussman & J. A. Lent (Eds.), *Transnational communications: Wiring the Third World* (pp. 234-262). Newbury Park, CA: Sage.

Penzias, A. (1989). *Ideas and information.* New York: Simon & Schuster.

Perrolle, J. A. (1987). *Computers and social change: Information, property, and power.* Belmont, CA: Wadsworth.

Perry, R. (1984). *Hidden power: The programming of the president.* New York: Beaufort.

Piller, C., & Wieman, L. (1992, August 7). America's computer ghetto. *New York Times,* p. A27.

Ponting, C. (1991). *A green history of the world: The environment and the collapse of great civilizations.* New York: Penguin.

Pool, I. de S. (1966). Communication and development. In M. Weiner (Ed.), *Modernization: The dynamics of growth* (pp. 105-116). Washington, DC: Voice of America Forum Lecture Series.

Pool, I. de S. (1974). Direct broadcast satellites and the integration of national cultures. In Aspen Institute Program on Communications and Society, *Control of the direct broadcast satellites: Values in conflict* (pp. 27-56). Palo Alto, CA: Aspen Institute.

Pool, I. de S. (1983a). *Forecasting the telephone.* Norwood, NJ: Ablex.

Pool, I. de S. (1983b). *Technologies of freedom.* Cambridge, MA: Harvard University Press.

Postman, N. (1993). *Technopoly: The surrender of culture to technology.* New York: Vintage.

Postman, N., & Powers, S. (1992). *How to watch TV news.* New York: Penguin.

Preston, W., Jr., Herman, E. S., & Schiller, H. I. (1989). *Hope and folly: The United States and UNESCO, 1945-1985.* Minneapolis: University of Minnesota Press.

Public Citizen. (1995, July/August). p. 7.

Puga, A. (1993, December 10). Violence in video games prompts Senate grilling. *Boston Globe,* p. 1.

Pursell, C. (1995). *The machine in America: A social history of technology.* Baltimore: Johns Hopkins University Press.

Quinn-Judge, P. (1994, May 28). Putting Aristide on radio debated. *Boston Globe,* pp. 1, 8.

Ramstad, E. (1995, August 1). It's no Mickey Mouse takeover. *The Oregonian,* pp. B13-B14.

Rebello, K. (1995, December 11). Up front. *Business Week,* p. 6.

Reed, B. (1995). The wealth of information: How media empires plan to amass it. In R. Albelda et al., *Real world micro* (5th ed., pp. 51-55). Somerville, MA: Dollars & Sense.

Regan, D. T. (1988). *For the record.* Orlando, FL: Harcourt Brace.

Reinecke, I. (1984). *Electronic illusions: A skeptic's view of our high-tech future.* New York: Penguin.

The Report of the South Commission. (1990). *The challenge to the South.* New York: Oxford University Press.

Reuters. (1994, May 29). Murdoch says Fox seeks more stations. *Boston Globe,* p. 4.

Reyes Matta, F. (1976). The information bedazzlement of Latin America. *Development Dialogue,* [Sweden], 2, pp. 29-42.

Rheingold, H. (1991). *Virtual reality.* New York: Simon & Schuster.

Risen, J. (1995, July 23). CIA moves into economic espionage. *San Jose Mercury News,* p. A10.

Ritzer, G. (1993). *The McDonaldization of society.* Newbury Park, CA: Pine Forge.

Roach, C. (Ed.). (1993). *Communication and culture in war and peace.* Newbury Park, CA: Sage.

Robin, V. (1994, September 13). Cairo delegates target lavish U.S. lifestyle. *The Oregonian,* p. B7.

Robins, K., & Webster, F. (1988). Cybernetic capitalism: Information, capitalism, everyday life. In V. Mosco & J. Wasko (Eds.), *The political economy of information* (pp. 44-75). Madison: University of Wisconsin.

Robinson, C. J. (1995). Mass media and the U.S. presidency. In J. Downing, A. Mohammadi, & M. Sreberny (Eds.), *Questioning the media: A critical introduction* (2nd ed., pp. 94-111). Thousand Oaks, CA: Sage.

Rogers, E., & Antola, L. (1985, Autumn). *Telenovelas:* A Latin American success story. *Journal of Communication,* 24-35.

Rose, F. (1984). *Into the heart of the mind: An American quest for artificial intelligence.* New York: Harper & Row.

Rosenberg, R. S. (1992). *The social impact of computers.* San Diego: Academic Press.

Ross, R. S. (1993). *American national government: Institutions, policy, and participation* (3rd ed.). Guilford, CT: Dushkin.

Rossi, J. P. (1985). A silent partnership? The U.S. government, RCA, and radio communications in East Asia, 1919-1928. *Radical History Review, 33,* 32-52.

Rowan, G. (1993, July 19). New technology propels air travel. *The Globe and Mail* [Toronto], p. B1.

Ryan, R. (1994, May 27). Legality of hidden mikes at businesses questioned. *Boston Globe,* pp. 1, 24.

Sagan, C. (1995, September 17). Where did TV come from? *Parade Magazine,* pp. 10, 12.

Saltus, R. (1993a, August 8). Computers not being taken personally. *Boston Globe,* pp. 1, 18.

Saltus, R. (1993b, August 9). Industry pins hopes on byte-sized machines. *Boston Globe,* pp. 1, 7.

Samarajiva, R., & Shields, P. (1989, May). *Value issues in telecommunication resource allocation in the Third World.* Paper presented at the International Communication Association Conference, San Francisco.

Saunders, R. J., Warford, J. J., & Wellenius, B. (1983). *Telecommunications and economic development.* Baltimore: Johns Hopkins University Press.

Schement, J. R., & Curtis, T. (1995). *Tendencies and tensions of the information age: The production and distribution of information in the United States.* New Brunswick, NJ: Transaction Books.

Schickel, R. (1968). *The Disney version: The life, times, art, and commerce of Walt Disney.* New York: Simon & Schuster.

Schiller, H. I. (1969). *Mass communications and American empire.* Boston: Beacon.

Schiller, H. I. (1973). *The mind managers.* Boston: Beacon.

Schiller, H. I. (1981). *Who knows: Information in the age of the Fortune 500.* Norwood, NJ: Ablex.

Schiller, H. I. (1993a, July 12). The information economy: Public way or private road? *The Nation,* pp. 64-66.

Schiller, H. I. (1993b). Not yet the post-imperialist era. In C. Roach (Ed.), *Communication and culture in war and peace* (pp. 97-116). Newbury Park, CA: Sage.

Schiller, H. I. (1996). *Information inequality: The deepening social crisis in America.* New York: Routledge.

Schor, J. B. (1992). *The overworked American: The unexpected decline of leisure.* New York: Basic Books.

Schreibman, V. (1993). Closing the "values gap": Industry plans "hostile takeover" of the NII. *FINS, 1*(2).

Schumpeter, J. A. (1942). *Capitalism, socialism, and democracy.* Sydney, Australia: Allen & Unwin.

Selvin, P. (1988, November 28). Campus hackers and the Pentagon. *The Nation,* pp. 563-566.

Shaiken, H. (1980). *Computer technology and the relations of power in the workplace.* Paper presented at the International Institute for Comparative Social Research, Berlin.

Shapiro, M. J. (1981). *Language and political understanding: The politics of discursive practices.* New Haven, CT: Yale University Press.

Shattuck, J., & Spence, M. M. (1988, April). The dangers of information control. *Technology Review,* 62-79.

Shenon, P. (1988, January 28). FBI papers show wide surveillance of Reagan critics. *New York Times,* pp. A1, A18.

Shenon, P. (1994, September 11). A repressed world says, "Beam me up." *New York Times,* p. E4.

Shenon, P. (1995, January 10). Singapore court finds a U.S. scholar and newspaper guilty of contempt. *New York Times,* p. A5.

Siegel, L. (1986, July/August). Microcomputers: From movement to industry. *Monthly Review,* 110-117.

Simon, S. (1995, January 1). Good morning, El Salvador! *New York Times Book Review,* p. 17.

Sims, C. (1994, April 28). Walt Disney reinventing itself. *New York Times,* pp. D1, D9.

Sington, D., & Weidenfeld, A. (1942). Broadcasting in the Third Reich. In A. Mattelart & S. Sieglaub (Eds.). (1979). *Communication and class struggle: Capitalism, imperialism* (pp. 272-277). New York: International General.

Sklair, L. (1991). *Sociology of the global system.* Baltimore: Johns Hopkins University Press.

Slack, J. D. (1984). *Communication technologies and society: Conceptions of causality and the politics of technological intervention.* Norwood, NJ: Ablex.

Sloane, L. (1992, September 12). Orwellian dream come true: A badge that pinpoints you. *New York Times,* p. 11.

Smythe, D. W. (1984, July). New directions for critical communications research. *Media, Culture & Society,* 205-217.

Southscan. (1990, June 17). *The Guardian,* p. 14.

Spero, J. E. (1985). *The politics of international economic relations* (3rd ed.). New York: St. Martin's.

Spigner, C. (1994, January). Black impressions: Television and film imagery. *Crisis,* 8-10, 16.

Spillane, M. (1994, November 21). Unplug it! *The Nation,* p. 600.

Starks, T. (1993, December 29). Few female heroes in video games. *San Jose Mercury News,* p. F1.

Stavrianos, L. S. (1981). *Global rift: The Third World comes of age.* New York: Morrow.

Stehman, J. W. (1925). *The financial history of the American Telephone and Telegraph Company.* Boston: Houghton Mifflin.

Steinbock, D. (1995). *Triumph and erosion in the American media and industries.* Westport, CT: Quorum Books.

Sterling, C. H. (1979). Television and radio broadcasting. In B. M. Compaine (Ed.), *Who owns the media: Concentration of ownership in the mass communications industry* (pp. 61-125). New York: Harmony.

Street, J. (1992). *Politics and technology.* New York: Guilford.

Streeter, T. (1987, June). The cable fable revisited: Discourse, policy, and the making of cable television. *Critical Studies in Mass Communication,* 174-200.

Sussman, G. (1982). Telecommunications technology: Transnationalizing the new Philippine information order. *Media, Culture & Society, 4,* 377-390.

Sussman, G. (1987, Spring). Banking on telecommunications: The World Bank in the Philippines. *Journal of Communication,* 90-105.

Sussman, G. (1991). The transnationalization of Philippine telecommunications: Postcolonial continuities. In G. Sussman & J. A. Lent (Eds.), *Transnational communications: Wiring the Third World* (pp. 125-149). Newbury Park, CA: Sage.

Sussman, G. (1993, September). Revolutionary communications in Cuba: Ignoring new world orders. *Critical Studies in Mass Communication,* 199-218.

Sussman, G., & Lent, J. A. (1991). Introduction: Critical perspectives on communication and Third World development. In G. Sussman & J. A. Lent (Eds.), *Transnational communications: Wiring the Third World* (pp. 1-26). Newbury Park, CA: Sage.

Tarr, J. A., Finholt, T., & Goodman, D. (1987, November). The city and the telegraph: Urban telecommunications in the pre-telephone era. *Journal of Urban History,* 38-80.

Taylor, P. M. (1992). *War and the media: Propaganda and persuasion in the Gulf War.* Manchester, UK: Manchester University Press.

Tenner, E. (1991, November-December). The impending information implosion. *Harvard Magazine,* 30-34.

Thompson, E. (1993, February 11). "Caribbean: Talk shows keep government on their toes" (Inter Press Service feature release).

Thompson, R. L. (1947). *Wiring a continent: The history of the telegraph industry in the United States, 1832-1866.* Princeton, NJ: Princeton University Press.

Toffler, A. (1980). *The third wave.* New York: Morrow.

Toffler, A. (1990). *Power shift: Knowledge, wealth, and violence at the edge of the 21st century.* New York: Bantam.

Tolan, S. (1996, July 16). Must NPR sell itself? *New York Times,* p. A11.

"Top 1000 ranked by industry." (1995, March 27). *Business Week,* pp. 104-149.

"Top 1000 ranked by industry." (1996, March 25). *Business Week,* pp. 106-145.

Traber, M., & Nordenstreng, K. (1992). *Few voices, many worlds: Toward a media reform movement.* London: World Association for Christian Communication.

Tuchman, G. (1978). *Making news: A study in the construction of reality.* New York: Free Press.

Tunstall, J. (1977). *The media are American.* New York: Columbia University Press.

Union of Concerned Scientists. (1993). *World scientists' warning to humanity* (Public statement signed by 1,670 scientists from 71 countries). Cambridge, MA: Author.

U.S. embarrassed by CIA role in nabbing Mandela. (1987, July/August). *Utne Reader,* p. 34.

U.S. Senate, Committee on Foreign Relations. (1966). *Hearings on new policies in Viet Nam* (89th Congress, 2nd Session).

U.S. Senate, Committee on Governmental Affairs. (1980, December). *Structure of corporate concentration* (Vol. 1). Washington, DC: Author.

U.S. Senate, Select Committee to Study Government Operations with Respect to Intelligence Activities. (1976). *Intelligence activities and the rights of Americans* (Books 2, 3, 4; 94th Cong., 2nd Session, Report No. 94-755). Washington, DC: Government Printing Office.

Volti, R. (1992). *Society and technological change* (2nd ed.). New York: St. Martin's.

Wasserman, G. (1988). *The basics of American politics* (5th ed.). Glenview, IL: Scott, Foresman.

Wasserman, H. (1972). *Harvey Wasserman's history of the United States.* New York: Harper & Row.

Watson, B. (1993, April). For a while, the Luddites had a smashing success. *Smithsonian,* 140-154.

Weiner, J. (1994, January 31). Tall tales and true. *The Nation,* pp. 133-135.

Weiner, T. et al. (1994, October 9). CIA spent millions to support Japanese right in 50's and 60's. *New York Times,* pp. A1, A11.

Weizenbaum, J. (1976). *Computer power and human reason: From judgment to calculation.* New York: Freeman.

Weizenbaum, J. (1981a). Once more, the computer revolution. In T. Forester (Ed.), *The microelectronics revolution: The complete guide to the new technology and its impact on society* (pp. 550-570). Cambridge: MIT Press.

Weizenbaum, J. (1981b). Where are we going?: Questions for Simon. In T. Forester (Ed.), *The microelectronics revolution: The complete guide to the new technology and its impact on society* (pp. 434-438). Cambridge: MIT Press.

Weizenbaum, J. (1985). The myths of artificial intelligence. In T. Forester (Ed.), *The information technology revolution* (pp. 84-94). Cambridge: MIT Press.

Wiener, N. (1948). *Cybernetics: Control and communication in the animal and the machine.* Cambridge: MIT Press.

Williams, R. (1975). *Television: Technology and cultural form.* New York: Schocken.

Wilson, M. (1995). The office farther back: Business services, productivity, and the offshore back office. In P. Harker (Ed.), *The service quality and productivity challenge* (pp. 203-224). Boston: Kluwer.

Winner, L. (1977). *Autonomous technology: Technics-out-of-control as a theme in political thought.* Cambridge: MIT Press.

Winner, L. (1993). Artifact/ideas and political culture. In A. Teich (Ed.), *Technology and the future* (6th ed., pp. 283-292). New York: St. Martin's.

Winston, B. (1986). *Misunderstanding media.* Cambridge, MA: Harvard University Press.

Winston, B. (1995). How are media born and developed? In J. Downing, A. Mohammadi, & A. Sreberny-Mohammadi (Eds.), *Questioning the media: A critical introduction* (2nd ed., pp. 54-74). Thousand Oaks, CA: Sage.

Wise, D. (1975). *The American police state: The government against the people.* New York: Random House.

Ziegler, D. W. (1990). *War, peace, and international politics* (5th ed.). Glenview, IL: Scott, Foresman.

Zimbalist, A. (1992). *Baseball and billions: A probing look inside the big business of our national pastime.* New York: Basic Books.

Zinn, H. (1980). *A people's history of the United States.* New York: Harper & Row.

Index

ABOUT THE AUTHOR

Gerald Sussman is Professor of Urban Studies and Communications at Portland State University, Portland, Oregon. His previous publications include *Transnational Communications: Wiring the Third World* (1991 and the forthcoming *Global Productions: Labor in the Making of the "Information Society"*), along with numerous journal articles and book chapters on communication technology and development.